AF588459

Sustainable Agriculture Reviews

Volume 33

Series Editor
Eric Lichtfouse
CEREGE, Aix Marseille Univ, CNRS, IRD, INRA, Coll France
Aix-en-Provence, France

Other Publications by Dr. Eric Lichtfouse

Books

Scientific Writing for Impact Factor Journals
https://www.novapublishers.com/catalog/product_info.php?products_id=42242

Environmental Chemistry
http://www.springer.com/978-3-540-22860-8

Sustainable Agriculture
Volume 1: http://www.springer.com/978-90-481-2665-1
Volume 2: http://www.springer.com/978-94-007-0393-3

Book series

Environmental Chemistry for a Sustainable World
http://www.springer.com/series/11480

Sustainable Agriculture Reviews
http://www.springer.com/series/8380

Journal

Environmental Chemistry Letters
http://www.springer.com/10311

Sustainable agriculture is a rapidly growing field aiming at producing food and energy in a sustainable way for humans and their children. Sustainable agriculture is a discipline that addresses current issues such as climate change, increasing food and fuel prices, poor-nation starvation, rich-nation obesity, water pollution, soil erosion, fertility loss, pest control, and biodiversity depletion.

Novel, environmentally-friendly solutions are proposed based on integrated knowledge from sciences as diverse as agronomy, soil science, molecular biology, chemistry, toxicology, ecology, economy, and social sciences. Indeed, sustainable agriculture decipher mechanisms of processes that occur from the molecular level to the farming system to the global level at time scales ranging from seconds to centuries. For that, scientists use the system approach that involves studying components and interactions of a whole system to address scientific, economic and social issues. In that respect, sustainable agriculture is not a classical, narrow science. Instead of solving problems using the classical painkiller approach that treats only negative impacts, sustainable agriculture treats problem sources.

Because most actual society issues are now intertwined, global, and fast-developing, sustainable agriculture will bring solutions to build a safer world. This book series gathers review articles that analyze current agricultural issues and knowledge, then propose alternative solutions. It will therefore help all scientists, decision-makers, professors, farmers and politicians who wish to build a safe agriculture, energy and food system for future generations.

More information about this series at http://www.springer.com/series/8380

Eric Lichtfouse
Editor

Sustainable Agriculture Reviews 33

Climate Impact on Agriculture

Editor
Eric Lichtfouse
CEREGE, Aix Marseille Univ, CNRS,
IRD, INRA, Coll France
Aix-en-Provence, France

ISSN 2210-4410 ISSN 2210-4429 (electronic)
Sustainable Agriculture Reviews
ISBN 978-3-319-99075-0 ISBN 978-3-319-99076-7 (eBook)
https://doi.org/10.1007/978-3-319-99076-7

Library of Congress Control Number: 2018957274

This Springer imprint is published by the registered company Springer Nature Switzerland AG
The registered company address is: Gewerbestrasse 11, 6330 Cham, Switzerland

Preface

Climate change is unavoidable but adaptation is possible. Climate change and agriculture are interrelated processes, both of which take place on a global scale[1]. Climate change affects agriculture through changes in average temperatures, rainfall and climate extremes; changes in pests and diseases; changes in atmospheric carbon dioxide; changes in the nutritional quality of some foods; and changes in sea level. Future climate change will likely negatively affect crop production in low latitude countries, while effects in northern latitudes may be positive or negative. Climate change will probably increase food insecurity for some vulnerable groups, such as the poor. Agriculture contributes to climate change both by anthropogenic emissions of greenhouse gases and by the conversion of non-agricultural land such as forests into agricultural land.

[1] https://en.wikipedia.org/wiki/Climate_change_and_agriculture

Soil erosion in wheat field, Pas de Calais, France, winter 1990. Copyright P. CHERY, INRA 1990

In order to adapt agriculture, there is actually an urgent need for management methods that will decrease negative impacts and allow food production on formerly sterile lands. This book reviews advanced knowledge and methods relevant to climate and agriculture. In the first chapter, Kulek reviews the agricultural nitrogen cycle, with focus on gas emissions of ammonia (NH_3), nitrous oxide (N_2O), commonly known as the laughing gas, and nitric oxide (NO) from animal husbandry and fertilisation. She found that camels emit much less ammonia and nitrous oxide than cattle, that the older the animal the higher the ammonia emission, and that fertilisation with calcium ammonium salts emits much less gases that urea fertilisation. In Chap. 2, Sarauskis evaluate the positive and negative effects of tillage; they found that sustainable tillage without ploughing reduces costs by 25–41%. Tsegaye reviews coffee production and climate change in Ethiopia, where the mean annual temperature has increased by 1.3°C between 1960 and 2006, and states that 'Africa can be easily converted into deserts', in Chap. 3.

Coastal agrosystems are particularly vulnerable to climate change and accelerated sea level rise. In Chap. 4, Banerjee et al. found that in some areas up to 40% of biodiversity has been lost; they propose adaptation practices such as agroforestry and salinity management. Singh et al. explain that wetland rice fields emit 15–20% of anthropogenic methane (CH_4) emissions; they list the various factors and practices controlling emissions in Chap. 5. In the same vein, Srivastava et al. review in Chap. 6 the factors that control carbon sequestration in soils, a practice which is foreseen to decrease CO_2 emissions; they found that dry tropical soils are far away from carbon saturation and thus have high potential for carbon sequestration.

In Chap. 7, Arora and Vanza present bacteria and fungi that can be used to decrease salt stress in plants; they found that wheat and corn yields can be increased by 10–12% under salinity stress. Bhaduri et al. review the types of degraded soils and the bioindicators of soil degradation, such as plant biomarkers and biosensors, in Chap. 8. Usman et al. discuss groundwater evolution in Pakistan, and consequence for irrigated agriculture, in Chap. 9. In the future, there will be more food production in closed systems due to climate changes and increasing urbanisation. Here, Hadavi and Ghazijahani review the types of closed systems used in agriculture, with inspiring experiments of food production in outer space, in Chap. 10. In the last Chap. 11, Zahedi presents biofuels such as bioethanol, biodiesel, crop residues and algae.

Aix-en-Provence, France Eric Lichtfouse

Contents

About the Editor

Eric Lichtfouse, PhD, born in 1960, is an environmental chemist working at the University of Aix-Marseille, France. He has invented carbon-13 dating, a method allowing to measure the relative age and turnover of molecular organic compounds occurring in different temporal pools of any complex media. He is teaching scientific writing and communication and has published the book *Scientific Writing for Impact Factor Journals*, which includes a new tool – the Micro-Article – to identify the novelty of research results. He is founder and chief editor of scientific journals and series in environmental chemistry and agriculture. He got the Analytical Chemistry Prize by the French Chemical Society, the Grand Prize of the Universities of Nancy and Metz, and a Journal Citation Award by the Essential Indicators.

Chapter 1
Impact of Human Activity and Climate on Nitrogen in Agriculture

Beata Kułek

Abstract High concentrations of gases containing nitrogen in the air and different nitrogen forms in soils, plants and water pose a threat both to the environment and to human health. Here I review the impact of various factors on the content of nitrate, nitrite, ammonium, organic and total nitrogen, urease and nitrate reductase in soils and plants. I also review impacts on ammonia, nitrous oxide, nitrogen dioxide and nitric oxide in the atmosphere. The strongest effect on concentrations of gases is the type of animals producing the gases. A weaker dependency is the distance from a farm, and the lowest effect is the type of plant species. The highest concentration of NH_3 and N_2O came from cattle (56.1 and 42.3 μg m^{-3}), whereas the lowest – from camels (0.3 and 0.5 μg m^{-3}), respectively. The following dependency prevailed: the longer the distance from animal farms, the lower the concentrations of ammonia.

Higher emissions of ammonia (92.0%) and nitrous oxide (74.8%) were found to come from urea in a crop field, whereas lower from calcium ammonium nitrate applied to grassland (1.6% of NH_3) and from ammonium salts used in a crop field (0.1% of N_2O). Similar tendencies were observed for NO. Total emission of ammonia was the highest when resulting from the spreading of waste (36%), whereas the lowest volatilization from grazing / outdoors (8%). The older the animals, the higher the NH_3 loss. The highest organic nitrogen concentration was noted after the application of pig slurry manure (3.5%) and the lowest after applying cattle and pig farmyard manure (FYM) (2.3%) above ryegrass field. The highest amounts of net nitrogen were found in *Melilotus alba*, whereas the lowest in *Poa pratensis*. A total nitrogen concentration also depended on the type of crops. Its level was higher in *Vicia faba* (48.71%) and the lowest in grained winter rye cereals (14.96%).

B. Kułek (✉)
Institute for Agricultural and Forest Environment, Polish Academy of Sciences, Poznań, Poland

Present Address: Horticultural and educational services Beata Kułek, Poznań, Poland
e-mail: B.Kulek@interia.eu

E. Lichtfouse (ed.), *Sustainable Agriculture Reviews 33*, Sustainable Agriculture Reviews 33, https://doi.org/10.1007/978-3-319-99076-7_1

Keywords Human activity · Climatic factors · Nitrogen forms · Agricultural ecosystems · Meadow · Shelterbelts · Crop fields · Water · Fertilizers · Animal husbandry · Organic residues

1.1 Introduction

Ammonia is deposited on vegetation, soil and water (Asman and Van Jaarsveld 1992). This deposition may cause acidification and eutrophication of natural ecosystems (Fangmeier et al. 1994). The highest amounts of NH_3-N were in spring and autumn corresponding to the largest main fertilization time – the application of organic fertilizers across the surrounding arable land. The large contribution of NH_3-N and N-NH_4^+ to the total N input poses a high eutrophication risk to peat moorland and changes in the biodiversity of the ecosystem (Hurkuck et al. 2014). Khonje et al. (1989) tested different fertilizers and the soil pH was very acidic when $(NH_4)_2SO_4$ and NH_4Cl were used (approx. 3.5), but when $NaNO_3$ was applied (6.5) and $Ca(NO_3)_2$ – nearly 6.0 (Bolan et al. 1991). About 5% of the total atmospheric greenhouse effect is attributed to N_2O from which 70% of the annual global anthropogenic emissions come from animal and crop productions (Mosier 2001). Nitrous oxide causes global warming, but NO_3^- ground water contamination (Lin et al. 2016). Over 80% of ammonia emissions originate from animal excreta and less than 20% of the total NH_3 emissions of agricultural origin in Europe from the use of fertilizers. A large variation in the amounts of this gas in the air above different countries was observed (Van der Hoek 1998). In areas of intensive animal husbandry, the NH_3 concentration was 50 μg m^{-3} (Asman et al. 1989), but in agricultural areas for several young (20–40 plant species) only 2–5 μg of NH_3 m^{-3} was found (Farquhar et al. 1980).

In an ecological farm system, higher urease activity, an accumulation of total nitrogen and lower concentrations of ammonium and nitrate ions compared with a conventional and integrated system were observed. Urease activity increased with increasing pH values and was higher in an ecological system than in other systems. Higher doses of nitrogen in fertilizers decreased the enzymatic activity (Meysner and Szajdak 2013). An amount of total nitrogen and urease activity were much higher in neutral mineral – organic soils than in very acid and acid mineral soils. An increase in an amount of rainfall and temperature was accompanied by an increase in an activity of urease (Szajdak and Matuszewska 2000). Stalenga and Kawalec (2008) showed that the emission of N_2O in an ecological crop production was more than two times lower than in two other systems (a conventional and an integrated one). Progressively decreasing urease activity was found with increasing depth when the greatest organic matter content existed and the most recent organic depositions were found (Myers and McGarity 1968).

A wide variation in a loss of NH_3 reported is due to a variety of factors, including: water pH, temperature, soil type, N-source and a dose and a method and time of application (Fenn and Hossner 1985). Small amounts of NH_3 loss were from decaying plant residues (Terman 1979) and maturing leaves (Hooker et al. 1980), but relatively large from urea and ammonium-based fertilizers, animal and sewage waste and submerged soils (Fenn and Hossner 1985). Factors which cause the NH_3 volatilization to the air are the following: the amount of urea applied, its rate of hydrolysis, an initial pH and pH buffer capacity of the soil, the level of soil moisture and the depth of application (Rachhpal and Nye 1988). Plants do not tolerate high amounts of ammonium and concentrations of 0.02–0.04% are toxic for them (Barker et al. 1967; Ajayi et al. 1970), but early growth was retarded when NH_3 and NH_4^+-N concentrations in soil reached 944 ppm at pH of 8.1 and was completely inhibited at concentrations of 1628 ppm at pH of 9.0. Roots tips of corn were brown and at 1000 ppm seeds resulted in significant stand reduction. High rates of NH_3 also caused NO_2^--N toxicity for plants (Colliver and Welch 1970). In light the emission of NH_3 from spring barley was high, but when the light was turned off it declined (Schjørring 1991).

Ammonia emissions to the atmosphere come from plant residues (Fenn and Hossner 1985) and mainly from farm animals and spreading of their manure, some also from application of mineral fertilizers, notably from surface application of urea. Plants act both as a source and sink of NH_3 (Farquhar et al. 1983; Sutton 1990; Schjørring 1991). High losses of ammonia occur when plants are diseased and during grain filling, but it was noted when the losses of nitrogen in barley crops were low (Schjørring et al. 1989; Holtan-Hartwig and Bøckman 1994). Greater amounts of ammonium were observed during senescence after proteolysis and deamination of amino compounds (Holtan-Hartwig and Bøckman 1994). Ammonium and nitrate are excreted through guttation and water percolate after rainfall or irrigation returns nitrogen to soil. Dew covers the plants during the night and early in the morning, then it evaporates from crop fields. What has not been taken up by plants from the air in these times will be re-emitted to the air as the vegetation dries (Sutton et al. 1992).

According to Wetselaar and Farquhar (1980), the following pathways of nitrogen losses from tops of plants can be distinguished: root exudates and losses from soil (by leaching and denitrification) of nitrogen transferred to roots, loss of pollen, flowers, fruits, leaves, plant material by insects, birds, microorganisms and excretions leaching from leaf surfaces by rain, dew dripping, sprinkler irrigation or spraying with pesticide and gaseous losses. These may be as: ammonia, amines, dinitrogen, nitric oxide, nitrogen dioxide and nitrous oxide (Dean and Harper 1986; Guanxiong et al. 1990).

An increase in the cation exchange capacity resulted in decreasing NH_3 losses. Ammonium sulfate produced higher soil pH values and NH_3 losses than ammonium nitrate, because one half of ammonium nitrate is in the NO_3^- form, so ammonia losses are much less than from ammonium sulfate at the same amount of nitrogen. A percent of NH_3-N losses and soil pH with ammonium nitrate decreased with increasing application dose, but with ammonium sulfate increased. When the fertilizer was present in deeper layers of soil, the NH_3-N loss was lower (Fenn and Kissel 1976).

Variations in temperature, precipitation, a content of clay and pH had significant effects on denitrification and N_2O emissions and stimulated higher values of nitrous oxide, but rainfall caused nitrate to limit emissions of this gas (Li et al. 1992). A significant influence on nitrous oxide and nitric oxide emissions was exerted by: (1) environmental factors (climate, soil organic C content, soil texture drainage and soil pH), (2) management-related factors (N application rate per the fertilizer type, a kind of crop, with major differences between grass, legumes and other annual crops) and (3) a length of measurement period and their frequency. The most important controls on nitric oxide emissions include the N application dose per the fertilizer type, soil organic C content and soil drainage. Global mean fertilizer-induced emissions of these gases amount to 0.9% and 0.7%, respectively of the N applied. The concentrations of these gases increased with the increasing dose of fertilizers. Nitrous oxide losses were lower from grassland than from croplands. Neutral to slightly acidic soils favor N_2O emissions. Urea gives the highest nitric oxide volatilization. Hence, ammonium-based fertilizers give high emissions of both gases and nitrate-based fertilizers increased NO, but decreased N_2O emissions. The lowest estimates for nitrous oxide are for animal manure, but ammonium nitrate and mixes of mineral fertilizers have the highest emissions. Contrary to nitrous oxide, the influence of climate was not significant for nitric oxide, differences in the soil texture had no effect on NO volatilization, because soil organic C content and drainage account for most of the variability (Bouwman et al. 2002).

Ammonia emissions increased with increasing soil temperatures, the time after this fertilizer application and at higher pH values, but decreased when the soil moisture was low and its dryness occurred (Ernst and Massey 1960). A total N loss at maturity (the grain filling period) was higher than during anthesis in winter wheat and increased with an increase in a fertilizer dose (Daigger et al. 1976). Many factors influence the concentrations of various nitrogen forms in the environment, as shown in Fig. 1.1.

Some of factors can act synergistically, because they overlap, and their effect is enhanced, but others have an inhibiting effect on the formation of nutrients in nature and gases in the atmosphere. Due to the complexity of relationships, which often overlap, the analysis of gas emissions and concentrations of different forms of nitrogen is needed and tests should be carried out directly and individually for each ecosystem, soil and plant. NOx, NH_3, NH_4^+ and NO_3^- lead to soil and plant community changes and to an eutrophication in semi-natural ecosystems. Ammonium nitrate, nitric oxide and soil NO_3^- can be harmful to human health and with ammonia also decrease biodiversity, but nitrous oxide causes changes in the climate (Erisman et al. 2003). Because of the harmful presence of excessive amounts of nitrogen forms in nature, investigates should be continually conducted to control the level of these forms and, if necessary, to limit their amounts e. g. through the use of phenylphosphorodiamidate, which inhibits urea volatilization by 54%, but the effect of this was in the lower soil pH (5.6), so when the pH was higher (7.2), N-(n-butyl) thiophosphoric triamide was applied and it limited ammonia emissions in 39% (Beyrouty et al. 1988). As a result, an appropriate method or an inhibitor of gas emissions should be selected for each environment and individual conditions. Trace elements like: Cu(I), Co(II), Cu(II), Fe(II), Ni(II), Pb(II), Zn(II), Al(III), As(III), Cr (III), Fe(III), V(IV), Mo(VI) and Se(VI) inhibited the nitrate reductase activity in acidic and neutral soils. The optimal pH for NR activity in soils is 7.0 - Fu and

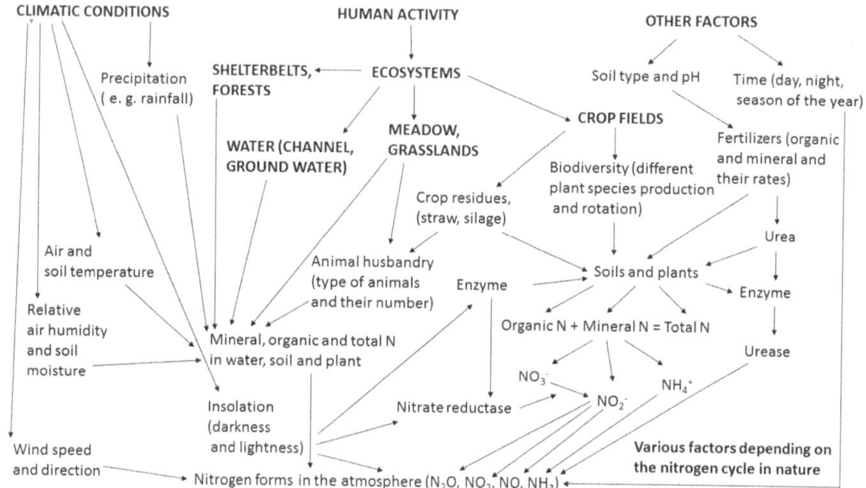

Fig. 1.1 Dependencies between human activity, climatic conditions and other factors on the content of different nitrogen forms in soils, plants, water and in the atmosphere in various agricultural ecosystems. The highest impact on gases emissions into the atmosphere was found to come from animal husbandry near meadow, crop residues and climatic conditions. Fertilizers visibly affect the amounts of different nitrogen forms in soils and plants growing in crop fields, especially manure and urea. The impact is associated with their doses and forms of application, but is less depended on a season of the year and the type of soil and plant species – contrary to a shelterbelt ecosystem. Insolation was found to affect nitrate reductase activity, whereas urease activity was stated to depend on the urea content in soil. It was also found that soil temperature and pH are important for the activity of both enzymes

Tabatabai 1989. Each enzyme has the highest activity at the optimal temperature. For the nitrate reductase it is below 40 °C (Abdelmagid and Tabatabai 1987) and it increases with time after incubation of soils (Binstock 1984). Higher rates of ammonium nitrate, higher N and the dry matter losses were from winter wheat leaves than from roots (Daigger et al. 1976). After anthesis, nitrogen was accumulated in grain in greater amounts. NO_x and NH_3 can cause acidification of soils and fresh waters. With a higher dose of urea prills, higher ammonia losses were observed for a longer time, but the amount of this gas decreases with an increasing dose of urea-liquid spray (Hargrove and Kissel 1979). A higher urea content in soil was related to higher activity of urease (Zantua and Bremner 1975). In sand, nitrate amounts were the lowest, but in clay loam – the highest (Groffman and Tiedje 1989). Nighttime volatilization rates of ammonia were only one-half of those observed by day for similar aqueous NH_3 concentrations. At 2–4 m s^{-1} wind speed, 3.5% of the ammonia loss was observed, but at 4–8 m s^{-1} 25%. Ammonia concentrations were higher at lower heights above ground (Denmead et al. 1982).

I investigated an effect of human activity on ammonia concentrations in the atmosphere using Gradko passive samplers from 5th September to 13th October 2008 and analyzes by ion chromatography (Fig. 1.2a–c).

Fig. 1.2 An impact of different activity: (**a**) 50 and 200 m from a cattle farm, (**b**) corn stubble after the application of POLIDAP fertilizer in a dose of 27 kg N ha^{-1} on 10th September 2008 and (**c**) straw ballots and silage on an amount of ammonia in the air. The source: own studies

Results of these investigations were illustrated in Fig. 1.3.

The highest concentration of ammonia was at a distance of 50 m from a cowshed (49.215 μg m^{-3}), but the lowest near straw ballots and silage (5.803 μg m^{-3}) (Fig. 1.3).

The effect of various factors on the activity of nitrate reductase was also analyzed. The presence of ammonium nitrate in soil caused stimulation of this enzyme activity in seedling leaves of barley, but reduced it in roots. In a presence of ammonium chloride in medium, no enzyme activity was found in leaves, but trace amounts were observed in roots (Skoczek 1992). Nitrate reductase activity was higher for wheat than for oat and barley and higher in light in leaves, especially for barley (Lillo and Henriksen 1984).

Soil pH and a kind of fertilizer have an effect on NH_3 losses, which were higher for ammonium nitrate at the pH level above 7.0 than at the value of 5.5 and lower from ammonium nitrate than from ammonium sulfate. At the pH above 5.5, emissions of this gas were higher from ammonium sulfate than from ammonium dihydrogen phosphate and ammonium nitrate, but lower than from diammonium hydrogen phosphate. Ammonia volatilization was also greater from diammonium hydrogen phosphate than from ammonium sulfate, but lower than from urea at the pH of 6.1. In very acidic soil (the pH of 3.7), ammonia emissions were very low only 1.4% of applied N with diammonium hydrogen phosphate to 0.6% with urea and to less than 0.1% with ammonium dihydrogen phosphate, ammonium sulfate and ammonium nitrate. Ammonium salts showed more variation in ammonia amounts from the pH of soils 5.5–7.4 than urea (Whitehead and Raistrick 1990).

The type of soil and a kind of fertilizer had also an impact on the ammonia emission. Ammonia volatilization from ammonium salts applied to calcareous soils was greatest from ammonium sulfate, about half as much from ammonium dihydrogen phosphate, diammonium hydrogen phosphate, ammonium nitrate and

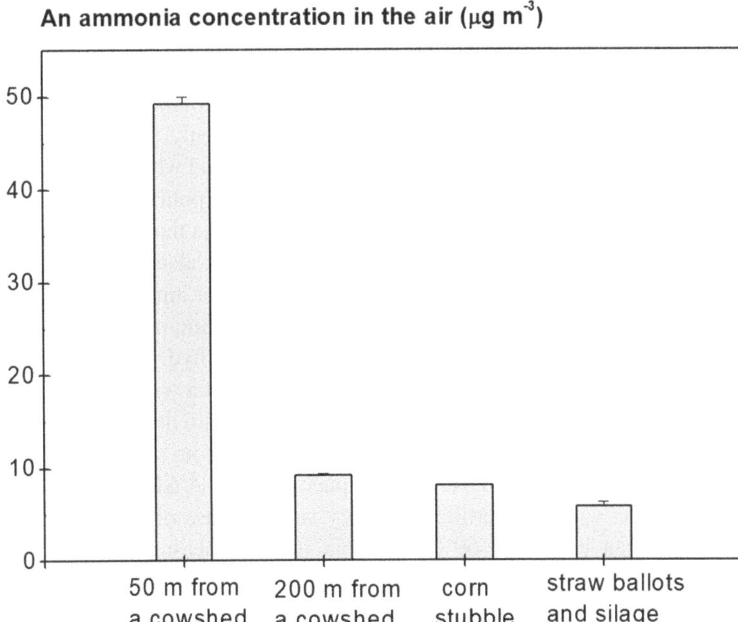

Fig. 1.3 The impact of different types of human activity on the content of ammonia in the air

negligible from magnesium ammonium phosphate. In acidic sandy soils treated with calcium carbonate the highest volatilization of ammonia was after using ammonium sulfate, but lower after using diammonium hydrogen phosphate, which acted in a similar manner to ammonium nitrate. Then the same soils were treated with barium carbonate. The highest ammonia emission was noticed after using ammonium nitrate, but lower after applying ammonium sulfate and the same for diammonium hydrogen phosphate. Further, for the soils treated with magnesium carbonate the highest loss of this gas was stated after using ammonium sulfate, lower after applying ammonium nitrate and the lowest after using diammonium hydrogen phosphate (Larsen and Gunary 1962). Diammonium hydrogen phosphate and ammonia introduce alkalinity into soil and a loss of ammonia can occur both from acidic and alkaline soils. The alkalinity of diammonium hydrogen phosphate is not great, so only small quantities of ammonia volatilization would be expected from acidic soils and the smallest – from acidic inorganic nitrogen such as: ammonium nitrate, ammonium sulfate, ammonium dihydrogen phosphate and ammonium chloride (Fenn and Hossner 1985). Quantities of NH_3 loss from surface-applied ammonium nitrate and ammonium chloride on calcareous soils are low. Higher amounts of urea, higher the NH_3 emission was. If an acidified N fertilizer is urea, little or no reduction of ammonia loss may occur, because urea does not hydrolyze immediately, but the acid is quickly neutralized. Ammonium sulfate and diammonium hydrogen phosphate caused high ammonia emission which occurs very rapidly (Fenn 1975). Dry potassium chloride and urea were not effective in reducing ammonia losses (Fenn and Hossner 1985). NH_4^+ added to urea is ineffective in replacing soil Ca to control ammonia losses (Fenn et al. 1982b). Ammonium nitrate with urea in solution reduces ammonia loss slightly and only due to the acidity of ammonium nitrate. No reduction of ammonia loss was found when only a content of urea was present. Even in acidic soil potassium nitrate and potassium chloride had a greater effect on reducing the ammonia emissions from urea than it was the case for ammonium nitrate. An application of urea to soil raises also a concentration of ammonium and soil pH, thus providing ideal conditions for ammonia volatilization (Rachhpal and Nye 1988). From Houston Black clay, maximum ammonia losses of 55–65% occurred from ammonium sulfate, diammonium hydrogen phosphate and ammonium fluoride at 22 °C. Losses of ammonia were higher with inorganic N salts. Organic urea is immobilized by microbes or by diffusion into the soil and maximum losses of ammonia from urea reached from 75% to 80% from sand and were lower from soils with higher the cation exchange capacity of soil. A 61% NH_3-N loss from ammonium sulfate was from fertilized Harkey sicl. The use of urea on calcareous sand resulted in an estimated 69% ammonia loss as measured by the residual fertilizer N available for growth of *Sorghum sudanense*. Ammonium sulfate in the same experiment resulted in an 80% NH_3-N loss of the applied nitrogen (Fenn and Miyamoto 1981). Ammonium dihydrogen phosphate and diammonium hydrogen phosphate in soil with high pH will produce a similar ammonia loss (El-Zahaby et al. 1982), but urea in acidic and alkaline soils similar to losses from reactive inorganic N compounds in calcareous soils. The NH_3 loss will be high both from acidic and calcareous soils, because urea is enzymatically hydrolyzed to ammonium carbonate (Fenn and Hossner 1985). Moving the fresh surface residues aside and applying urea to the mineral surface will result in reduced NH_3 losses, because lower urease activity limits the urea hydrolysis rate. The cation exchange capacity does not control ammonia losses even at the highest values of the CEC (Fenn et al. 1984; Touchton and Hargrove 1982). Hydrolysis of urea is a function of microbial activity and when the temperature is reduced to a point where microbial activity essentially ceases, the ammonia loss is stopped. Calcium nitrate does not produce ammonia losses. Ammonium sulfate and diammonium hydrogen phosphate react with calcium carbonate to produce an increase in the soil solution pH and the ammonia loss (Fenn and Hossner 1985). Twenty percent of ammonia emitted from surface applied urea to a sugarcane crop was captured by the crop leaves (Denmead et al. 1993). The highest volatilization of ammonia was noticed from urea applied on the surface of grass, but the lowest – from calcium ammonium nitrate and a little higher – from ammonium dihydrogen phosphate. The impact of fertilizers was higher than the pH of soils – Franco et al. (1979).

A dose of fertilizer had also an effect on the ammonia concentration in the air. A higher rate of fertilizer meant a higher amount of ammonia in the air from wheat. No increase in % N lost occurred with increasing rates of surface application of ammonium nitrate from Houston Black clay (Maheswari et al.

1992). Higher doses of ammonium sulfate mean a higher loss of ammonia and higher value of the soil pH (Fenn and Kissel 1974).

The method of urea application and a dose of fertilizer were important for the amount of NH_3 lost. The application of urea ammonium nitrate liquid sprayed onto non-tilled soil surface resulted in less corn grain yield at higher dose of this fertilizer and a 76% loss of NH_3 of the applied N was reached (Fenn and Hossner 1985).

At an apoplastic optimum pH of 7.0 plants emit NH_3 and produce ammonium in the highest amounts (Holtan-Hartwig and Bøckman 1994). Maximum content of ammonium in oldest leaves occurred slightly before anthesis in barley (Schjoerring et al. 1993b).

Climatic conditions also had an effect on the ammonia emission and the activity of urease. If the rainy season occurs in the winter, then optimum moisture conditions for ammonia loss can exist at the time of spring fertilization and plant growth initiation. It was concluded that a significant loss of NH_3 will occur only as a result of the presence of urea, whether synthetic or of animal origin. Broad application of ammonium nitrate will produce maximum ammonia loss under all conditions (Fenn and Hossner 1985). Rainfall would move urea into the soil and lower losses of ammonia were also in cooler temperatures (Fisher and Parks 1958). Urease activity is the highest at an optimal temperature of 37 °C (Gould et al. 1973), but also it is optimal in the range of 60 °C (for an Indian Vertisol) to 70 °C (for Alfisol) and increases with an increase in the moisture content, but was not detected in soil samples collected in late summer, when the soil moisture was below a pressure of 15 bar (Sahrawat 1984). The total quantity of soil moisture present in soil is not important to the ammonia loss from urea, if the surface is air dry (Fenn and Hossner 1985). An addition of urea will stimulate ureolytic microbes and if organic residues do not limit the production of urease, a maximum ammonia loss can occur (Paulson and Kurtz 1969).

Other scientists examined the effect of residues and heavy metals on ammonia emissions to the atmosphere. Fenn et al. (1984) stated that an addition of fresh organic residues can double ammonia losses, especially at lower doses of urea. Soils that are low in fresh organic residues could be surface fertilized with urea with reduced risk of ammonia loss. Toxic heavy metals and organic compounds to block the hydrolysis of urease reduced the ammonia loss (Bremner and Douglas 1971b).

Also time was another factor that had an effect on the ammonia volatilization. The more hours after an application of manure passed, the higher of cumulative NH_3-N loss was. Daily ammonia concentration in ppm decreased with increasing air temperature, but nitrous oxide increased – Franco et al. (1979).

In the shelterbelt, NH_4^+ amounts were 2–4 times higher down to the depth of 85 cm. In sandy loam and Luisiana clay and peat soils, the NO_3^- amounts were very high in contrast to NH_4^+ – Franco et al. (1979).

An activity of nitrate reductase correlated positively with yield of bean (*Phaseolus vulgaris* L.) seeds – Franco et al. (1979) and with a content of nitrogen in wheat of cv. Ottawa, but not in the case of cv. Gage – Eilrich and Hageman (1973).

I described processes and reactions taking place in the nitrogen cycle and methods of detecting its different forms in soils, plants and in the air and in addition I presented my new and very sensitive method for the determination of trace amounts of ammonia in the atmosphere using passive samplers and a spectrophotometer (Kułek 2015). This review presents the influence of various factors on the content of nitrogen forms in agricultural ecosystems.

1.2 Type of Human Activity

The percentage of total ammonia emissions for each type of farm management is shown in Fig. 1.4.

The highest volatilization of ammonia was from the spreading waste, but the lowest – from grazing or outdoors (Fig. 1.4).

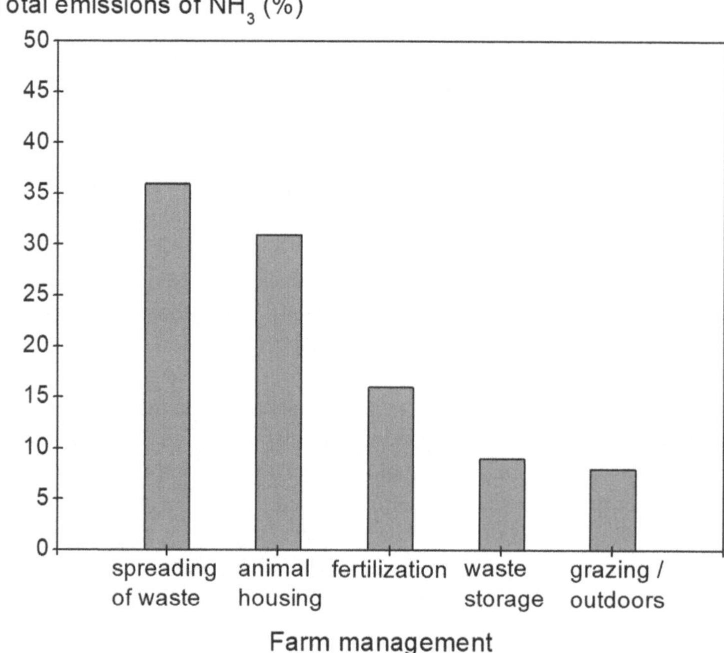

Fig. 1.4 The effect of the kind of human activity on the total emission of ammonia. These data originate from an article by Pain et al. (1998)

1.2.1 Emissions of Gases from Animal Husbandry

1.2.1.1 Types of Animals

Emissions of ammonia and nitrous oxide from various types of livestock kept in the same conditions were presented in Fig. 1.5a–b.

The highest emissions of both gases were observed from cattle (Fig. 1.5a–b), but the lowest values of ammonia were noticed from camels and a little higher – from mules and donkeys (Fig. 1.5a), but in the case of nitrous oxide lower volatilization of this gas was from camels and the lowest – from mules and donkeys (Fig. 1.5b).

The volatilization of ammonia from other animals is shown in Table 1.1.

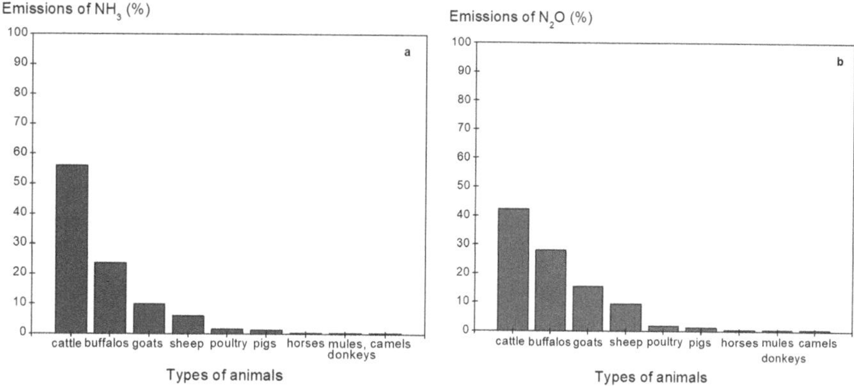

Fig. 1.5 The volatilization of ammonia (**a**) and nitrous oxide (**b**) expressed in % from different categories of animals, according to Aneja et al. (2012)

Table 1.1 Emissions of ammonia from various types of animals

Types of animals	Ammonia emissions (%)	References
Heifers	32.7	Chai et al. (2014)
Steers	31.5	Chai et al. (2014)
Dairy cows	28.5	Van der Hoek (1998)
Suckling cows	14.3	Van der Hoek (1998)
Cows	11.5	Chai et al. (2014)
Bulls	16.2	Aneja et al. (2008)
Suckling calves	3.9	Chai et al. (2014)
Ducks	0.9	Van der Hoek (1998)
Geese	0.9	Van der Hoek (1998)
Turkeys	0.9	Van der Hoek (1998)
Lying hens	0.4	Van der Hoek (1998)
Broilers	0.3	Van der Hoek (1998) and Aneja et al. (2008)
Deer	0.1	Pain et al. (1998)

The highest losses of ammonia were observed from heifers, but the lowest – from deer (Table 1.1). Zbieranowski and Aherne (2012) stated higher emissions of ammonia from cattle than from pigs. Nitrous oxide and ammonia concentrations in ppm were higher from gestating sows than from fattening pigs (Philippe et al. 2015).

1.2.1.2 Number of Animals

Sometimes the content of ammonia in the air is not correlated with the number of animals. The ammonia loss resulting from 12,000 animals was higher than that coming from 6000 but also from 25,000 animals (McGinn et al. 2003), which shows that the type of animals had a stronger effect than their number.

1.2.1.3 Age of Animals

Total emission of ammonia also depended on the age of animals and was expressed in another unit (kg NH_3 animal^{-1} year^{-1}) and achieved the following values for different livestock: dairy cows (28.5), sows (for female adult animals, the emissions of gas from the young animals are included in the given values) – 16.43, other cattle (young cattle, beef cattle, sucking cows) 14.3, horses, mules and donkeys 8.0, fattening pigs 6.39, fur animals 1.69, sheep and goats 1.34, other poultry (ducks, geese, turkeys) 0.92, laying hens and parents 0.37 and broilers and parents 0.28 (Van der Hoek 1998).

1.2.2 Mineral and Organic Fertilization

1.2.2.1 Type of Fertilizer

Ammonia emissions to the atmosphere from fertilizers presents in Table 1.2.

The highest volatilization of ammonia was noticed from urea, applied in a crop field (Aneja et al. 2012), but the lowest – from monoammonium phosphate used to grassland (Whitehead and Raistrick 1990). Emissions of this gas from different fertilizers depended on: the type of ecosystem, the method of using of these fertilizers and the time of their application (of a month or a season) – Table 1.2.

A very important effect has the type of fertilizer on this process. Calcium ammonium nitrate caused higher ammonia and nitrous oxide volatilization than slurry (Bourdin et al. 2014). Higher ammonia values were achieved from urea than from ammonium nitrate (Sutton et al. 1995; Pain et al. 1998) and it was also confirmed in this review. More amounts of this gas were obtained also after the application of urea than slurry (Salazar et al. 2014). It depended on the kind of slurry – more cattle slurry was related to a higher ammonia content in the air, but to a lesser extent, when pig slurry was used (Svensson 1994). Ammonium nitrate with urea in

Table 1.2 The percentage of ammonia volatilization to the air from different fertilizers, which were used in many ecosystems, various times and ways of their application. Explanation of abbreviations

The type of fertilizer, its application in different ecosystems and seasons	NH_3 emissions (%)	References
Urea in a crop field	92	Aneja et al. (2012)
Urea on native prairie grasses	60	Power (1979)
Solid urea, urea solution, urea-KCl in the spring in a field	27–41	Lightner et al. (1990)
Solid urea, urea solution, urea-KCl solution in the summer in a field	12–37	Lightner et al. (1990)
Urea on arable land	30	Pain et al. (1998)
Urea on one meadow	28	Pain et al. (1998)
Urea on grassland	25	Hyde et al. (2003)
Urea on grassland	23	Sommer and Jensen (1994)
Urea in June on temperate grassland	20.8	Watson et al. (1994)
Urea on pasture surface	15–20	Henzell (1971)
Urea on grassland	16.5	Whitehead and Raistrick (1990)
Urea on crops and grasslands	15	Van der Hoek (1998)
Urea in May on temperate grassland	5.5	Watson et al. (1994)
UAN in soil with corn	67	Hargrove et al. (1983)
AS on a surface	55	Hargrove et al. (1977)
AS in a crop field	15	Aneja et al. (2012)
AS on grassland	9.9	Whitehead and Raistrick (1990)
AS to crops and grasslands	8	Van der Hoek (1998)
AS on grassland	< 5	Sommer and Jensen (1994)
AS on a surface of grassland	3.3	Van der Hoek (1998)
Spreading of livestock manure	47.5	Xu et al. (2015)
Manure from livestock housing	15	Hendriks et al. (2016)
DAP on grassland	14	Sommer and Jensen (1994)
DAP to crops and grasslands	5	Van der Hoek (1998)
DAP on a surface of grassland	5	Whitehead and Raistrick (1990)
DAP on grassland	4.9	Whitehead and Raistrick (1990)
DAP on a crop field	4.6	Aneja et al. (2012)
Urine with dicyandiamide in a temperate grassland	12.9	Fischer et al. (2016)
Urine in a temperate grassland	11.2	Fischer et al. (2016)
Urea and AN to crops and grasslands	10	Whitehead and Raistrick (1990)
Urea and AN on a grassland surface	8	Van der Hoek (1998)

(continued)

Table 1.2 (continued)

The type of fertilizer, its application in different ecosystems and seasons	NH_3 emissions (%)	References
AN on arable land	10	Pain et al. (1998)
AN on meadow	7	Pain et al. (1998)
AN to grassland	2.5	Whitehead and Raistrick (1990)
AN to crops and grasslands	2	Van der Hoek (1998)
MAP on a surface of grassland	5	Whitehead and Raistrick (1990)
MAP to crops and grasslands	2	Van der Hoek (1998)
MAP to grassland	1.5	Whitehead and Raistrick (1990)
Cattle dung in a temperate grassland	3.9	Fischer et al. (2016)
CAN on a surface of grassland	2	Van der Hoek (1998)
		Whitehead and Raistrick (1990)
CAN on a surface of grassland, CAN to crops and grasslands	< 2	Sommer and Jensen (1994)
CAN to grassland	1.6	Hyde et al. (2003)

AN ammonium nitrate, *AS* ammonium sulfate, *MAP* mono ammonium phosphate, *CAN* calcium ammonium nitrate, *DAP* diammonium phosphate, *UAN* urea ammonium nitrate

solution reduces the ammonia loss slightly and only due to acidity of ammonium nitrate. No reduction in ammonia loss was found, where only the urea content of the urea-ammonium nitrate solution was present. Potassium nitrate and potassium chloride even on an acidic soil much more reduced ammonia loss from urea than did the acidic ammonium nitrate (Fenn and Hossner 1985). The daily total NH_3-N loss (in kg N ha^{-1}) from fertilizers on a surface applied at 200 kg N ha^{-1} to soil with *Dactylis glomerata* L. in May was the highest from urea granules (62.5), an urea solution (62.3), urea-potassium chloride solution (62.2), urea-calcium chloride solution (49.6), urea-urea phosphate (49.4), ammonium nitrate (only 8.2) and without N (6.3), so it depended on the way of application of fertilizers and their mixtures (Lightner et al. 1990). Ammonia losses are well-known from surface applications of urea on both acidic and alkaline soils (Terman 1979). Calcium and potassium salts added to urea reduced ammonia volatilization (Lightner et al. 1990). A net flux of ammonia from the soil-plant system greatly increased within 4–48 h after urea application on a grazed, tropical pasture depending on initial soil water content (Harper et al. 1983). Ammonia emissions depended also on the ecosystem and was higher from urea application to arable land use than from ammonium nitrate used for grassland (Pain et al. 1998). Higher urea lower ammonium in soils was observed (Tabatabai and Bremner 1972).

Nitrous oxide emissions to the atmosphere from fertilizers are presented in Table 1.3.

Table 1.3 Percentage of nitrous oxide volatilization to the air from different fertilizers, which were used in different ecosystems

Fertilizers in various ecosystems	N_2O emissions (%)	References
Urea in a crop field	74.8	Aneja et al. (2012)
DAP in a crop field	18.8	Aneja et al. (2012)
AS in a crop field	6.2	Aneja et al. (2012)
Anhydrous ammonia in a crop field	2.3	Bouwman (1996)
Cattle slurry and NH_4^+ fertilizers for grasslands	2.0	Velthof et al. (1997)
CAN to grasslands	0.6	Velthof et al. (1997)
AN in crop fields	0.3	Bouwman (1996)
Nitrate (NO_3^-) salts in a crop field	0.2	Bouwman (1996)
Ammonium (NH_4^+) salts and urea in crop fields	0.1	Bouwman (1996)

Explanation of abbreviations: *DAP* diammonium phosphate, *AS* ammonium sulfate, *CAN* calcium ammonium nitrate, *AN* ammonium nitrate

The highest level of nitrous oxide in the air was noticed after the application of urea in a crop field (Aneja et al. 2012), but the lowest – after using of ammonium salts with urea also in the same environment (Bouwman 1996) – Table 1.3.

1.2.2.2 Fertilizer Dose

Ammonia emissions from fertilizers also depended on the rate of nutrients and not always increased with an increase in the dose – for example a higher rate of AN, lower losses of ammonia were observed from the Huston Black clay (Fenn and Kissel 1974). Contrary results were obtained by Felix et al. (2014). A concentration of ammonia increased with an increasing dose of urea ammonium nitrate applied in a corn field. The highest dose of nitrogen in corn grain was after an application of ammonium nitrate, lower – after urea ammonium nitrate and the lowest – after urea (Touchton and Hargrove 1982). Higher rate of pig slurry from fattening pigs at sowing higher ammonia volatilization was and was higher than after the ammonium nitrate application at a lower dose (Bosch-Serra et al. 2014). The highest ammonia concentration was after the application of ammonium nitrate, lower after ammonium chloride and the lowest after ammonium sulfate using at the highest rate and the amounts of ammonia increased with an increasing dose of each fertilizer (Mennen et al. 1996). Higher doses of slurry resulted in higher levels of ammonia emissions. At 12 o'clock the volatilization of this gas was maximal (Menzi et al. 1998).

1.2.3 Storage of Crop Residues, Straw Ballots and Silage

Nitrous oxide concentrations in the air were the highest above a dung heap, a little lower, when a tractor was in the field, high above an organic farm, lower at a poultry farm and the lowest above a compost heap (Hensen et al. 2013).

The highest activity of urease in soil was after the application of manure, a little lower – after the use decomposed leaves, lower after dried undecomposed grass and the lowest when organic matter was absent (Kumar and Wagenet 1984). Losses of NH_3 were the highest from an uncovered field, lower from: straw, oil, PVC foil. The highest ammonia concentrations were above stored separately cattle and pig slurries. These losses correspond to 12% of the total N and 21% of the total ammonia nitrogen in cattle slurry. Corresponding values for pig slurry are 8% of total N and 12% of the total ammonia nitrogen (Somers et al. 1983). The ammonia emissions in Switzerland were higher from: animals, liquid manure (42 Kt of NH_3-N $year^{-1}$) than from artificial fertilizers (3.3), traffic (2.4), cattle on alpine pastures (2.0), industry (1.7), arable land (1.6), people and pets (1.4), wild animals (0.4), an application of sludge (0.9), wild animals (0.4) and natural sources (0.3) – from grassland (Menzi et al. 1997; Buwal 1995). The addition of fresh organic residues can double ammonia losses in many cases, especially at lower doses of urea addition. Soils that are low in fresh organic residues could be surface fertilized with urea with a reduced risk of NH_3 loss (Fenn et al. 1984). Ammonia volatilization from very acidic soils would not occur (Fenn and Hossner 1985). Small amounts of ammonia loss from decaying plant residues (Terman 1979) and maturing plants (Hooker et al. 1980) were noticed. A barley crop recovered 29–41% of ^{15}N added to soil in ryegrass shoots and it was not related to the soil type. Soils fertilized with $^{15}NH_4{}^{15}NO_3$ exhibited the highest ^{15}N recovery (from 61% to 77%) in sandy loam soil and this fertilizer was assimilated by barley at the earing phase – Thomsen (1993).

The application of tannins to compost reduced cumulative emission of nitrous oxide by 17.0% and volatilization of ammonia by 51.0% compared to control, when the emission of nitrous oxide achieved 66.6%, but ammonia 33.4% (Jordan et al. 2015).

1.2.4 Cultivation of Different Plant Species and Cultivars – Biodiversity

The content of mineral N in soil in relation to the growing of winter rape and clover-grass mixture was favorable to the accumulation of mineral N forms in the soil. Less advantageous were potatoes, and the smallest amount of mineral N was found in the soil with cereals and corn grown for green forage (Mazur 1987). The highest amount of net nitrogen = mineral (NH_4^+-N and NO_3^--N) in soil was in *Melilotus alba*, high in *Medicago lupulina*, lower in *Poa pratensis* and the lowest – in control soils (Magid et al. 2001). An ammonia compensation point (nmol NH_3 mol of air^{-1}) was the highest in barley cv. Golf during tilling, but in cv. *Laevigatum* during grain filling and for both cultivars was the lowest after anthesis – depending on the cultivar of barley and the stage of their growth and the ammonia exchange concentration increased with increasing leaf temperature (Schjoerring et al. 1998; Harper et al. 1989). The compensation point for wheat increases from about 13 mg NH_3-N m^{-3} at early grain filling to 23 at late grain filling achieving 40 during senescence (Morgan and Parton 1989). In this case, also the stage of plant's growth had an important significance.

1.3 Climatic Conditions

1.3.1 Air and Soil Temperature

The ammonia concentration increased with increasing air temperatures over a spruce forest (Huber and Kreutzer 2002). Grassland and forest soils have less negative effects on the environment by decreasing the production of NO_3^- and the risk of NO_3^- leaching under increasing temperature by global warming (Lang et al. 2010). Higher air temperatures also meant higher amounts of ammonia in the atmosphere (Blunden and Aneja 2008; Barthelmie and Pryor 1998; Hu et al. 2014; Shen et al. 2011; Phillips et al. 2004; Robarge et al. 2002; Wichink Kruit et al. 2007) and were 5 μg m^{-3} at 20–25 °C, but at −5 °C only above 2 μg m^{-3} (Huber and Kreutzer 2002). Higher air temperatures, higher the nitrous oxide concentrations in the air also occurred (Bai et al. 2014; Abalos et al. 2016).

Higher soil temperatures near roots of ryegrass higher concentrations of ammonium and nitrate in plants occurred (Clarkson and Warner 1979) and more ammonia (Svensson 1994) and higher the activity of urease (Kumar and Wagenet 1984). The highest urease activity was at 70 °C when the highest release of ammonium from plants occurred (Frankenberger and Tabatabai 1982). Ammonia efflux after the urea application was more intense under high soil temperatures than under lower ones and more rapid with higher initial soil water contents (Harper et al. 1983). The higher temperature may have contributed to the high ammonia volatilization from dry soil (Sommer and Christensen 1992). The largest initial losses occurred at the highest temperature, with slower ammonia losses at the lower temperatures (Fenn and Hossner 1985).

Fenn and Kissel (1974) showed that 8–10% of soil calcium carbonate content was necessary for maximum ammonia losses, which depended on the ammonium sulfate application dose. Application of fertilizer in large granules will create high local concentrations of reaction products which can result in a maximum pH increase. An application of ammonium sulfate as a liquid at the lowest possible level and the soil pH would show the least increase and the lowest ammonia loss. If diammonium hydrogen phosphate reacts to form calcium hydrogen phosphate, the solubility of the calcium reaction product is much lower than the solubility of calcium sulfate and it should produce a slightly higher rate of ammonia loss (Fenn and Kissel 1973). Urea will produce ammonia losses in acidic and alkaline soils similar to losses of the reactive inorganic N compounds will produce in calcareous soils. Since urea is hydrolyzed to ammonium carbonate by urease, ammonia loss will be high both on acidic and calcareous soils (Fenn and Hossner 1985). Urea-Ca and urea-KCl products greatly reduce NH_3 losses (Fenn et al. 1981, 1982b). The retentive capacity of the soil for cations will exert limited control on ammonia losses from surface-applied N fertilizers (Fenn and Hossner 1985). An addition of calcium with urea to reduce the loss of ammonia was found to be most effective in soils of low cation exchange capacity (Fenn et al. 1982a). A loss of ammonia from the applied urea was 5.5% in May and 20.8% – in June. N-(n-butyl) thiophosphoric triamide (nBTPT) inhibited in 97% the volatilization of NH_3 from urea and increased the dry matter yield by 9% as compared to an application of urea alone and increased the shot recovery of ^{15}N to 80.9%.

Total ^{15}N recovery in the soil – plant system was 17% over by urea with nBTPT. It has economic significance and for market grassland (Watson et al. 1994) and is also beneficial for the purity of the atmosphere.

1.3.2 Relative Humidity of Air and Soil

Higher relative air humidity was, the lower the ammonia content was e. g. over spruce forest (Huber and Kreutzer 2002; Barthelmie and Pryor 1998). A moisture loss and a loss of ammonia from slurry were noticed. An increased content of NH_3 at the slurry surface is observed in drying conditions (Sommer et al. 1991). This dependence is similar in the case of relative air humidity and soil moisture. Water content in soil as soil moisture is important. Ammonia volatilization increased as soil water content increased. The extent of NH_3 volatilization is determined by the soil pH, texture, temperature, moisture, exchangeable cations, fertilizer source, and rate of application (Fenn and Hossner 1985; Ferguson and Kissel 1986). A loss of ammonia from granular urea applied to calcareous soils was related to the initial moisture content and increased by 8% for every 10% increase in initial soil moisture. Higher cumulative water evaporation and higher cumulative ammonia volatilization were also observed (Al-Kanani et al. 1991). Nitrogen losses from urea and an urea ammonium nitrate solution through ammonia volatilization were significantly influenced by the initial soil moisture content. The maximum ammonia loss occurred at initial soil water potential of – 0.01 MPa, regardless of drying or water replenishment. Increased clay content of the soil reduced an NH_3 emission possibly due to adsorption of ammonium (Al-Kanani et al. 1991). Ammonia losses from dry and wet soils were 20% and 50% of injected ammonia. From a dry soil losses of gaseous ammonia took place within the first hours after injection, but from a wet soil ammonia was lost more gradually between 6 h and 6 day (Sommer and Christensen 1992).

1.3.3 Wind Speed and Direction

At east and north-northeast wind directions ammonia concentrations were the highest and at 4–5 m s^{-1} wind speed and then decreased over forests with spruce (Huber and Kreutzer 2002). High ammonia amounts were observed when the direction of wind was south-southwest and west (Viguria et al. 2015), when was wind direction of 278° and to 1.5 m s^{-1} wind speed (Robarge et al. 2002). The highest ammonia concentration volatilized occurred for wind direction 320° and 4 m s^{-1} wind speed, but the lowest at 300° and 6.3, respectively (Hovmand et al. 1998). Schjoerring (1995) achieved the opposite results – the highest ammonia concentrations above crop fields were at wind speeds of 5–6 m s^{-1}, but the lowest near 2–3 m s^{-1}. Higher ammonia concentrations were at lower height than at higher – above wheat, pea and oil-seed rape, but not in the case of barley. It depended on species of plants. Ammonia concentrations increased only to 4–5 m s^{-1} of wind speed and then decreased at higher wind speeds and at north-northeast and east wind directions were the highest (Huber and Kreutzer 2002). Contrary results were received by Welch et al. (2005), the highest ammonia amounts were at approx. 6–7 m s^{-1} wind speed and at the wind direction of 360°. The higher the wind speed was the lower the concentrations of ammonia were observed (Sommer et al. 1991; Sutton et al. 2003). At the height of 1.5 m less ammonia than at 0.5 m was observed (Denmead et al. 2008). From 210° to 250° of the wind direction maximal ammonia concentrations occurred. At wind direction of 278° and wind speed of 1.5 m s^{-1}, at increasing air temperatures higher amounts of ammonia were noticed (Robarge et al. 2002). At 180° wind direction the ammonia concentrations were the highest above Auchencorth Moss (Fowler et al. 1998). The highest ammonia concentration in December was at 330° wind direction and 8 m s^{-1} wind speed, but in May at approx. 240° and 10 m s^{-1} (Hu et al. 2014). When the wind speed was the highest (7 m s^{-1}) a NH_3 loss was low at 12 o'clock, but approx. 9 p.m. the NH_3-N concentration was the highest when the wind speed was low (near 3 m s^{-1}) – McGinn et al. (2003). A concentration of ammonia in the air not only depended on the wind speeds and directions, but sometimes more on a type of ecosystem, species of plants and the height above a ground and season.

1.3.4 Precipitation

The amount and distribution of rain after the urea application on a grazed tropical pasture appeared to control the total ammonia loss from urea. Rain reduced ammonia efflux by dispersing the urea in the soil and thus limiting the development of very high NH_3 and NH_4^+ concentrations in a soil solution around urea prills. The ammonia efflux from such sites would be very rapid (Harper et al. 1983). Rainfall would move the unhydrolyzed urea into the soil where it would hydrolyze with less possibility of ammonia loss. Lower losses of ammonia for cooler temperature may be partially due to rainfall terminating the NH_3 loss conditions (Fisher and Parks 1958). Fox and Hoffmann (1981) found that if 10 mm rainfall fell within 3 days, less than 10% of NH_3 loss from a surface-applied urea occurred. However, if 3–5 mm fell within 5 days or 7–9 mm within 9 days, ammonia losses were from 10% to 30%. No rainfall within 6 days after an surface application resulted in a loss of ammonia higher than 30%. Ernst and Massey (1960) speculated that when soil was dried within 4–5 days after the urea application hydrolysis and the subsequent ammonia loss were inhibited and the NH_3 loss was directly related to the initial soil moisture content. The data are similar for inorganic N fertilizer compounds. When 12 mm of precipitation was received within 96 h after urea application, losses of ammonia were reduced in the semi-open system from 18% to 9% in a 50 year old Douglas fir stand with a dominant tree height of 26 m (Marshall and Debell 1980).

With higher precipitation, lower nitrous oxide concentrations in the air occurred (Abalos et al. 2016). In the wet season, the emissions of nitrous oxide from soil were higher than in the dry season (Bai et al. 2014).

1.3.5 Insolation

The highest ammonia fluxes occurred during the period of highest solar radiation over grazed tropical pasture. Net ammonia transport over this ecosystem is affected by soil and microclimate conditions (Harper et al. 1983).

The nitrous oxide emission in different ecosystems and plant species was the highest after plants treatment with UV-A, a little lower after UV-B, lower after PAR and when plants were exposed on sunlight, but the lowest when plants were in darkness. Nitrous oxide concentrations are also higher when air temperature is increasing (Bruhn et al. 2014).

Light has an effect also on the activity of nitrate reductase. In light the enzyme activity increases, but in darkness decreases (Gniazdowska-Skoczek 1988a, b; Nicholas et al. 1976; Somers et al. 1983).

1.4 Other Factors

1.4.1 pH and Type of Soil

The pH values of ground water under a shelterbelt and adjoining cultivated fields were neutral and ranged from 6.65 to 7.91 (Jaskulska and Szajdak 2010). New adjoining cultivated fields had lower pH than older near *Robinia pseudoacacia* and *Crataegus*, but the new shelterbelt had higher pH than the shelterbelt with *Crataegus* and *Robinia pseudoacacia* (Jaskulska and Szajdak 2010). The highest pH of ground water was in a new shelterbelt (7.05–7.57) and the lowest in the *Robinia pseudoacacia* shelterbelt (6.69–7.39). The highest pH was in the *Crataegus* adjoining cultivated field and the lowest in the *Robinia pseudoacacia* adjoining cultivated field (Jaskulska and Szajdak 2010).

A higher pH of soil, lower ammonia values were observed (Blunden and Aneja 2008). Ammonium salts increased the pH from 5.5 to 7.4 of different soils, 23 times more ammonia from soil number 5 and 2, but after the application of urea only 2 times less. From soils numbers 2 and 3 at pH of 5.5 to 6.1 diammonium hydrogen phosphate increased the pH of soil number 3. Ammonium carbonate increased pH of the soils to 7.5 and 7.9. The higher ammonia emissions occurred at pH of 8.0 than 5.0 after using the mixture of ammonium dihydrogen phosphate, diammonium hydrogen phosphate, ammonium sulfate, ammonium nitrate and urea with 4 soils. Soils number 1 and 2 are Batcombe, number 3 – Hucklesbrook, 4 – Frilsham and 5 – Andover (Whitehead and Raistrick 1990). The impact of fertilizers on ammonia losses (in % of the N applied) was the following for the soils with numbers:

1. pH 3.7 from diammonium hydrogen phosphate (1.4%), urea (0.6%) and ammonium dihydrogen phosphate, ammonium sulfate and ammonium nitrate (< 1%),
2. pH 5.5 from diammonium hydrogen phosphate a NH_3 loss was greater than from urea (23.5%), ammonium sulfate, ammonium nitrate (< 1%) and ammonium dihydrogen phosphate (< 0.5%),
3. pH 6,1 from urea, diammonium hydrogen phosphate a NH_3 loss was higher (4.0%) than from ammonium sulfate, ammonium nitrate (< 1%) and ammonium dihydrogen phosphate (< 0.5%),
4. pH 7.1 from ammonium sulfate a gas loss was higher (31.9%) than from diammonium hydrogen phosphate and urea, from ammonium nitrate (8%), from ammonium dihydrogen phosphate (< 0.5%),
5. pH 7.4 from urea (43.1%) a NH_3 loss was higher than from: diammonium hydrogen phosphate, ammonium sulfate and ammonium dihydrogen phosphate (34.6%), but from ammonium nitrate – the lowest (10%).

The activation of fertilizers depended on the soil pH and type. The ammonia loss from surface-applied ammonium nitrate and ammonium chloride is low on calcareous soils (Fenn and Hossner 1985).

The higher pH of soil (about 9.0), the higher the urease activity, but lower and higher than 10.0 caused lower the activity of this enzyme (Tabatabai and Bremner 1972). The activity of urease was the highest at the pH of buffer 7.5 (Frankenberger Jr and Tabatabai 1982).

Soils under a shelterbelt without the addition of urea were very acidic and pH ranged from 3.48 to 3.70. While under adjoining cultivated fields to the *Robinia pseudoacacia* shelterbelt were weakly acidic to neutral and the pH ranged from 6.17 to 7.08. After the addition of urea in these both soils, the pH increased under a shelterbelt from 5.72 to 7.48 and under an adjoining cultivated field pH values were from 7.22 to 8.21.

The biggest differences in the nitrate concentration were found in the deeper soil layers in the field with wheat. The soil solution at the depth of 55 cm had a several time higher concentration of nitrate in comparison with the soil under a shelterbelt. In the shelterbelt, NH_4^+ amounts were 2–4 times higher down to the depth of 85 cm. A soil solution from the deeper layers of the soil profile of the arable land was 2–4 times richer in nitrate in comparison with the concentration of the NH_4^+ form (Bartoszewicz 2000).

Losses of N and N_2O were greater in the mineral soil than in the peat soils, with losses of 3% and less than 1% of N applied, respectively after 100 days. The NO_3^--N amount was detected only in leachates from mineral soils after the urine application (Clough et al. 1996). Urea formaldehyde, sulfur coated urea and coated calcium nitrate in incubated sphagnum moss peat released 3 and 20% of the applied N in 6 weeks. The losses from NPK fertilizers were 20 times more than from slow release N, except of Osmocote® and Duna, which released 30–40% of the applied N as mineral nitrogen. It was noted that temperature did not influence on nutrients release from NPK fertilizers (Engelsjord et al. 1997). Ammonium sulfate and diammonium hydrogen phosphate react with calcium carbonate to produce an increase in soil solution pH and an ammonia loss. The inorganic fertilizers are not susceptible to gaseous ammonia loss in acidic soils. Urea is biologically hydrolyzed and ammonium carbonate is formed. For all types of soil high ammonia losses may occur as a result. For potential control of ammonia losses are the use of acids urease inhibitors, and addition of Ca, Mg, K salts with urea (Fenn and Hossner 1985).

1.4.2 Time: Day, Night, Season of the Year

Ammonia concentrations (in μg m^{-3}) depended on season of the year and were in: spring (12.3), summer (30.5), autumn (13.3) and winter (6.2), so the highest amount of ammonia was in summer, but the lowest in winter – Shen et al. (2011) and also Thöni et al. (2004). More time of incubation, more ammonium was in soil in dependence on its type (the highest in Ida soil, lower in Marshall and the lowest in Edina soils) – Tabatabai and Bremner (1972). On day 21, after the application of urea the highest volatilization of ammonia was observed (Rachhpal and Nye 1988). More hours, since manure application more cumulative NH_3-N loss was and the lowest from the compost. It was higher from surface ammonia emissions and were higher than from tilled 15 cm depth. Similar trends were obtained also for inorganic nitrogen (McGinn and Sommer 2007).

Amounts of nitrate during the vegetation season remained on a considerably higher level than in the autumn-winter period (Bartoszewicz 2000).

Mean values of ammonium in the ground water of area planted with trees and in the case of cultivated field during winter and summer time were almost identical (Bartoszewicz 2000). More days after manuring with slurry higher ammonia and ammonium losses were above and from winter wheat (Dosch and Gutser 1996).

At daytime, in the late morning the ammonia concentrations were higher than at nighttime (Phillips et al. 2004; Huber and Kreutzer 2002). The highest N_2O-N flux concentration was approx. at 2 p.m. hour above fields with wheat-corn and cotton, when the air temperature was also the highest (Liu et al. 2014). The smallest nitric oxide fluxes were found in the early morning hours from 4 to 8 a.m., whilst peak emissions were found at noon (from 12 to 2 p.m.) – Medinets et al. (2016).

NO-N emission depended on plant species and was 0.5% for barley (Laville et al. 2011) and cotton (Cruvinel et al. 2011), 0.14 and 1.46% for wheat and corn (Cruvinel et al. 2011; Liu et al. 2011; Cui et al. 2012) and 0.6% for sugarcane (Paton-Walsh et al. 2011). The optimum for nitric oxide emission was found at the soil temperature range of 10–20 °C and dissolved inorganic nitrogen concentrations

(15–18 mg N kg^{-1} sdm) for a wide range of soil moisture levels (approx. 25–80%) – Medinets et al. (2015). The highest nitric oxide emission rates were observed during summer and the lowest – during the winter (Gasche and Papen 2002).

Mean NH_3-N losses from calcium ammonium nitrate were 85% lower than from urea. A maleic and itaconic acid polymer did not decrease the NH_3-N emission, but N-(n-butyl) thiophosphoric triamide caused a 78.5% reduction and then combined with dicyandiamide a 74% reduction compared with urea alone. Mean spring and summer losses were similar, but in spring more variable in Irish temperate grassland (Forrestal et al. 2016). Daily ammonia concentrations in ppm decreased with increasing air temperature, but nitrous oxide increased (Stinn et al. 2014).

1.4.3 Spatial Changes in the Concentration of Nitrogen Forms – Horizontal and Vertical Gradients

A horizontal gradient of ammonia concentration was observed. These amounts of gas decreased with a greater distance from the animal farm and when 1100 cows were in a barn near a farm fence the ammonia concentration in μg m^{-3} was 40,100 m Westerly the farm 22, at a forest margin 12 and a southwest wind direction was. When 400 cows were in a stable: 6, 11 and 8 μg of NH_3 m^{-3} were noticed and southwest and southeast wind directions were, respectively. More amounts of ammonia were above uncovered than covered wind-hoods in stables (Adema et al. 1993). The concentration of ammonia decreased with an increasing distance from the dairy barn at 50 m was above 40 μg m^{-3} and at 200 m 20 μg m^{-3} (Felix et al. 2014). Pogány et al. (2012) horizontal and vertical gradients of the concentration of ammonia observed near a cattle farm with 425 animals. Manure from buildings was transported to an uncovered farmyard manure stack, but liquid fraction of cattle manure was stored in an uncovered slurry tank, both near the farm buildings. The highest ammonia concentration in the air was 46 m from this farm (60 μg m^{-3}) when the west wind was blowing from the farm. The smallest amount of this gas was far from the farm above a field with winter wheat (2 μg m^{-3}). The concentration of ammonia was the highest at the height of 2.4 m above ground and the lowest at the height of 0.5 m. Hovmand et al. (1998) measured at different climatic conditions the N-NH_3 concentration in μg m^{-3} and the highest (0.76) was at 334° wind direction and at a 4 m s^{-1} wind speed, between 3–6 hours p.m. and at 15.5 °C, but the lowest (0.04) was at: 300, 6.3, between 9–12 h a.m. and at 13.4 °C, respectively. Amounts of this gas decreased with an increasing distance from forest edge.

A vertical gradient of ammonia concentrations was also observed. When the height above ground was higher (1.5 m), a higher ammonia concentration in μg of this gas m^{-3} (20) was detected for barley than at a lower height of 0.5 m (15). Contrary results were obtained for wheat, pea and oilseed rape above plant canopy, so it depended also on plant species. The wind speed was greater at a higher height of

1.5 m, but ammonia concentrations were lower in this case (Schjoerring 1995). At lower heights, the ammonia loss was higher (Denmead et al. 1982). It was also confirmed by Husted et al. (2000) above oilseed rape canopy. In other ecosystems, reverse results were obtained by different authors. The ammonia flux was measured at height of 30 m in forest (with a value of 1860) – Duyzer et al. 1994, at 4 m in heathland (810) – Erisman et al. (1994) and at a height of 1.5 m in heathland (550) – Duyzer et al. (1989). Genermont et al. (1998) contrary results obtained and stated that a normalized ammonia concentration decreased with higher height, but after slurry spreading under field conditions.

The nitrogen dioxide concentration decreased with a higher height: at 3 m the lowest amount of this gas occurred, but at 1 m – the highest (Felix and Elliott 2014).

At the layer of soil 90 cm nitrate nitrogen was in higher concentration than when was deeper (130 cm) below the ground surface (Karlen et al. 1998). This was also confirmed by Abbasi and Adams (1999) and for ammonium nitrogen, too in grassland soils. Contrary for urea – in deeper layers of soil less of this fertilizer was found (Rodgers and Pruden 1984). In the case of oilseed rape canopy, NH_4^+ was at higher concentrations when at higher depths of soil it was measured (Husted et al. 2000).

1.5 Different Gases Occurring in the Atmosphere

1.5.1 Ammonia

The highest concentrations of ammonia (2.7–6.3 μg NH_3 m^{-3}) occurred in agricultural areas with intensive livestock farming, but low – in forests (0.6–1.6) – Thöni et al. (2004). The highest amounts originated from the application of manure and from a cattle grazing. In agricultural areas, ammonia concentrations achieved values in the range of values from 2 to 5 μg of NH_3 m^{-3} (Farquhar et al. 1980). No detectable loss of ammonia occurred from the ammonium nitrate fertilizer applied in March and April 1993 and also in 1994 to winter wheat. In the period from mid-May until August during the generative stage of growth the main part of ammonia was emitted. After anthesis, wheat plants, which had received 25% of reduced N supply started to emit more ammonia than plants fertilized with the standard N amount and had a relatively high shoot dry matter N concentrations during the grain filled period of the plants at a reduced N supply. In 1994, the total ammonia emission from winter wheat was about 5 kg of NH_3-N ha^{-1} or more than double that in 1993. Approximately half of the ammonia loss took place before anthesis in the second half of June. The pre-anthesis ammonia emission peak was 0.45 kg of NH_3-N $week^{-1}$ in May 1993 and 2.5 kg of NH_3-N ha^{-1} in 1994 prior to stem elongation. From anthesis to maturity in 1993 the ammonia emission was approx. 1.0 kg of NH_3-N ha^{-1} (Schjoerring and Mattsson 2001). The first application of urea in March resulted in 1993 in a fertilizer – derived ammonia loss of 5 kg N ha^{-1} (to 8% of the applied N) while in 1994 only about 1.5 kg N ha^{-1} (3% of the applied N was lost), because in

1994 was 13 mm of rainfall immediately when the urea was applied, so that the urea was transported into the soil. The second round of urea application in late April resulted in a loss of 7–8% of the applied N in both years. Total amounts of urea – derived NH_3 – N lost were 12 and 9 mg N ha^{-1} in 1993 and 1994, respectively, corresponding to about 7% of the applied urea – N in both years. In 1994, there was a post – anthesis foliar – derived ammonia loss from urea – fertilized wheat amounting to approx. 3 kg NH_3-N ha^{-1}. During flowering of winter oilseed rape and after the application of fertilizer in April 1993 1–2 kg of NH_3-N ha^{-1} was volatilized (0.5–1.0% of the ammonium nitrate) applied late March 1993. Later, during flowering, the rapid increase in ammonia emissions from another field in 1993 (4 kg of NH_3-N ha^{-1}) was due to volatilization of about 7% of the liquid mixture of ammonium nitrate, urea and organic N applied early in May. In April 1993 a plant – based emission of 1 kg NH_3 ha^{-1}. In 1994, there was no loss from the ammonium nitrate applied late March, while about 1% (approx. 1 kg of NH_3-N ha^{-1}) of ammonium nitrate and liquid ammonium sulfate applied in the beginning of May was lost. Also in 1994 in field-IV, the application of Roundup resulted in a 4-fold increase in atmospheric ammonia concentrations above the crop. In late April and in May 1993 a net deposition of approx. 0.1 kg NH_3-N ha^{-1} $week^{-1}$ occurred, when the field had received 13 kg of N ha^{-1} in ammonium nitrate as part of the NPK fertilizer applied on April 17th in 1993. In early June, due to drought and N-deficiency accelerated anthesis, ammonia emission increased to about 0.3 kg of NH_3-N ha^{-1} $week^{-1}$. After rainfall, ammonia losses achieved 0.5 kg of NH_3-N ha^{-1} $week^{-1}$ at the end of July and declined until barley plants were fully senescent in mid-August.

The total amount of NH_3-N loss from the spring barley crop in 1993 was about 3 kg ha^{-1}. The plants emitted about 1 kg of NH_3-N ha^{-1} in growth stages prior to stem elongation and during anthesis 0.8 kg of NH_3-N ha^{-1} $week^{-1}$. The rate of ammonia emission declined towards maturity except for pea at the beginning of August, where the plants were spray-killed with Roundup. The total amount of NH_3-N lost from spring barley in 1994 was 5 kg of this ha^{-1}. The weekly rate of ammonia emissions from the pea crops amounting to 0.3 kg of NH_3-N ha^{-1} $week^{-1}$ in 1993 between flowering and maturity, while in 1994 the whole growing period was induced and a substantial ammonia emission followed spray-killing of the crop in the end of July 1994. The calculated loss amounted to 2 and 5 kg of NH_3-N ha^{-1} in 1993 and 1994, respectively. For wheat, oilseed rape and barley the highest of ammonia emissions were measured in the year where the above ground N content at anthesis was the highest (in 1993 for oilseed rape and in 1994 – for the cereal crops). There were no indications that ammonia losses were correlated with crop N status at maturity. There was no relation between the amount of N fertilizer applied on the accumulated ammonia loss over the whole growth cycle (Schjoerring and Mattsson 2001). Relatively small emissions (1–2 kg of NH_3-N ha^{-1}) per season were measured from spring barley growing at 3 different rates of nitrogen application (Schjoerring et al. 1993).

Animal types in the total ammonia emissions from livestock in China (in %) were: poultry (34), pigs, cattle (29), sheep (4) and rabbits, mules, donkeys and horses separately (1) – Xu et al. (2015). The total ammonia emission form a livestock achieved 80–95% and from fertilizers 3–21% in Europe (Van der Hoek 1998). The largest ammonia loss occurring during a 20 day period after the ammonium nitrate fertilizer application (11.4% of the applied fertilizer) from the soil and plants. Additional ammonia losses were observed from the wheat plants between anthesis and harvest (9.8%). During the plant senescence 11% of ammonia was from stems and leaves. Plant N after anthesis – about half of the grain N was from the redistribution within the plant with the balance assimilated directly from the soil. Concentrations of nitrogen compounds in leaves and stems were higher, but in roots lower. The large proportion of N in the grain came directly from the soil after anthesis from mineralized organic N, which may have management potential for increasing grain N content. Nitrogen was lost from wheat after the application of fertilizer (about 21% equivalent of the applied fertilizer – ammonium nitrate) and 1% – from the plant absorption during the senescence period. Total plant N was in the highest amount in grain, a little less in leaves, lower in stems and the lowest in roots (Harper et al. 1987). Some ammonia in soils is utilized by plants, most of it is first oxidized to nitrite and to nitrate by soil microorganisms. When the cation exchange capacity is low, less NH_4^+ will be absorbed, a greater amount of ammonia will be in soil solution, and toxicity will be increased to *Nitrobacter*. When the pH of soil increases, ammonia may accumulate to level toxic to the nitrite oxidizing organisms – from 12.8 ppb (Smith 1964). Factors favoring ammonia loss from flooded soils and well-drained soils include high urease activity, solution pH and temperature, elevated rates of an urea application, a surface application of urea and low cation exchange capacity (Freney and Simpson 1981). Maximal conversion of urea to ammonia occurred at about pH 8.0 under flooded soils (Delaune and Patrick 1970). Volatile ammonia loss ranges from approx. 3% to 10% for ammonium sulfate and from 5% to 50% when urea was the source of nitrogen (MacRae and Ancajas 1970; Mikkelsen et al. 1978; Vlek and Stumpe 1978).

Vlek and Craswell (1979) measured losses of up to 50% of the applied urea as ammonia over a 2–3 week period from flooded soils with 4–5 cm of standing water. They found approx. 60% losses of the applied N from ammonium sulfate and over 50% of the urea-N in the floodwater immediately following the application of these fertilizers. Floodwater pH increased from 6.45 to 7.15 in the control and to near 8.0 for the urea treatment. When urea fertilizer is added to the forest floor, the soil pH is high and loss of fertilizer N as ammonia also (Fenn and Hossner 1985). Losses from forest soils of ammonia are less than 5% (Overrein 1968) to more than 20 and 40% (Acquaye and Cunningham 1965; Watkins et al. 1972).

Chai et al. (2014) the following annual ammonia volatilization (in kg N animal^{-1} year^{-1}) gave for various animals types: steers, heifers (38), bulls (14), heifers (10), beef cows (9), calves (5).

Total NH_3-N losses in kg ha^{-1} season^{-1} were higher from urea than urea ammonium nitrate from the sugarcane field under retained than burned residues (Dattamudi et al. 2016). Cumulative NH_3-N volatilization losses in kg ha^{-1} was higher from urea in fall and a little lower in spring than from dairy slurry and higher from dairy slurry in summer and the lowest in winter. The highest losses form urea were in fall, but the lowest – in winter. More important was the type of fertilizer than the season (Salazar et al. 2014). The fertilizer N-input higher, the NH_3-N loss higher (Misselbrook et al. 2000). Ammonia emissions from: urea, diammonium hydrogen phosphate, ammonium sulfate and calcium ammonium nitrate were: 25, 14, less than 5 and less than 2%, respectively after 15–20 days (Sommer and Jensen 1994). Acetylene inhibited ammonia emissions from field plots grown with pea and barley (Bertelsen and Jensen 1992).

1.5.2 Nitrous Oxide

Global annual emissions of N_2O-N from a fertilized cropland were higher than from grassland (Stehfest and Bouwman 2006). The higher N deposition was correlated with higher nitrous oxide emission – the highest above deciduous forest, higher over coniferous forest and lower and similar above grassland and wetlands. It depended on the type of ecosystem and species of plants (Bühlmann et al. 2015). The highest amounts of nitrous oxide were above artificial lands, lower above croplands and the lowest in forests (Nicolini et al. 2013). Total N_2O-N losses were higher in beech forests than spruce forests (Butterbach-Bahl et al. 2002).

Nitrous oxide emission from anhydrous ammonia was the highest (2.3%), from ammonium nitrate high (0.3%) and from urea – the lowest (0.1%) – Bouwman (1996).

Abalos et al. (2016) also confirmed that the higher volatilization of this gas was from ammonium nitrate than urea. The higher the dose of rapeseed cake manure was in kg N h^{-1} $year^{-1}$ the higher the loss of nitrous oxide was (a positive correlation was observed) during the 3 cycles from April 2011 to April 2014 above rice (Liu et al. 2015). Nitrous oxide cumulative emissions depended on a type of manure and were the highest for a layer of manure, high – for broiler litter, low – for pig farmyard manure – FYM and the lowest – for cattle FYM at Drayton in 2003 and Gleadthorpe sites (Webb et al. 2014). Nitrous oxide losses were the highest (1.81 kg ha^{-1}) when coming from fertilizer urea at 180 kg N ha^{-1} dose, a little lower (1.29) from urea in the rate of 144 kg ha^{-1} and organic manure (60 kg above a rice field ha^{-1}) and the lowest (0.55) from organic manure at a rate of 300 kg ha^{-1} (Zhao et al. 2015). It is a mean from 2 years 2012 and 2013 in comparison with control (0.29) – untreated with fertilizers. In intensively managed grasslands the nitrous oxide emission was the highest after the application of NH_4^+ fertilizers and cattle slurry and was 2%, from calcium ammonium nitrate 0.6% and from NH_4^+ and NO_3^- only 0.1% of the applied N. During cold and dry conditions in early spring an emission from both N forms was small – less than 0.1% and large after the application to a poorly drained sand soil during the wet spring. A total of 5–12% and 8–14% of the applied N was lost as nitrous oxide by denitrification. This gas emission depends on the N fertilizer, the type and moisture of soil (Velthof et al. 1997).

Biochar decreased nitrous oxide emissions from urea up to 54% and 53% during the rice and wheat seasons and increased grain yield and biomass and the production of rice and wheat by 12 and 17% and also increased soil retention. Biochar increased crop production and decreased the nitrous oxide emissions (Wang et al. 2012). The same inhibitor decreased the total nitrous oxide emission from the corn field and poor calcareous loamy soil and limited the total global warming potential, increased a total N in soil and had no effect on a soil mineral N contents (Zhang et al. 2012). The emission of nitrous oxide was the highest from urea, lower from ammonium nitrate and the lowest from the ammonium sulfate. Inhibitors: dicyandiamide, nitrapyrin, encapsulated calcium carbide caused lower and mean values. An emission of nitrous oxide was higher from urea above corn than wheat, but after adding inhibitors to urea above wheat (Mosier et al. 1996).

A very high emission of nitrous oxide above sugarcane occurred at high N inputs and high rainfall and high temperatures over summer and was temporal and spatial variability (Reeves et al. 2016).

Total nitrous oxide production depended in 100% on soil characteristics and a content of calcium carbonate, water soluble carbon and an amount of sand and increased when nitrate and ammonium increased (Vermoesen et al. 1996).

1.5.3 Nitrogen Dioxygen

Total NO_2-N losses were higher in beech forests than in spruce forests (Butterbach-Bahl et al. 2002). Felix and Elliott (2014) gave the following nitrogen dioxide concentration (in ppb) in dependence on a source: vehicle exhaust (50.8), a field with corn – fertilized soil with urea ammonium nitrate (19.8), poultry facility – Turkey waste (7.1), dairy barn – cow waste (5.5) and cattle – cow waste (4.0). The highest nitrogen dioxide concentration (approx. 13 ppb) was when south-southwest wind direction was (Mouzourides et al. 2015) and the highest were in winter, but the lowest – in summer in South-East Mediterranean climatic conditions. The nitrogen dioxide concentration decreased with increasing wind speed, increased with increasing air temperature and in winter was the highest, but in summer – the lowest. Low wind speeds (below 2 m s^{-1}) were associated with high levels of NO_x (Grundström et al. 2015). An accumulation of NO_3^- or NO_2 in the needles and the addition of nitrate to the soil could cause NO_x and NO_2 emissions from boreal Scots pine forests.

1.5.4 Nitric Oxide

Total NO-N losses were higher in beech forests than in spruce forests (Butterbach-Bahl et al. 2002). Global annual emissions of NO-N from a fertilized cropland were higher than from another ecosystem (grassland) – Stehfest and Bouwman (2006). The highest nitric oxide fluxes were at night and at high humidity (Joensuu et al. 2015). The nitric oxide emissions from a green manure (mustard) used before oilseed rape growth were higher than from fallow plots and were related with the soil ammonium concentration (Vos et al. 1994). The direct volatilization factors of nitric oxide increased nonlinearly with increasing N rates in fertilizers above winter rice (Zhao et al. 2015). Nitric oxide emissions increased with higher forest floor temperatures (Fowler et al. 2009). The highest amounts of NO (in µg m^{-3}) in the air were in winter, but the lowest – in summer. The highest cumulative stomatal flux of nitrogen dioxide and nitric oxide were in spring, but the lowest in autumn for *Eucalyptus citriodora*, *Acacia auriculiformis* and *Schima superba* (Hu et al. 2016). The nitric oxide emission was the highest above forest, lower above grassland and the lowest above wetland in Switzerland, so it depended on the kind of ecosystem (Bühlmann et al. 2015). The production of total nitric oxide depended on the content of calcium carbonate, NH_4^+ concentration in soil and soil pH and nitrification for 97% on the soil characteristics and 3% on the quantity of nitrogen added.

At low pH of soil the nitric oxide amount was the highest from nitrate, at higher pH values – from ammonium. A low pH has a negative effect on nitrification. The nitric oxide emission also increased with increasing nitrate and ammonium applications (Vermoesen et al. 1996).

1.6 Nitrogen Forms Occurring in Plants and Soils in Various Ecosystems: Water, Meadow, a Shelterbelt and a Crop Field

1.6.1 Nitrate

The lowest values of nitrate were observed in ground water under the *Robinia pseudoaccacia* shelterbelt (3.0–3.5 mg L^{-1}) and the highest under adjoining cultivated fields located near new shelterbelts (19.78) and the *Crataegus* shelterbelt (18.51). All shelterbelts had less N-NO_3^- than adjoining cultivated fields. The highest amounts were in a new shelterbelt (15.82 mg L^{-1} N-NO_3^-), lower in the *Crataegus* shelterbelt (14.11) and the lowest in the *Robinia pseudoacacia* shelterbelt (3.35). The same trends were in the case of adjoining cultivated field and the values were: 19.78, 18.51 and 8.43, respectively. Nitrate in ground water from adjoining cultivated fields were in the highest amounts. The content of N-NO_3^- in ground water under the *Robinia pseudoacacia* shelterbelt decreased to 60% than in an adjoining cultivated field (Jaskulska and Szajdak 2010). The quantity of nitrate in soils under a shelterbelt with *Robinia pseudoacacia* was 10.3 times higher than in an adjoining cultivated field (Szajdak and Gaca 2010).

The concentration of nitrate in the water under tree plantings were low, but under the field high (Bartoszewicz 2000). A significant decrease in the nitrate concentration of the nitrogen form in the soil solution and ground water was observed (Bartoszewicz 2000).

The nitrate accumulation was maximum, when the cation exchange capacity was maximum (Smith 1964). A soil nitrate of fertilized with urea plots was similar or even higher under monocropping than under crop rotation, where corn better grew, especially in deeper soil layers and at the end of the cropping period (Horst and Härdter 1994). In soil with corn less nitrate concentrations were observed in a deeper layer of soil (Dou and Fox 1995). Nitrate sorption in the profile of an acid tropical soil was found to increase with depth to 25–50% in the 90–120 cm layer at water and NO_3^- under field condition (Cahn et al. 1992). A soil nitrate content was higher after the urea application than urea with inhibitors – encapsulated calcium carbide or nitrapyrin – Mosier et al. (1996). Losses of NO_3^- during the drainage period from grazed grassland swards increased with increasing soil N-NO_3^- levels in the soil profile.

Thus, as an activity of nitrate reductase increased, decreased leaching of mineral nitrogen occurred. A decrease was observed also with an increasing dose of fertilizer (Jarvis and Barraclough 1991). Higher N rates, higher NO_3^- concentrations were in soil, especially in upper layers and in kernel more than in straw of plants (Hera 1996). Urea increased the level of NO_3^- in soil much than ammonium sulfate, which is an acidic fertilizer (Singh and Yadav 1981). The concentration of nitrate was the highest after the ammonium nitrate application, lower after ammonium sulfate and the lowest after urea using to soil (Naseem and Nasrallah 1981). The amounts of N-NO_3^- formed in soils was highly positively correlated with soil pH, but not higher than 6.0 and was not significantly correlated with the organic N of total N content of the soils (Sahrawat 1982). In sandy loam and Luisiana clay and peat soils, the NO_3^- concentrations were in trace amounts, but in silty loamy Pila Clay, Lipa Loam Maahas clay alkalized were very high in the contrast to NH_4^+ conditions. Reverse results received for NH_4^+ contribution (Sahrawat 1982). Nitrate leaching is an economic loss and causes eutrophication and a health hazard, but gaseous emissions (NH_3, N_2O, NO and NO_2) may prove to be the most serious environmentally (Jenkinson 2001). In the objects fertilized with manure and NPK, the N-NO_3^- leaching was lower than in the control soil. The nitrate nitrogen participation in total N was higher that of ammonium nitrogen only in soils without fertilization and with animal slurry in a dose I (Mazur and Budzyńska 1994).

1.6.2 Nitrite

After the urea application to the soil the highest amounts of NH_4^+, lower of NO_3^- and the lowest of NO_2^- in each soil were and a biuret added inhibited nitrification (Sahrawat 1980). Nitrite is an intermediate product in the aerobic nitrification and anaerobic denitrification processes. Because of its low stability in acid conditions, nitrite can be a key compound in N loss, because nitric oxide and nitrogen dioxide cannot escape from the medium promote the production of some nitrite (Van Cleemput and Baert 1984). The NO_2-N concentration was the highest after the application of urea, lower after ammonium sulfate and the lowest after ammonium nitrate using. The gaseous loss of N from urea was greater than the loss of N from ammonium sulfate or ammonium nitrate, because NO_2-N, $(NH_4)_2CO_3$ and NH_3 are present in the urea-soil system (Naseem and Nasrallah 1981).

In the soil of the former lake Texcoco (Mexico), concentrations of NO_2^- and NH_4^+ increased with increased salinity and availability of NO_3^-. The larger amount of NO_3^- was observed. In alkaline-saline soil higher concentrations of NO_2^- and smaller NH_4^+ were noticed (Dendooven et al. 2006). The highest concentration of urea was accompanied by an accumulation of NO_2^-. The emission of nitrous oxide was delayed at the higher urea level. The pH and NH_4^+ produced inhibitory concentration of ammonia at the highest urea concentration and it was negative effect of urea both on nitrifying and denitrifying bacteria (Petersen et al. 2004). The highest emissions of nitrous oxide from forest and agricultural light textured soils were induced by the addition of NO_2^- to the soil and it had a biological origin (a very active denitrifying population was present), because no significant nitrous oxide emissions were measured, when the soil was autoclaved (Castaldi and Smith 1998). Nitrite nitrogen was leaching in higher amounts to water after the application of slurry in a rate II (Mazur and Budzyńska 1994).

1.6.3 Ammonium

The highest concentration for ammonium was observed in ground water under an adjoining cultivated field to the *Crataegus* shelterbelt (4.48 mg L^{-1}). The lowest values of ammonium were determined under an adjoining cultivated field (about 29%) in comparison with ground water under the *Robinia pseudoacacia* shelterbelt. No significant differences in ground water under an adjoining cultivated field and a new shelterbelt were found. The N-NH_4^+ concentrations were: 1.44, 1.74 and 2.32 mg L^{-1} in: a new shelterbelt, the *Crataegus* shelterbelt and the *Robinia pseudoacacia* shelterbelt, respectively, but the concentration of ammonium nitrogen was the highest in a *Crataegus* adjoining cultivated field (4.48) and the lowest in a new adjoining cultivated field (1.43). Ground water under the *Robinia pseudoacacia* shelterbelt showed higher concentration of ammonium than from the adjoining cultivated field (Jaskulska and Szajdak 2010). Mean concentration values of ammonium nitrogen in ground waters under the field and tree plantings were identical (Bartoszewicz 2000). Ammonium nitrogen was leaching in larger amounts to water after the application of slurry in a dose II. More ammonium nitrogen was leached from soil fertilized with NPK than with manure and animal slurry in rate I (Mazur and Budzyńska 1994).

The concentrations of ammonium in soils under shelterbelt were approx. Two times higher than under an adjoining cultivated field (Szajdak and Gaca 2010). After the application of urea and ammonium nitrate with increasing doses increased also ammonium concentration in humus and mineral soils in higher amounts from ammonium nitrate than urea in pine (Nômmik et al. 1994). The content of NH_4^+-N was the highest when soil was treated with ammonium sulfate, lower – with urea and the lowest – with ammonium nitrate (Naseem and Nasrallah 1981).

In beech the highest ammonium uptake rates around noon and in the afternoon and minima at midnight (50% less), but in spruce ammonium uptake was constant during the day. It depended more on plant species (Geβler et al. 2002). The uptake of N as ammonium was higher than as nitrate, which indicated a preference for ammonium uptake by wheat (Crawford and Chalk 1993) – it also depended on species of plants.

1.6.4 Mineral Nitrogen

The proportion of NH_4^+-N in total nitrogen increased and that of NO_3^- decreased. Applications of slurry rate II and NPK caused an increase of mineral nitrogen by 48–49%, of manure by 35% and of slurry dose I by 27% as compared with control. The application of a supplemental mineral fertilizer increased soil mineral nitrogen only at the slurry rate of I. Average proportions of NH_4^+-N and NO_3^--N in the total mineral nitrogen were 59% and 41%, respectively. The highest amount of soil mineral nitrogen was found in April and the lowest in September and August (Mazur 1987).

Incorporation of oilseed rape stems residues into the soil resulted in an increase in gross N mineralization rate (24%) in the case of this plant, but in winter wheat straw residues increase in gross N mineralization achieved 12% (Watkins and Barraclough 1996).

1.6.5 Organic Nitrogen

The content of organic nitrogen increased to 36% in ground water under an adjoining cultivated field to the *Crataegus* shelterbelt and to 22% in comparison with ground water under the *Robinia pseudoaccacia* shelterbelt. Amounts of organic N were the highest in the *Crataegus* shelterbelt and an adjoining cultivated field and the lowest in a new shelterbelt and an adjoining cultivated field (Jaskulska and Szajdak 2010).

The highest organic N concentration (in %) was after the application pig slurry manure (3.5), a little lower after poultry litter (3.4), lower after layer manure (2.9), much lower after cattle slurry (2.5) and the lowest after cattle farmyard manure – FYM and pig FYM (2.3) above ryegrass field. Up to 70% of the organic N was mineralized from pig slurry and layer manure, compared to 10–30% from the cattle slurry and straw – based farmyard manures – FYMs (Bhogal et al. 2016).

Concentrations of organic N and total mineral N decreased in deeper layers of soils from harvest of oilseed rape before sowing of oats when was ploughing and green manure (Vos et al. 1994).

1.6.6 Total Nitrogen

Default N contents in different material – in: fertilizers (1000 kg N mg^{-1} of FW), soybean oil cake (70.2), soybean (56.4), feed milk (56.3), rape cake (49.3), meet – live animals (46.0), whey (35.0), eggs (18.1), alfalfa (18.0), silage clover grass (9.1), silage – grass (8.5), silage – whole crop (6.0), whole crop fresh (5.8), straw (5.4), corn (3.9), silage beef pulp (3.8), wool (3.0) – were listed by Dalgaard et al. (2012).

Nitrogen contents (in kg N mg^{-1}) were different for the following types of crops: 48.71 – for faba beans (*Vicia faba* L.), 31.45 – for oilseed rape (*Brassica napus*), 18.79 – for winter wheat (*Triticum aestivum*), 14.96 – for grain of winter rye cereals (*Secale cereale*) – Dalgaard et al. (2012). The same author presented default N contents (in kg mg^{-1}) for different manure types: solid poultry manure (21.0), pig farmyard manure (8.8), mixed FYM (8.6), cattle FYM and separately sheep or goat FYM (8.4), solid fraction of pig manure – sows and piglets (8.1), horse FYM (7.5), sewage sludge (6.0), solid fraction of pig – fatteners or mixed manure (5.9), mixed slurry and liquid fraction of cattle manure and separately pig slurry – fattening pigs (5.4), liquid fraction of mixed manure (5.0), pig slurry – sows and piglets (4.6), degassed cattle slurry and liquid fraction of pig manure (4.0), degassed cattle slurry (3.9) and other organic fertilizer (e. g. bone meal) or a composted manure – compost from other materials (2.0).

The total nitrogen in soil under a shelterbelt was 3.5 times higher than in the adjoining cultivated field (Szajdak and Gaca 2010).

The highest total N content in wheat was on day 153 (leaf senescence) in grain (93.4) and the lowest in roots (20.7). During anthesis (day 125) the highest amounts of total N were observed in leaves (61.0) and the smallest in roots (18.6). The highest absorption of ^{15}N (in mg $plant^{-1}$) was in the plant growth stage – harvesting, smaller – during fruiting, then at flowering and the lowest in seedling. The highest amount of ^{15}N was after the application of nitrate, lower – after ammonium using and the lowest after the urea application (Tan et al. 2000). It was depended on the phase of plant growth (Harper et al. 1987) and on the nitrogen form applied in fertilizers (Tan et al. 2000).

In the wheat field using the same dose (112 kg N ha^{-1}) 33.3% loss of ^{15}N received in straw (Myers and Paul 1971) and 25% and 6% in clay wet and dry, respectively (Craswell and Martin 1975). In barley crop, when 90 kg N ha^{-1} was applied as NH_4OH the ^{15}N loss achieved 30% and 31% in the case of $(NH_4)_2SO_4$ (Koren'kov et al. 1975). In the same N dose on Sandy loam in pearl millet field, when urea was applied the ^{15}N loss was higher (46.7%) – Ganry et al. (1978). Yields of pearl millet were similar whether urea or calcium ammonium nitrate were used, but ^{15}N uptake from calcium ammonium nitrate was higher by plants and ammonia emission was also higher from urea from soils and in years of strong rainfall (Christiansen et al. 1990).

The amount of N leached from soil depended on the kind of organic fertilization and animal slurry dose. After manure application the largest amounts of total-N, organic-N, N-NH_4^+ and N-NO_2^- and after the application of animal slurry used of N-NO_3^- were leached from the soil fertilized with rates of animal slurry, manure and NPK doses balanced in nitrogen. The animal slurry dose equivalent to manure in terms of organic carbon caused the strongest leaching of all nitrogen forms and the increase was significant also in comparison with manure fertilization (Mazur and Budzyńska 1994).

1.7 Very Important Enzymes Related to the Nitrogen Cycle

Main enzymes occurring in plants and soils are nitrate reductase and urease.

1.7.1 Nitrate Reductase

The activity of nitrate reductase (NR) depended on the part of plant. The highest was in upper leaves, lower in lower leaves and in stems and the lowest – in roots of *Verbascum* (Güleryüz and Arslan 1999). The activity of this enzyme in roots and leaf petioles in *Rumex obtusifolius* L. was low, but the highest in leaf blades (Gebauer et al. 1984).

The activity of nitrate reductase correlated positively with grain yield – Eilrich and Hageman (1973), yield of been (*Phaseolus vulgaris* L.) seeds – 59% over control (Franco et al. 1979) and an accumulation of grain protein and total reduced nitrogen in a vegetative material at maturity (Deckard et al. 1973).

The pH of intact tissue had an effect on the activity of nitrate reductase. The level of this enzyme was the highest, when the pH ranged from 7.5 to 8.0 and the lowest near pH of 6.0 in the case of soybean leaves (Jaworski 1971).

The activity of nitrate reductase decreased when soybeans were exposed to dark. The subsequent light stimulation increased the activity of this enzyme (Nicholas et al. 1976). Low levels of NR activity were noticed in leaf extracts from plants grown on nitrate in darkness. When barley etiolated plants were transferred to the light the activity of nitrate reductase increased (Somers et al. 1983). The same results were obtained for corn (Rao et al. 1981) and barley (Gniazdowska-Skoczek 1988a, b) leaves. Osmotic stress caused dehydration of barley leaves and decreased the activity of nitrate reductase (Bandurska 1993).

The activity of nitrate reductase depended on forest trees species: the higher was in *Populus deltoides*, lower in *Glechoma hederacea* much lower in *Quercus robur* and the lowest in *Prunus padus* after the ammonium chloride application (Pearson and Soares 1998).

Another factor which was the effect on the activity of nitrate reductase was the level of soil profile. In deeper layers of soil the activity of this enzyme was higher (Jarvis and Barraclough 1991).

1.7.2 Urease

The activity of urease depended on plant species. The highest activity of enzyme was found in leaves of *Poa pratensis*, lower in *Botriochloa caucasica* and the lowest in *Bromus inermis* (Frankenberger Jr and Tabatabai 1982).

The activity of urease also depended on the depth in soil. In tropical flooded soils, unplanted and planted to rice in the subsurface of soil (1–2 cm) the enzyme activity was higher than in a soil subsurface (20–25 cm). In deeper layers of soils lower the activity of enzyme was detected. Progressively decreasing urease activity was found with increasing depth (Fenn and Hossner 1985).

The longer time of the soil incubation was the higher the activity of enzyme was noticed. Also, the stage of plant growth was important.

The highest urease activity was in panicle stage of the crop, but the lowest during maturity stage. During rice growth at the time of seedling and tillering the activity of urease was higher after the application of green manure (*Sesbania rostrata*) at 66.4 kg N ha^{-1}, lower after urea (as pilled urea) using at 60 kg N ha^{-1} and the lowest – after the application of green manure plus urea (both at a 30 kg N ha^{-1} dose). Contrary results were obtained at panicle initiation and maturity stages of this crop growth (Pattnaik et al. 1999).

The activity of urease, which is activated by nickel (Ni) was hardly detectable in different plants (rye, wheat, soybean, rape, zucchini and sunflower) grown without supplementary Ni, but on urea-based media. As a consequence plants showed a reduced dry matter production and the soluble amino acid N concentration and reduced total N amounts, which was illustrated by the chlorotic appearance of these plants (Gerendás and Sattelmacher 1997).

The addition of urea stimulated the ureolytic microbes and if organic residues did not limit the production of urease, the maximum ammonia loss could occurred (Paulson and Kurtz 1969). Urease gradually loses its hydrolytic activity at pH values below 4.0 (Bremner and Douglas 1971a). Urease activity did increase with increasing soil pH values and gave better correlation to activity at neutrality (McGarity and Myers 1967; Van Slyke and Zacharias 1914). Much lower NH_3 losses were achieved using N-(n-butyl) thiophosphoric triamide to inhibit urea hydrolysis by urease owing to the gradually accumulation of NH_4^+ that limited the rise in pH, produced more NO_3^- with very small amounts of NO_2^- and immobilization was reduced (Guimarães et al. 2016). Gibson (1930) reported that the hydrolysis of urea in forest soils was more rapid than in cultivated soils. Strongly acid peat soils (pH 3.1–3.3) hydrolyzed from 0.44% to 0.86% of their own dry weights of urea in 24 h at 22–23 °C.

A sunlight, high temperature and drying may denature the urease enzyme produced in soil (Fenn and Hossner 1985).

1.8 Conclusions

1. The highest ammonia and nitrous oxide emissions were from cattle, but the lowest ammonia volatilization was from deer, but the nitrous oxide emission – from camels. From all animals total ammonia emissions were the highest from the spreading of waste, but the lowest from grazing or outdoors and not always were higher when the number of animals was higher. Older and dairy cows volatilized more ammonia to the air than younger ones, so the age of animals was also important.
2. A high ammonia concentration was above manure. The highest ammonia emission was noted from urea, but the lowest from calcium ammonium nitrate applied to grassland. The lowest emissions of ammonia and nitrous oxide was from slurry. The content of NH_4^+ in soils was the highest when ammonium sulfate was used, lower after urea and the lowest after ammonium nitrate applications. When slurry was applied, higher ammonium and nitrite concentrations were in ground water. To conclude slurry, manure and organic residues – oilseed rape stems and winter wheat straw increased the level of mineral nitrogen in soils. In total, N mineral is more often as nitrate than ammonium. Amounts of organic nitrogen also depended on the type of organic fertilizers and the highest were after the application of pig slurry manure and the lowest after using manure from cattle and pigs separately. For most types of fertilizers, ammonia emissions were higher with higher doses, but this did not always apply to the ammonium nitrate fertilizer. The highest nitrous oxide and ammonia emissions were noticed after urea application, but the lowest – after using ammonium nitrate or ammonium sulfate in crop fields. High concentrations of these gases were above a heap of dung and manure, but low above compost and were higher from organic than mineral fertilizers. Greater doses of nitrogen in fertilizers mean more nitrate in soil. There is a positive correlation between a high level of nitrate and organic and total nitrogen concentrations in soils and a pH value of 6.0 at low cation exchange capacity in soils. Urea increased the level of NO_3^- and NO_2^- in acidic soil and NO_2 in the air much than ammonium nitrate and ammonium sulfate, but contrary results were obtained for alkaline soils, except when the last soils were under salinity stress. Nitrate and total N were higher when mineral fertilizers were without organic manure and slurry.
3. Emissions of ammonia from urea can be reduced by potassium nitrate, potassium chloride, ammonium nitrate, calcium chloride, and calcium with urea at high cation exchange capacity, in soil, but not always. Also, calcium, magnesium, potassium salts with urea, tannins, acetylene and N-(n-butyl) thiophosphoric triamide limited the ammonia volatilization and the activity of urease.

4. Ammonia and nitrous oxide concentrations were higher, when the air temperature was higher. Ammonia emissions are high, when the relatively humidity of air is low, but the soil water content is high. Higher temperatures of soil, higher urease activity and more ammonium and nitrate occurred in soils, especially dry ones and with the higher level of NH_3. Ammonia concentrations in the air are high, when the wind speed ranges from 2 to max. 4 m s^{-1} and not higher, often at south-southwest, west to east and north-northeast wind directions, respectively, but not always. It also depended on the type of ecosystem and fertilizer. At higher heights above the ground, less ammonia was noticed. Roundup increased the level of ammonia in soil.
5. Rainfall decreased volatilization of ammonia, but more NH_3 and NH_4^+ in soils occurred. Also higher precipitation caused lower nitrous oxide amounts, but in the wet season the emissions of N_2O from soil were higher than in the dry season in contrast to NH_3. Ammonia and nitrous oxide concentrations are the lowest in winter and at nighttime and in the early morning and the highest in summer, when it is warm and at daytime, especially at noon. In beech, the highest ammonium uptake rates occurred at noon and in the afternoon and the lowest uptake rates were observed at midnight, but the content of NH_4^+ in spruce was constant all day. It depended on the type of plant.
6. Nitrous oxide and ammonia emissions were higher when the intensity of light was higher. Also, the activity of enzymes (nitrate reductase and urease) was higher when it was in light than in darkness. Higher pH of soil, lower ammonia values in the air were found, but sometimes results were contrary, because the effect of fertilizers was higher than the pH value of soil. Higher pH of soil also caused a high level of urease activity in the soil.
7. The emission of ammonia decreased with an increasing distance from an animal farm (a horizontal gradient of NH_3 was observed). At higher heights above the ground, it was higher for barley than at lower heights, but it more depended on plant species and for other plants it was contrary – similar to the emission of nitrogen dioxide. Concentrations of nitrate and ammonium in soils depended on their depth and were in higher amounts in deeper layers of the soil at high cation exchange capacity, but not always. Sometimes it depended on the type of fertilizer and was higher in upper layers of the soil and in kernel more than in straw. The level of NO_3^- was higher than NH_4^+, but the highest amounts of ammonia in the air were from the surface than from deeper layers of soil (vertical gradients of NH_3 were observed). Concentrations of organic and total nitrogen were lower in deeper layers of soil when ploughing and green manure was and the harvest of oilseed rape before sowing oats. In deeper layers of soil, the activity of nitrate reductase was higher, but the activity of urease was lower.
8. A loss of ammonia was lower from calcareous soils than from acidic ones and higher when cation exchange capacity was low in soils. Urea increases the pH of soils. Unfertilized soil under a shelterbelt was very acidic, while under adjoining cultivated fields weakly acidic to neutral. The loss of N_2O was greater from mineral than from peat soils. From slow release fertilizers, losses of gases and

various mineral nutrients were lower. Inorganic fertilizers did not cause high emissions of ammonia in acidic soils. Also, amounts of NH_4^+ in soils depended on their type.

9. From animal farms the emissions of ammonia and nitrous oxide were higher from crop fields than from grasslands, shelterbelts and forests. It depended on the type of ecosystem.
10. Higher concentrations of ammonia were observed during anthesis, but the highest during plant senescence – from stems and leaves. Total plant N was in higher amounts in grain, a little less in leaves, lower in stems and the lowest in roots similar as other nitrogen forms. It depended on the stage of plants' growth and the type of their organs.
11. Nitrous oxide emissions depended on soil characteristics: the calcium carbonate content, water soluble carbon in sand and increased when the content of NH_4^+ and NO_3^- increased. Inhibitors of nitrous oxide emissions include: dicyandiamide, nitrapyrin, encapsulated calcium carbide and Biochar.
12. Nitrogen dioxide concentrations were higher from an animal farm than from fertilizers. The highest were when a south-southwest wind direction was, low wind speeds and in winter and increased with increasing air temperatures.
13. Nitric oxide values were also higher in winter than in summer and at night and at high relative humidity of air and were higher with increasing forest floor temperatures. Nitric oxide emissions were the highest above forests than above grasslands and wetlands. At a low pH value of soil the nitric oxide amount was higher from nitrate, but at higher pH value – from ammonium and increased when doses of these forms were higher.
14. Nitrate and ammonium concentrations were the highest in ground waters, lower in soils from crop fields and the lowest from shelterbelts. More NH_4^+, but less of NO_3^- concentrations were in soils and waters from the *Robinia pseudoacacia* shelterbelt than *Crataegus* and new shelterbelts. It depended on the age and plant species.
15. The content of mineral nitrogen depended on the plant species – for example it was higher for white sweet clover and lower for winter rape and on the stomatal compensation point and the cultivar of plant. Higher amounts of ammonia from barley cv. Golf than cv. *Laevigatum* were during grain filling. Nitrous oxide emissions also depended on the plant species (higher values were observed above a deciduous than a coniferous forest). Also, nitric oxide amounts in the air depended on the type of plant species. Encapsulated calcium carbide and nitrapyrin inhibit the amount of nitrate in soils.
16. A total nitrogen amount depended on the material and the plant species and was the highest in fertilizers, soybean oil cake and the lowest in wool and silage of beef pulp. Higher concentrations of total nitrogen were observed in solid poultry manure, but the lowest – in a composted manure. More of this form was in faba beans than in grains of winter rye. The stage of plant growth also had the effect on the amount of total nitrogen and was the highest during harvesting and the lowest in seedling. The total nitrogen was in the highest values after NO_3^- applications, lower after an NH_4^+ using and the lowest after urea. It also

depended on the type of ecosystem – its higher amounts were in soils under shelterbelts than under adjoining cultivated fields. A part of the plant also has significance. The highest concentrations of total nitrogen occurred in grains during leaf senescence and in leaves during anthesis, but the lowest in roots. It also depended on the type of fertilizer – the highest amount of total nitrogen was observed after manure application. Animal slurry caused leaching of all nitrogen forms in high amounts to ground water.

17. When the time of a soil incubation was longer higher activities of nitrate reductase and urease were detected. The highest nitrate reductase activity occurred in leaves, the lowest – in roots and it was positively correlated with grain protein accumulation, grain yield and the total reduced nitrogen in vegetative material (in which the highest level of enzymes was observed) than at maturity. An optimum pH value of intact plants for nitrate reductase activity was between 7.5 and 8.0 similar to urease activity at 7.0. Drying and higher temperatures stimulated the activity of these enzymes. The optimal value for nitrate reductase was approximately 35 °C, but for urease 37 °C and from 60 °C to 70 °C. An enzymatic activity also depended on plant species. Osmotic stress decreased the activity of nitrate reductase. Heavy metals inhibit nitrate reductase activity in acidic and neutral soils.
18. The lack of nickel inhibited urease activity, whereas total nitrogen and organic residues did not limit the production of urease. At the same time the maximum ammonia loss occurred. Hydrolysis of urea is higher and more rapid in forest soils than in cultivated soils taken from fields.

Acknowledgments The investigations were conducted at the Institute for Agricultural and Forest Environment, Polish Academy of Sciences in Poznań, Poland within the framework of the NitroEurope Integrated Project No. 017841 titled "The nitrogen cycle and its influence on the European greenhouse gas balance", which was a part of the Sixth Framework Programme for Research and Technological Development of the European Union.

References

Abalos D, Cardenas LM, Wu L (2016) Climate change and N_2O emissions from South West England grasslands: a modelling approach. Atmos Environ 132:249–257. https://doi.org/10.1016/j.atmosenv.2016.03.007

Abbasi MK, Adams WA (1999) Assessment of the contribution of denitrification to N losses from compacted grassland soil by NO_3^- disappearance and N_2O production during anaerobic incubation. Can J Soil Sci 79:57–64. https://doi.org/10.4141/S98-022

Abdelmagid HM, Tabatabai MA (1987) Nitrate reductase activity of soils. Soil Biol Biochem 19 (4):421–427

Acquaye DK, Cunningham RK (1965) Losses of nitrogen by ammonia volatilization from surface-fertilized tropical forest soils. Trop Agric Trinidad 42:281–292

Adema EH, Mejstřík V, Binek B (1993) The determination of NH_3-concentration gradients in a spruce forest using a passive sampling technique. Water Air Soil Pollut 69:321–335

Ajayi O, Maynard DN, Barker AV (1970) The effects of potassium on ammonium nutrition of tomato (*Lycopersicon esculentum* Mill.). Agron J 62(6):818–821

Al-Kanani T, MacKenzie AF, Barthakur NN (1991) Soil water and ammonia volatilization relationships with surface-applied nitrogen fertilizer solutions. Soil Sci Soc Am J 55:1761–1766

Aneja VP, Blunden J, Roelle PA, Schlesinger WH, Knighton R, Niyogi D, Gilliam W, Jennings G, Duke CS (2008) Workshop on agricultural air quality: state of the science. Atmos Environ 42 (14):3195–3208. https://doi.org/10.1016/j.atmosenv.2007.07.043

Aneja VP, Schlesinger WH, Erisman JW, Behera SN, Sharma M, Battye W (2012) Reactive nitrogen emissions from crop and livestock farming in India. Atmos Environ 47:92–103. https://doi.org/10.1016/j.atmosenv.2011.11.026

Asman WAH, van Jaarsveld HA (1992) A variable-resolution transport model applied for NH_x in Europe. Atmos Environ A Gen Top 26:445–464

Asman WAH, Pinksterboer EF, Maas HFM, Erisman JW, Waijers-Ypelaan A, Slanina J, Horst TW (1989) Gradients of ammonia concentration in a nature reserve – model results and measurements. Atmos Environ 23:2259–2265

Bai Z, Yang G, Chen H, Zhu Q, Chen D, Li Y, Wang X, Wu Z, Zhou G, Peng C (2014) Nitrous oxide fluxes from three forest types of the tropical mountain rainforests on Hainan Island, China. Atmos Environ 92:469–477. https://doi.org/10.1016/j.atnosenv/2014.04.059

Bandurska H (1993) *In vivo* and *in vitro* effect of proline on nitrate reductase activity under osmotic stress in barley. Acta Physiol Plant 15(2):83–88

Barker AV, Maynard DN, Lachman WH (1967) Induction of tomato stem and leaf lesions, and potassium deficiency, by excessive ammonium nutrition. Soil Sci 103(5):319–327

Barthelmie RJ, Pryor SC (1998) Implications of ammonia emissions for fine aerosol formation and visibility impairment – a case study from the Lower Fraser Valley, British Columbia. Atmos Environ 32(3):345–352. https://doi.org/10.1016/S1352-2310(97)83466-8

Bartoszewicz A (2000) Effect of the change of soil utilization on the concentration of nitrogen mineral forms in soil and ground waters. Polish J Soil Sci XXXIII(2):13–20

Bertelsen F, Jensen ES (1992) Gaseous nitrogen losses from field plots grown with pea (*Pisum sativum* L.) or spring barley (*Hordeum vulgare* L.) estimated by ^{15}N mass balance and acetylene inhibition techniques. Plant Soil 142:287–295. https://doi.org/10.1007/BF00010974

Beyrouty CA, Sommers LE, Nelson DW (1988) Ammonia volatilization from surface-applied urea as affected by several phosphoroamide compounds. Soil Sci Soc Am J 52:1173–1178

Bhogal A, Williams JR, Nicholson FA, Chadwick DR, Chambers KH, Chambers BJ (2016) Mineralization of organic nitrogen from farm manure applications. Soil Use Manag 32:32–43. https://doi.org/10.1111/sum.12263

Binstock DA (1984) Potential denitrification in an acid forest soil: dependence on wetting and drying. Soil Biol Biochem 16(3):287–288. https://doi.org/10.1016/0038-0717(84)90019-1

Blunden J, Aneja VP (2008) Characterizing ammonia and hydrogen sulfide emissions from a swine waste treatment lagoon in North Carolina. Atmos Environ 42(14):3277–3290. https://doi.org/10.1016/j.atmosenv.2007.02.026

Bolan NS, Hedley MJ, White RE (1991) Processes of soil acidification during nitrogen cycling with emphasis on legume based pastures. Plant Soil 134(1):53–63. https://doi.org/10.1007/BF00010717

Bosch-Serra AD, Yagüe MR, Teira-Esmatges MR (2014) Ammonia emissions from different fertilizing strategies in Mediterranean rainfed winter cereals. Atmos Environ 84:204–212. https://doi.org/10.1016/j.atmosenv.2013.11.044

Bourdin F, Sakrabani R, Kibblewhite MG, Lanigan GJ (2014) Effect of slurry dry matter content, application technique and timing on emissions of ammonia and greenhouse gas from cattle slurry applied to grassland soils in Ireland. Agric Ecosyst Environ 188:122–133. https://doi.org/10.1016/j.agee.2014.02.025

Bouwman AF, Boumans LJM, Batjes NH (2002) Modeling global annual N2O and NO emissions from fertilized fields. Global Biogeochem Cycles 16(4):28-1-28-9. https://doi.org/10.1029/2001GB001812

Bouwman AF (1996) Direct emission of nitrous oxide from agricultural soils. Nutr Cycl Agroecosyst 46:53–70. https://doi.org/10.1007/BF00210224

Bremner JM, Douglas LA (1971a) Decomposition of urea phosphate in soils. Soil Sci Soc Am Proc 35:575–578

Bremner JM, Douglas LA (1971b) Inhibition of urease activity in soils. Soil Biol Biochem 3:299–307

Bruhn D, Albert KR, Mikkelsen TN, Ambus P (2014) UV-induced N_2O emission from plants. Atmos Environ 99:206–214. https://doi.org/10.1016/j.atmosenv.2014.09.077

Butterbach-Bahl K, Gasche R, Willibald G, Papen H (2002) Exchange of N-gases at the Höglwald forest – a summary. Plant Soil 240:117–123. https://doi.org/10.1023/A:1015825615309

Buwal (1995) Vom Menschen verursachte Luftschadstoff-Emissionen in der Schweiz von 1900 bis 2010. In: Buwal (ed.) Bern, Schriftenreihe Umwelt – Luft, 256:121

Bühlmann T, Hiltbrunner E, Körner C, Rihm B, Achermann B (2015) Induction of indirect N_2O and NO emissions by atmospheric nitrogen deposition in (semi-)natural ecosystems in Switzerland. Atmos Environ 103:94–101. https://doi.org/10.1016/j.atmosenv.2014.12.037

Cahn MD, Bouldin DR, Cravo MS (1992) Nitrate sorption in the profile of an acid soil. Plant Soil 143:179–183

Castaldi S, Smith KA (1998) The effect of different N substrates on biological N_2O production from forest and agricultural light textured soils. Plant Soil 199:229–238. https://doi.org/10.1023/A:1004383015778

Chai L, Kröbel R, Janzen HH, Beauchemin KA, McGinn SM, Bittman S, Atia A, Edeogu I, MacDonald D, Dong R (2014) A regional mass balance model based on total ammoniacal nitrogen for estimating ammonia emissions from beef cattle in Alberta Canada. Atmos Environ 92:292–302. https://doi.org/10.1016/j.atmosenv.2014.04.037

Christiansen CB, Bationo B, Henao J, Vlek PLG (1990) Fate and efficiency of N fertilizers applied to pearl millet in Niger. Plant Soil 125:221–231

Clarkson DT, Warner AJ (1979) Relationships between root temperatures and the transport of ammonium and nitrate ions by italian and perennial ryegrass (*Lolium multiflorum* and *Lolium perenne*). Plant Physiol 64:557–561

Clough TJ, Sherlock RR, Oameron KC, Ledgard SF (1996) Fate of urine nitrogen on mineral and peat soils in New Zealand. Plant Soil 178:141–152. https://doi.org/10.1007/BF00011172

Colliver GW, Welch LF (1970) Toxicity of preplant anhydrous ammonia to germination and early growth of corn: II. Laboratory studies. Agron J 62:346–348

Craswell ET, Martin AE (1975) Isotopic studies of the nitrogen balance in a cracking clay. II Recovery of nitrate ^{15}N added to columns of packed soil and microplots growing wheat in the field. Aust J Soil Res 13:53–61

Crawford DM, Chalk PM (1993) Sources of N uptake by wheat (*Triticum aestivum* L.) and N transformations in soil treated with a nitrification inhibitor (nitrapyrin). Plant Soil 149:59–72. https://doi.org/10.1007/BF00010763

Cruvinel EBF, Bustamante MM d C, Kozovits AR, Zepp RG (2011) Soil emissions of NO, N_2O and CO_2 from croplands in the savanna region of central Brazil. Agric Ecosyst Environ 144:29–40. https://doi.org/10.1016/j.agee.2011.07.016

Cui F, Yan G, Zhou Z, Zheng X, Deng J (2012) Annual emissions of nitrous oxide and nitric oxide from a wheat-maize cropping system on a silt loam calcareous soil in the North China Plain. Soil Biol Biochem 48:10–19. https://doi.org/10.1016/j.soilbio.2012.01.007

Daigger LA, Sander DH, Peterson GA (1976) Nitrogen content of winter wheat during growth and maturation. Agron J 68:815–818. https://doi.org/10.2134/agronj1976.00021962006800050033x

Dalgaard T, Bienkowski JF, Bleeker A, Dragosits U, Drouet JL, Durand P, Frumau A, Hutchings NJ, Kedziora A, Magliulo V, Olesen JE, Theobald MR, Maury O, Akkal N, Cellier P (2012) Farm nitrogen balances in six European landscapes as an indicator for nitrogen losses and basis for improved management. Biogeosciences 9:5303–5321. https://doi.org/10.5194/bg-9-5303-2012

Dattamudi S, Wang JJ, Dodla SK, Arceneaux A, Viator HP (2016) Effect of nitrogen fertilization and residue management practices on ammonia emissions from subtropical sugarcane production. Atmos Environ 139:122–130. https://doi.org/10.1016/j.atmosenv.2016.05.035

Dean JV, Harper JE (1986) Nitric oxide and nitrous oxide production by soybean and winged bean during the *in Vivo* nitrate reductase assay. Plant Physiol 82:718–723. https://doi.org/10.1104/pp.82.3.718

Deckard EL, Lambert RJ, Hageman RH (1973) nitrate reductase activity in corn leaves as related to yields of grain and grain protein. Crop Sci 13(3):343–350. https://doi.org/10.2135/cropsci1973.0011183X001300030017x

Delaune RD, Patrick WH (1970) Urea conversion to ammonia in waterlogged soils. Proc Soil Sci Soc Am 34(4):603–607

Dendooven L, Vega-Jarquin C, Cruz-Mondragon C, Van Cleemput O, Marsch R (2006) Dynamics of inorganic nitrogen in nitrate and glucose-amended alkaline-saline soil. Plant Soil 279:243–252. https://doi.org/10.1007/s11104-005-1359-8

Denmead OT, Freney JR, Simpson JR (1982) Dynamics of ammonia volatilization during furrow irrigation of maize. Soil Sci Soc Am J 46:149–155

Denmead OT, Freney JR, Dunin FX, Jackson AV, Reyenga W, Saffina PG, Smith JWB, Wood AW (1993) Effect of canopy development on ammonia uptake and loss from sugarcane fields fertilized with urea. In: Proceedings of Australian Society of Sugar Cane Technologists, pp. 285–292

Denmead OT, Chen D, Griffith DWT, Loh ZM, Bai M, Naylor T (2008) Emissions of the indirect greenhouse gases NH_3 and NO_x from Australian beef cattle feedlots. Aust J Exp Agric 48:213–218. https://doi.org/10.1071/EA07276

Dosch P, Gutser R (1996) Reducing N losses (NH_3, N_2O, N_2) and immobilization from slurry through optimized application techniques. Fertil Res 43:165–171. https://doi.org/10.1007/BF00747697

Dou Z, Fox RH (1995) Using NCSWAP to simulate seasonal nitrogen dynamics in soil and corn. Plant Soil 177:235–247. https://doi.org/10.1007/BF00010130

Duyzer JH, Verhagen HLM, Erisman JW (1989) De depositie van verzurende stoffen op de Asselse heide. Report No. R 89/29, TNO/RIVM, Delft, The Netherlands

Duyzer JH, Verhagen HLM, Weststrate JH, Bosveld FC, Vermetten AWM (1994) The dry deposition of ammonia onto a Douglas-fir forest in the Netherlands. Atmos Environ 28:1241–1253. https://doi.org/10.1016/1352-2310(94)90271-2

Eilrich GL, Hageman RH (1973) Nitrate reductase activity and its relationship to accumulation of vegetative and grain nitrogen in wheat (*Triticum aestivum* L.). Crop Sci 13:59–66. https://doi.org/10.2135/cropsci1973.0011183X001300010018x

El-Zahaby EM, Chien SH, Savant NK, Vlek PLG, Mokwunye AV (1982) Effect of pyrophosphate on phosphate sorption and ammonia volatilization by calcareous soils treated with ammonium phosphates. Soil Sci Soc Am J 46:733–740

Engelsjord ME, Fostad O, Singh BR (1997) Effects of temperature on nutrient release from slow-release fertilizers. *I. Commercial and experimental products.* Nutr Cycl Agroecosyst 46:179–187. https://doi.org/10.1007/BF00420552

Erisman JW, Van Elzakker BG, Mennen MG, Hogenkamp J, Zwart E, Van den Beld L, Römer FG, Bobbink R, Heil G, Raessen M, Duyzer JH, Verhage H, Wyers GP, Otjes RP, Möls JJ (1994) The Elspeetsche Veld experiment on surface exchange of trace gases: summary of results. Atmos Environ 28(3):487–496. https://doi.org/10.1016/1352-2310(94)90126-0

Erisman JW, Grennfelt P, Sutton M (2003) The European perspective on nitrogen emission and deposition. Environ Int 29:311–325. https://doi.org/10.1016/S0160-4120(02)00162-9

Ernst JW, Massey HF (1960) The effects of several factors on volatilization of ammonia formed from urea in the soil. Soil Sci Soc Am J 24:87–90. https://doi.org/10.2136/sssaj1960.03615995002400020007x

Fangmeier A, Hadwiger-Fangmeier A, van der Eerden L, Jaeger HJ (1994) Effects of atmospheric ammonia on vegetation – a review. Environ Pollut 86:43–82. https://doi.org/10.1016/0269-7491(94)90008-6

Farquhar GD, Firth PM, Wetselaar R, Weir B (1980) On the gaseous exchange of ammonia between leaves and the environment: determination of the ammonia compensation point. Plant Physiol 66(4):710–714

Farquhar GD, Wetselaar R, Weir B (1983) Gaseous nitrogen losses from plants. In: Freney JR, Simpson JR (eds) Gaseous loss of nitrogen from plant-soil systems. Martinus Nijhoff / Dr. W. Junk Publ., The Hague / Boston / Lancaster, pp. 159–180

Felix JD, Elliott EM (2014) Isotopic composition of passively collected nitrogen dioxide emissions: vehicle, soil and livestock source signatures. Atmos Environ 92:359–366. https://doi.org/10.1016/j.atmosenv.2014.04.005

Felix JD, Elliott EM, Gish T, Maghirang R, Cambal L, Clougherty J (2014) Examining the transport of ammonia emissions across landscapes using nitrogen isotope ratios. Atmos Environ 95:563–570. https://doi.org/10.1016/j.atmosenv.2014.06.061

Fenn LB (1975) Ammonia volatilization from surface applications of ammonium compounds on calcareous soils: III. Effects of mixing low and high loss ammonium compounds. Soil Sci Soc Am Proc 39:366–368

Fenn LB, Hossner LR (1985) Ammonia volatilization from ammonium or ammonium-forming nitrogen fertilizers. In: Stewart BA (ed) Advances in soil science, vol 1. Springer, New York, pp. 123–169. https://doi.org/10.1007/978-1-4612-5046-3_4

Fenn LB, Kissel DE (1973) Ammonia volatilization from surface applications of ammonium compounds on calcareous soils: I. General theory. Soil Sci Soc Am Proc 37:855–859

Fenn LB, Kissel DE (1974) Ammonia volatilization from surface applications of ammonium compounds on calcareous soils: II. Effects of temperature and rate of ammonium nitrogen application. Soil Sci Soc Am Proc 38:206–210

Fenn LB, Kissel DE (1976) The influence of cation exchange capacity and depth of incorporation on ammonia volatilization from ammonium compounds applied to calcareous soils. Soil Sci Soc Am J 40:394–398. https://doi.org/10.2136/sssaj1976.03615995004000030026x

Fenn LB, Miyamoto S (1981) Ammonia loss and associated reactions of urea in calcareous soils. Soil Sci Soc Am J 45:537–540

Fenn LB, Taylor RM, Matocha JE (1981) Ammonia losses from surface-applied nitrogen fertilizer as controlled by soluble calcium and magnesium: general theory. Soil Sci Soc Am J 45:777–781

Fenn LB, Matocha JE, Wu E (1982a) Soil cation exchange capacity effects on ammonia loss from surface-applied urea in the presence of soluble calcium. Soil Sci Soc Am J 46:78–81

Fenn LB, Matocha JE, Wu E (1982b) Substitution of ammonium and potassium for added calcium in reduction of ammonia loss from surface-applied urea. Soil Sci Soc Am J 46:771–776

Fenn LB, Malstrom HL, Wu E (1984) Ammonia losses from surface applied urea as affected by added calcium and fresh plant residues. Soil Sci 137:270–279

Ferguson RB, Kissel DE (1986) Effects of soil drying on ammonia volatilization from surface applied urea. Soil Sci Soc Am J 50:485–490

Fischer K, Burchill W, Lanigan GJ, Kaupenjohann M, Chambers BJ, Richards KG, Forrestal PJ (2016) Ammonia emissions from cattle dung, urine and urine with dicyandiamide in a temperate grassland. Soil Use Manag 32:83–91. https://doi.org/10.1111/sum.12203

Fisher WB, Parks WL (1958) Influence of soil temperature on urea hydrolysis and subsequent nitrification. Soil Sci Soc Am Proc 22:247–248

Forrestal PJ, Harty M, Carolan R, Lanigan GJ, Watson CJ, Laughlin RJ, McNeill G, Chambers BJ, Richards KG (2016) Ammonia emissions from urea, stabilized urea and calcium ammonium nitrate: insights into loss abatement in temperate grassland. Soil Use Manag 32:92–100. https://doi.org/10.1111/sum.12232

Fowler D, Flechard CR, Sutton MA, Storeton-West RL (1998) Long term measurements of the land-atmosphere exchange of ammonia over moorland. Atmos Environ 32(3):453–459. https://doi.org/10.1016/S1352-2310(97)00044-7

Fowler D, Pilegaard K, Sutton MA, Ambus P, Raivonen M, Duyzer J, Simpson D, Fagerli H, Fuzzi S, Schjoerring JK, Granier C, Neftel A, Isaksen ISA, Laj P, Maione M, Monks PS, Burkhardt J, Daemmgen U, Neirynck J, Personne E, Wichink-Kruit R, Butterbach-Bahl K, Flechard C, Tuovinen JP, Coyle M, Gerosa G, Loubet B, Altimir N, Gruenhage L, Ammann C, Cieslik S, Paoletti E, Mikkelsen TN, Ro-Poulsen H, Cellier P, Cape JN, Horváth L, Loreto F, Nijnemets Ü, Palmer PI, Rinne J, Misztal P, Nemitz E, Nilsson D, Pryor S, Gallagher MW,

Vesala T, Skiba U, Brüggemann N, Zechmeister-Boltenstern S, Williams J, O'Dowd C, Facchini MC, de Leeuw G, Flossman A, Chaumerliac N, Erisman JW (2009) Atmospheric composition change: Ecosystems-Atmosphere interactions. Atmos Environ 43(33):5193–5267. https://doi.org/10.1016/j.atmosenv.2009.07.068

Fox RH, Hoffmann LE (1981) The effect of N fertilizer source on grain yield, N uptake, soil pH, and lime requirement in no-till corn. Agron J 73:891–895

Franco AA, Pereira JC, Neyra CA (1979) Seasonal patterns of nitrate reductase and nitrogenase activities in *Phaseolus vulgaris* L. Plant Physiol 63:421–424. https://doi.org/10.1104/pp.63.3.421

Frankenberger WT Jr, Tabatabai MA (1982) Amidase and urease activities in plants. Plant Soil 64:153–166. https://doi.org/10.1007/BF02184247

Freney JR, Simpson JR (1981) Ammonia volatilization. In: Clark FE, Rosswall T (eds) Terrestrial nitrogen cycles, Ecological bulletin, vol 33. Swedish Natural Science Research Council, Stockholm, pp. 291–302

Fu MH, Tabatabai MA (1989) Nitrate reductase activity in soils: effects of trace elements. Soil Biol Biochem 21(7):943–946. https://doi.org/10.1016/0038-0717(89)90085-0

Ganry F, Guiraud G, Dommergues Y (1978) Effect of straw incorporation on the yield and nitrogen balance in the sandy soil-pearl millet cropping system of Senegal. Plant Soil 50(3):647–662

Gasche R, Papen H (2002) Spatial variability of NO and NO_2 flux rates from soil of spruce and beech forest ecosystems. Plant Soil 240:67–76. https://doi.org/10.1023/A:1015860322656

Gebauer G, Melzer A, Rehder H (1984) Nitrate content and nitrate reductase activity in *Rumex obtusifolius* L.: I. Differences in organs and diurnal changes. Oecologia 63:136–142. https://doi.org/10.1007/BF00379795

Geβler A, Kreuzwieser J, Dopatka T, Rennenberg H (2002) Diurnal courses of ammonium net uptake by the roots of adult beech (*Fagus sylvatica*) and spruce (*Picea abies*) trees. Plant Soil 240:23–32. https://doi.org/10.1023/A:1015831304911

Genermont S, Cellier P, Flura D, Morvan T, Laville P (1998) Measuring ammonia fluxes after slurry spreading under actual field conditions. Atmos Environ 32(3):279–284. https://doi.org/10.1016/S1352-2310(97)00007-1

Gerendás J, Sattelmacher B (1997) Significance of Ni supply for growth, urease activity and the concentrations of urea, amino acids and mineral nutrients of urea-grown plants. Plant Soil 190:153–162. https://doi.org/10.1023/A:1004260730027

Gibson T (1930) The decomposition of urea in soils. J Agric Sci 20:549–558

Gniazdowska-Skoczek H (1988a) Effect of nitrates and sucrose on the induction of nitrate reductase in etiolated seedling leaves from selected barley genotypes in darkness. Acta Physiol Plant 20:363–368. https://doi.org/10.1007/s11738-9980021-5

Gniazdowska-Skoczek H (1988b) Effect of light and nitrates on nitrate reductase activity and stability in seedling leaves of selected barley genotypes. Acta Physiol Plant 20:155–160. https://doi.org/10.1007/s11738-998-0007-3

Gould WD, Cook FD, Webster GR (1973) Factors affecting urea hydrolysis in several Alberta soils. Plant Soil 38(2):393–401

Groffman PM, Tiedje JM (1989) Denitrification in north temperate forest soils: spatial and temporal patterns at the landscape and seasonal scales. Soil Biol Biochem 21(5):613–620. https://doi.org/10.1016/0038-0717(89)90053-9

Grundström M, Hak C, Chen D, Hallquist M, Pleijel H (2015) Variation and co-variation of PM_{10}, particle number concentration, NO_x and NO_2 in the urban air – relationships with wind speed, vertical temperature gradient and weather type. Atmos Environ 120:317–327. https://doi.org/10.1016/j.atmosenv.2015.08.057

Guanxiong CS, Shuhui Y, Kewei Y, Adong W, Yujie Jie W (1990) Investigation on the emission of nitrous oxide by plant. J Appl Ecol 1:94–96

Guimarães GGF, Mulvaney RL, Khan SA, Cantarutti RB, Silva AM (2016) Comparison of urease inhibitor N-(n-butyl) thiophosphoric triamide and oxidized charcoal for conserving urea-N in soil. J Plant Nutr Soil Sci 179(4):520–528. https://doi.org/10.1002/jpln.201500622

Güleryüz G, Arslan H (1999) Nitrate Reductase Activity in *Verbascum* L. (*Scrophulariaceae*) species from the eastern mediterranean in dependence on altitude. Tr J Botany 23:89–96

Hargrove WL, Kissel DE (1979) Ammonia volatilization from surface applications of urea in the field and laboratory. Soil Sci Soc Am J 43:359–363

Hargrove WL, Kissel DE, Fenn LB (1977) Field measurements of ammonia volatilization from surface applications of ammonium salts to calcareous soil. Agron J 69:473–476

Hargrove WL, Rawniker RA, Bock BR (1983) Ammonia volatilization from urea in no-tillage. Agron Abst, Ann Meetings, pp 170–171

Harper LA, Catchpoole VR, Davis R, Weir KL (1983) Ammonia volatilization: soil, plant, and microclimate effects on diurnal and seasonal fluctuations. Agron J 75(2):212–218. https://doi.org/10.2134/agronj1983.00021962007500020014x

Harper LA, Sharpe RR, Langdale GW, Giddens JE (1987) Nitrogen cycling in a wheat crop: soil, plant, and aerial nitrogen transport. Agron J 79(6):965–973

Harper LA, Giddens JE, Langdale GW, Sharpe RR (1989) Environmental effects on nitrogen dynamics in soybean under conservation and clean tillage systems. Agron J 81(4):623–631

Hendriks C, Kranenburg R, Kuenen JJP, Van den Bril B, Verguts V, Schaap M (2016) Ammonia emission time profiles based on manure transport data improve ammonia modelling across north western Europe. Atmos Environ 131:83–96. https://doi.org/10.1016/j.atmosenv.2016.01.043

Hensen A, Skiba U, Famulari D (2013) Low cost and state of the art methods to measure nitrous oxide emissions. Environ Res Lett 8(2):1–10. https://doi.org/10.1088/1748-9326/8/2/025022

Henzell EF (1971) Recovery of nitrogen from four fertilizers applied to Rhodes grass in small plots. Aust J Exp Agric Anim Hus 11:420–430

Hera C (1996) The role of inorganic fertilizers and their management practices. Fertil Res 43:63–81. https://doi.org/10.1007/BF00747684

Holtan-Hartwig L, Bøckman OC (1994) Ammonia exchange between crops and air. Norw J Agric Sci Suppl 14:5–41

Hooker ML, Sander DH, Peterson GA, Daigger LA (1980) Gaseous N losses from winter wheat. Agron J 72:789–792

Horst WJ, Härdter R (1994) Rotation of maize with cowpea improves yield and nutrient use of maize compared to maize monocropping in an alfisol in the northern Guinea Savanna of Ghana. Plant Soil 160:171–183. https://doi.org/10.1007/BF00010143

Hovmand MF, Andersen HV, Løfstrøm P, Ahleson H, Jensen NO (1998) Measurements of the horizontal gradient of ammonia over a conifer forest in Denmark. Atmos Environ 32 (3):423–429. https://doi.org/10.1016/S1352-2310(96)00351-2

Hu Q, Zhang L, Evans GJ, Yao X (2014) Variability of atmospheric ammonia related to potential emission sources in downtown Toronto, Canada. Atmos Environ 99:365–373. https://doi.org/10.1016/j.atmosenv.2014.10.006

Hu Y, Zhao P, Niu J, Sun Z, Zhu L, Ni G (2016) Canopy stomatal uptake of NO_X, SO_2 and O_3 by mature urban plantations based on sap flow measurement. Atmos Environ. Part A 125:165–177. https://doi.org/10.1016/j.atmosenv.2015.11.019

Huber C, Kreutzer K (2002) Three years of continuous measurements of atmospheric ammonia concentrations over a forest standat the Höglwald site in southern Bavaria. Plant Soil 240:13–22

Hurkuck M, Brümmer C, Mohr K, Grünhage L, Flessa H, Kutsch WR (2014) Determination of atmospheric nitrogen deposition to a semi-natural peat bog site in an intensively managed agricultural landscape. Atmos Environ 97:296–309. https://doi.org/10.1016/j.atmosenv.2014.08.034

Husted S, Schjoerring JK, Nielsen KH, Nemitz E, Sutton MA (2000) Stomatal compensation points for ammonia in oilseed rape plants under field conditions. Ammonia exchange special issue. Agric For Meteorol 105(4):371–383. https://doi.org/10.1016/S0168-1923(00)00204-5

Hyde BP, Carton OT, O'Toole P, Misselbrook TH (2003) A new inventory of ammonia emissions from Irish agriculture. Atmos Environ 37(1):55–62. https://doi.org/10.1016/S1352-2310(02)00692-1

Jarvis SC, Barraclough D (1991) Variation in mineral nitrogen under grazed grassland swards. Plant Soil 138:177–188. https://doi.org/10.1007/BF00012244

Jaskulska R, Szajdak LW (2010) Impact of the shelterbelts on the nitrogen and phosphorus concentration in ground water. In: Szajdak LW, Karabanov AK (eds) Physical, chemical and biological processes in soils. Prodruk, Poznań, pp 249–258

Jaworski EG (1971) Nitrate reductase assay in intact plant tissue. Biochem Biophys Res Commun 43(6):1274–1279. https://doi.org/10.1016/S0006-291X(71)80010-4

Jenkinson DS (2001) The impact of humans on the nitrogen cycle, with focus on temperate arable agriculture. Plant Soil 228:3–15. https://doi.org/10.1023/A:1004870606003

Joensuu J, Raivonen M, Kieloaho A-J, Altimir N, Kolari P, Sarjala T, Bäck J (2015) Does nitrate fertilization induce nox emission from scots pine (*p. sylvestris*) shoots? Plant Soil 388:283–295. https://doi.org/10.1007/s11104-014-2328-x

Jordan G, Predotova M, Ingold M, Goenster S, Dietz H, Joergensen RG, Buerkert A (2015) Effects of activated charcoal and tannin added to compost and to soil on carbon dioxide, nitrous oxide and ammonia volatilization. J Plant Nutr Soil Sci 178:218–228. https://doi.org/10.1002/jpln.201400233

Karlen DL, Kramer LA, Logsdon SD (1998) Field-scale nitrogen balances associated with long-term continuous corn production. Agron J 90:644–650

Khonje DJ, Varsa EC, Klubek B (1989) The acidulation effects of nitrogenous fertilizers on selected chemical and microbiological properties of soil. Commun Soil Sci Plant Anal 20:1377–1395

Koren'kov DA, Romanyuk LI, Varyushkina NM, Kirpaneva LI (1975) Use of the stable N^{15} isotope to study the balance of fertilizer nitrogen in field lysimeters on sod-podzolic sandy loam soil. Soviet Soil Sci 7:244–249

Kułek B (2015) Nitrogen transformations in soils, agricultural plants and the atmosphere. In: Lichtfouse E (ed.) Sustainable Agriculture Reviews 18. Springer International Publishing, Cham, pp. 1–44. https://doi.org/10.1007/978-3-319-21629-4_1

Kumar V, Wagenet RJ (1984) Urease activity and kinetics of urea transformation in soils. Soil Sci 137(4):263–269

Lang M, Cai Z-C, Mary B, Hao X, Chang SX (2010) Land-use type and temperature affect gross nitrogen transformation rates in Chinese and Canadian soils. Plant Soil 334:377–389. https://doi.org/10.1007/s11104-010-0389-z

Larsen S, Gunary E (1962) Ammonia loss from ammoniacal fertilizers applied to calcareous soils. J Sci Food Agric 13:566–572

Laville P, Lehuger S, Loubet B, Chaumartin F, Cellier P (2011) Effect of management, climate and soil conditions on N_2O and NO emissions from an arable crop rotation using high temporal resolution measurements. Agric For Meteorol 151(2):228–240. https://doi.org/10.1016/j.agrformet.2010.10.008

Li C, Frolking S, Frolking TA (1992) A model of nitrous oxide evolution from soil driven by rainfall events: 1. Model structure and sensitivity. J Geophys Res 97(D9):9759–9776. https://doi.org/10.1029/92JD00509

Lightner JW, Mengel DB, Rhykerd CL (1990) Ammonia volatilization from nitrogen fertilizer surface applied to orchardgrass sod. Soil Sci Soc Am J 54:1478–1482

Lillo C, Henriksen A (1984) Comparative studies of diurnal variations of nitrate reductase activity in wheat, oat and barley. Physiol Plant 62:89–94. https://doi.org/10.1111/j.13993054.1984.tb05928.x

Liu YT, Li YE, Wan YF, Chen DL, Gao QZ, Li Y, Qin XB (2011) Nitrous oxide emissions from irrigated and fertilized spring maize in semi-arid northern China. Agric Ecosyst Environ 141 (3–4):287–295. https://doi.org/10.1016/j.agee.2011.03.002

Liu C, Yao Z, Wang K, Zheng X (2014) Three-year measurements of nitrous oxide emissions from cotton and wheat-maize rotational cropping systems. Atmos Environ 96:201–208. https://doi.org/10.1016/j.atmosenv.2014.07.040

Liu Y, Zhou Z, Zhang X, Xu X, Chen H, Xiong Z (2015) Net global warming potential and greenhouse gas intensity from the double rice system with integrated soil-crop system management: a three-year field study. Atmos Environ 116:92–101. https://doi.org/10.1016/j.atmosenv.2015.06.018

Lin HCH, Huber JA, Gerl G, Hülsbergen KJ (2016) Nitrogen balance and nitrogen-use efficiency of different organic and conventional farming systems. Nutr Cycl Agroecosyst 105:1–23. https://doi.org/10.1007/s10705-016-9770-5

MacRae IC, Ancajas R (1970) Volatilization of ammonia from submerged tropical soils. Plant Soil 33:97–103

Magid J, Henriksen O, Thorup-Kristensen K, Mueller T (2001) Disproportionately high N-mineralisation rates from green manures at low temperatures – implications for modeling and management in cool temperate agro-ecosystems. Plant Soil 228:73–82. https://doi.org/10.1023/A:1004860329146

Maheswari M, Nair TVR, Abrol YP (1992) Ammonia metabolism in the leaves and ears of wheat (*Triticum aestivum* L.) during growth and development. J Agron Crop Sci 168:310–317. https://doi.org/10.1111/j.1439-037X.1992.tb01014.x

Marshall VG, Debell DS (1980) Comparison of four methods of measuring volatilization losses of nitrogen following urea fertilization on forest soils. Can J Soil Sci 60:549–563

Mazur T (1987) The content of mineral nitrogen forms in the soil treated with slurry, manure and NPK on the basis of fourteen-year field experiment. Polish J Soil Sci XX(2):47–52

Mazur T, Budzyńska D (1994) Nitrogen leaching from soil fertilized with animal slurry, manure and NPK. Polish J Soil Sci 37(2):151–157

McGarity JW, Myers MG (1967) A survey of urease activity in soils of northern New South Wales. Plant Soil 27(2):217–238. https://doi.org/10.1007/BF01373391

McGinn SM, Sommer SG (2007) Ammonia emissions from land-applied beef cattle manure. Can J Soil Sci 87:345–352. https://doi.org/10.4141/S06-053

McGinn SM, Janzen HH, Coates T (2003) Atmospheric ammonia, volatile fatty acids, and other odorants near beef feedlots. J Environ Qual 32:1173–1182. https://doi.org/10.2134/jeq2003.1173

Medinets S, Skiba U, Rennenberg H, Butterbach-Bahl K (2015) A review of soil NO transformation: associated processes and possible physiological significance on organisms. Soil Biol Biochem 80:92–117. https://doi.org/10.1016/j.soilbio.2014.09.025

Medinets S, Gasche R, Skiba U, Medinets V, Butterbach-Bahl K (2016) The impact of management and climate on soil nitric oxide fluxes from arable land in the Southern Ukraine. Atmos Environ 137:113–126. https://doi.org/10.1016/j.atmosenv.2016.04.032

Mennen MG, Van Elzakker BG, Van Putten EM, Uiterwijk JW, Regts TA, Van Hellemond J, Wyers GP, Otjes RP, Verhage AJL, Wouters LW, Heffels CJG, Römer FG, Van den Beld L, Tetteroo JEH (1996) Evaluation of automatic ammonia monitors for application in an air quality monitoring network. Atmos Environ 30(19):3239–3256. https://doi.org/10.1016/1352-2310(96)00079-9

Menzi H, Katz PE, Fahrni M, Neftel A, Frick R (1998) A simple empirical model based on regression analysis to estimate ammonia emissions after manure application. Atmos Environ 32(3):301–307. https://doi.org/10.1016/S1352-2310(97)00239-2

Menzi H, Frick R, Kaufmann R (1997) Ammoniak-Emissionen in der Schweiz: Ausmass und technische Beurteilung des Reduktionspotentials. Eidgenössische Forschungsanstalt für Agrarökologie und Landbau (FAL), Zürich-Reckenholz, Schriftenreihe der FAL 26, 107 pp.

Meysner T, Szajdak LW (2013) Changes content of the nitrogen forms and urease activity in the ecological, conventional and integrated farming systems. ProEnvironment 6:124–129

Mikkelsen DS, De Datta SK, Obceniea WN (1978) Ammonia losses from flooded rice soils. Soil Sci Soc Am J 42:725–730

Misselbrook TH, Van Der Weerden TJ, Pain BF, Jarvis SC, Chambers BJ, Smith KA, Phillips VR, Demmers TGM (2000) Ammonia emission factors for UK agriculture. Atmos Environ 34(6):871–880. https://doi.org/10.1016/S1352-2310(99)00350-7

Morgan JA, Parton WJ (1989) Characteristics of ammonia from spring wheat. Crop Sci 29:726–731

Mosier AR, Duxbury JM, Freney JR, Heinemeyer O, Minami K (1996) Nitrous oxide emissions from agricultural fields: assessment, measurement and mitigation. Plant Soil 181:95–108. https://doi.org/10.1007/BF00011296

Mosier AR (2001) Exchange of gaseous nitrogen compounds between agricultural systems and the atmosphere. Plant Soil 228:17–27. https://doi.org/10.1023/A:1004821205442

Mouzourides P, Kumar P, Neophytou MK-A (2015) Assessment of long-term measurements of particulate matter and gaseous pollutants in South-East Mediterranean. Atmos Environ 107:148–165. https://doi.org/10.1016/j.atmosenv.2015.02.031

Myers M, McGarity JW (1968) The urease activity in profiles of five great soil groups from northern New South Wales. Plant Soil 28:25–37

Myers RJK, Paul EA (1971) Plant uptake and immobilization of ^{15}N-labelled ammonium nitrate in a field experiment with wheat. In: Nitrogen-15 in soil-plant studies. IAEA, Vienna, pp 55–64

Naseem MG, Nasrallah AK (1981) The effect of sulfur on the response of cotton to urea under alkali soil conditions in pot experiments. Plant Soil 62:255–263. https://doi.org/10.1007/BF02374089

Nicholas JC, Harper JE, Hageman RH (1976) Nitrate reductase activity in soybeans (*Glycine max* [L.] Merr.). II. Energy limitations. Plant Physiol 58(6):736–739

Nicolini G, Castaldi S, Fratini G, Valentini R (2013) A literature overview of micrometeorological CH_4 and N_2O flux measurements in terrestrial ecosystems. Atmos Environ 81:311–319. https://doi.org/10.1016/j.atmosenv.2013.09.030

Nômmik H, Pluth DJ, Larsson K, Mahendrappa MK (1994) Isotopic fractionation accompanying fertilizer nitrogen transformations in soil and trees of a Scots spine ecosystem. Plant Soil 158:169–182. https://doi.org/10.1007/BF00009492

Overrein LN (1968) Lysimeter studies on tracer nitrogen in forest soil. I. Nitrogen losses by leaching and volatilization after addition of urea-N^{15}. Soil Sci 106:280–290

Pain BF, Van der Weerden TJ, Chambers BJ, Phillips VR, Jarvis SC (1998) A new inventory for ammonia emissions from U.K. agriculture. Atmos Environ 32:309–313. https://doi.org/10.1016/S1352-2310(96)00352-4

Paton-Walsh C, Wilson SR, Naylor T, Griffith DWT, Denmead OT (2011) Transport of NO_X emissions from sugarcane fertilisation into the great barrier reef lagoon. Environ Model Assess 16(5):441–452. https://doi.org/10.1007/s10666-011-9260-8

Pattnaik P, Mallick K, Ramakrishnan B, Adhya TK, Sethunathan N (1999) Urease activity and urea hydrolysis in tropical flooded soil unplanted or planted to rice. J Sci Food Agric 79:227–231

Paulson KN, Kurtz LT (1969) Locus of urease activity in soil. Soil Sci Soc Am Proc 33:897–901

Pearson J, Soares A (1998) Physiological responses of plant leaves to atmospheric ammonia and ammonium. Atmos Environ 32(3):533–538. https://doi.org/10.1016/S1352-2310(97)00008-3

Petersen SO, Stamatiadis S, Christofides C (2004) Short-term nitrous oxide emissions from pasture soil as influenced by urea level and soil nitrate. Plant Soil 267:117–127. https://doi.org/10.1007/s11104-005-4688-8

Philippe F-X, Laitat M, Wavreille J, Nicks B, Cabaraux J-F (2015) Effects of a high-fibre diet on ammonia and greenhouse gas emissions from gestating sows and fattening pigs. Atmos Environ 109:197–204. https://doi.org/10.1016/j.atmosenv.2015.03.025

Phillips SB, Arya SP, Aneja VP (2004) Ammonia flux and dry deposition velocity from near-surface concentration gradient measurements over a grass surface in North Carolina. Atmos Environ 38:3469–3480. https://doi.org/10.1016/j.atmosenv.2004.02.054

Pogány A, Weidinger T, Bozóki Z, Mohácsi Á, Bieńkowski J, Józefczyk D, Eredics A, Bordás Á, Gyöngyösi AZ, Horváth L, Szabó G (2012) Application of a novel photoacoustic instrument for ammonia concentration and flux monitoring above agricultural landscape – results of a field measurement campaigh in Choryń, Poland. IDŐJÁRÁS 116(2):93–107

Power JF (1979) Use of slow release N fertilizers on native mixed prairie. Agron J 71:446–449

Rachhpal S, Nye PH (1988) Processes controlling ammonia losses from fertilizer urea. In: Jenkinson DS, Smith KA (eds) Nitrogen efficiency in agricultural soils. Elsevier Applied Science, London / New York, pp. 246–255

Rao LVM, Rajasekhar VK, Sopory SK, Guha-Mukherjee S (1981) Plant Cell Physiol 22 (3):577–582

Reeves S, Wang W, Salter B, Halpin N (2016) Quantifying nitrous oxide emissions from sugarcane cropping systems: optimum sampling time and frequency. Atmos Environ 136:123–133. https://doi.org/10.1016/j.atmosenv.2016.04.008

Robarge WP, Walker JT, McCulloch RB, Murray G (2002) Atmospheric concentrations of ammonia and ammonium at an agricultural site in the southeast United States. Atmos Environ 36(10):1661–1674. https://doi.org/10.1016/S1352-2310(02)00171-1

Rodgers GA, Pruden G (1984) Field Estimation of ammonia volatilisation from ^{15}N-labelled urea fertiliser. J Sci Food Agric 35:1290–1293. https://doi.org/10.1002/jsfa.2740351204

Sahrawat KL (1980) On the criteria for comparing the ability of compounds for retardation of nitrification in soil. Plant Soil 55:487–490. https://doi.org/10.1007/BF02182707

Sahrawat KL (1982) Nitrification in some tropical soils. Plant Soil 65:281–286. https://doi.org/10.1007/BF02374659

Sahrawat KL (1984) Effects of temperature and moisture on urease activity in semi-arid tropical soils. Plant Soil 78:401–408. https://doi.org/10.1007/BF02450373

Salazar F, Martínez-Lagos J, Alfaro M, Misselbrook T (2014) Ammonia emission from a permanent grassland on volcanic soil after the treatment with dairy slurry and urea. Atmos Environ 95:591–597. https://doi.org/10.1016/j.atmosenv.2014.06.057

Schjoerring JK (1995) Long-term quantification of ammonia exchange between agricultural cropland and the atmosphere – I. Evaluation of a new method based on passive flux samplers in gradient configuration. Atmos Environ 29(8):885–893. https://doi.org/10.1016/1352-2310(95)00020-Y

Schjoerring JK, Mattsson M (2001) Quantification of ammonia exchange between agricultural cropland and the atmosphere: measurements over two complete growth cycles of oilseed rape, wheat, barley and pea. Plant Soil 228:105–115. https://doi.org/10.1023/A:1004851001342

Schjoerring JK, Husted S, Mattsson M (1998) Physiological parameters controlling plant-atmosphere ammonia exchange. Atmos Environ 32(3):491–498. https://doi.org/10.1016/S1352-2310(97)00006-X

Schjørring JK, Nielsen NE, Jensen HE, Gottschau A (1989) Nitrogen losses from field-grown spring barley plants as affected by rate of nitrogen application. Plant Soil 116:167–175

Schjørring JK (1991) Ammonia emission from the foliage of growing plants. In: Sharkey TD, Holland EA, Mooney HA (eds) Trace gas emission by plants. Academic, San Diego, pp. 267–292

Schjoerring JK, Kyllingsbaek A, Mortensen JV, Byskov-Nielsen S (1993a) Field investigations of ammonia exchange between barley plants and the atmosphere. I. Concentration profiles and flux densities of ammonia. Plant Cell Environ 16:161–167. https://doi.org/10.1111/j.1365-3040.1993.tb00857.x

Schjoerring JK, Kyllingsbaek A, Mortensen JV, Byskov-Nielsen S (1993b) Field investigations of ammonia exchange between barley plants and the atmosphere. II. Nitrogen reallocation, free ammonium content and activities of ammonium-assimilating enzymes in different leaves. Plant Cell Environ 16:169–178. https://doi.org/10.1111/j.1365-3040.1993.tb00858.x

Shen J, Liu X, Zhang Y, Fangmeier A, Goulding K, Zhang F (2011) Atmospheric ammonia and particulate ammonium from agricultural sources in the North China Plain. Atmos Environ 45:5033–5041. https://doi.org/10.1016/j.atmosenv.2011.02.031

Skoczek H (1992) Nitrate reductase activity and protein content in leaves and roots of two doubled haploid barley lines depending on seedling age and nitrogen source. Acta Physiol Plant 14 (2):77–84

Smith JH (1964) Relationships between soil cation-exchange capacity and the toxicity of ammonia to the nitrification process. Soil Sci Soc Am Proc 28:640–644

Somers DA, Kuo TM, Kleinhofs A, Warner RL, Oaks A (1983) Synthesis and degradation of barley nitrate reductase. Plant Physiol 72:949–952. https://doi.org/10.1104/pp.72.4.949

Sommer SG, Christensen BT (1992) Ammonia volatilization after injection of anhydrous ammonia into arable soils of different moisture levels. Plant Soil 142:143–146. https://doi.org/10.1007/BF00010184

Sommer SG, Jensen C (1994) Ammonia volatilization from urea and ammoniacal fertilizers surface applied to winter wheat and grassland. Fertil Res 37:85–92. https://doi.org/10.1007/BF00748549

Sommer SG, Olesen JE, Christensen BT (1991) Effects of temperature, wind speed and air humidity on ammonia volatilization from surface applied cattle slurry. J Agric Sci 117:91–100

Stalenga J, Kawalec A (2008) Emission of greenhouse gases and soil organic matter balance in different farming systems. Int Agrophys 22(3):287–290

Stehfest E, Bouwman L (2006) N_2O and NO emission from agricultural fields and soils under natural vegetation: summarizing available measurement data and modeling of global annual emissions. Nutr Cycl Agroecosyst 74:207–228. https://doi.org/10.1007/s10705-006-9000-7

Singh M, Yadav DS (1981) Transformation of urea and ammonium sulphate in different soils. Plant Soil 63:511–515. https://doi.org/10.1007/BF02370053

Stinn JP, Xin H, Shepherd TA, Li H, Burns RT (2014) Ammonia and greenhouse gas emissions from a modern U.S. swine breeding-gestation-farrowing system. Atmos Environ 98:620–628. https://doi.org/10.1016/j.atmosenv.2014.09.037

Sutton MA (1990) The surface / atmosphere exchange of ammonia. Ph.D. Thesis, University of Edinburgh, Edinburg, p 194

Sutton MA, Moncrieff JB, Fowler D (1992) Deposition of atmospheric ammonia to moorlands. Environ Pollut 75(1):15–24. https://doi.org/10.1016/0269-7491(92)90051-B

Sutton MA, Place CJ, Eager M, Fowler D, Smith RI (1995) Assessment of the magnitude of ammonia emissions in the United Kingdom. Atmos Environ 29:1393–1411. https://doi.org/10.1016/1352-2310(95)00035-W

Sutton MA, Asman WAH, Ellermann T, Van Jaarsveld JA, Acker K, Aneja V, Duyzer J, Horvath L, Paramonov S, Mitosinkova M, Tang YS, Achermann B, Gauger T, Bartniki J, Neftel A, Erisman JW (2003) Establishing the link between ammonia emission control and measurements of reduced nitrogen concentrations and deposition. Environ Monit Assess 82(2):149–185. https://doi.org/10.1023/A:1021834132138

Svensson L (1994) Ammonia volatilization following application of livestock manure to arable land. J Agric Eng Res 58:241–260. https://doi.org/10.1006/jaer.1994.1054

Szajdak LW, Gaca W (2010) The influence of nitrogen on denitrification processes in soil under shelterbelt and adjoining cultivated field. In: Szajdak LW, Karabanov AK (eds) Physical, chemical and biological processes in soils. "Prodruk" Publishing–Printing House, Poznań, pp. 225–235

Szajdak L, Matuszewska T (2000) Reaction of woods in changes of nitrogen in two kinds of soil. Polish J Soil Sci XXXIII(1):9–17

Tabatabai MA, Bremner JM (1972) Assay of urease activity in soils. Soil Biol Biochem 4 (4):479–487. https://doi.org/10.1016/0038-0717(72)90064-8

Tan XW, Ikeda H, Oda M (2000) The absorption, translocation, and assimilation of urea, nitrate or ammonium in tomato plants at different plant growth stages in hydroponic culture. Scientia Hortic 84:275–283. https://doi.org/10.1016/S0304-4238(99)00108-9

Terman GL (1979) Volatilization losses of nitrogen as ammonia from surface-applied fertilizers, organic amendments and crop residue. Adv Agron 31:189–223

Thomsen IK (1993) Nitrogen uptake in barley after spring incorporation of ^{15}N-labelled Italian ryegrass into sandy soils. Plant Soil 150:193–201. https://doi.org/10.1007/BF00013016

Thöni L, Brang P, Braun S, Seitler E, Rihm B (2004) Ammonia monitoring in Switzerland with passive samplers: patterns, determinants and comparison with modelled concentrations. Environ Monit Assess 98:93–107. https://doi.org/10.1023/B:EMAS.0000038181.99603.6e

Touchton JT, Hargrove WL (1982) Nitrogen sources and methods of application for no-tillage corn production. Agron J 74:823–826

Van Cleemput O, Baert L (1984) Nitrite: a key compound in N loss processes under acid conditions? Plant Soil 76:233–241. https://doi.org/10.1007/BF02205583

Van der Hoek KW (1998) Estimating ammonia emission factors in Europe: summary of the work of the UNECE ammonia expert panel. Atmos Environ 32(3):315–316. https://doi.org/10.1016/S1352-2310(97)00168-4

Van Slyke DD, Zacharias G (1914) The effect of hydrogen ion concentration and of inhibiting substances on urease. J Biol Chem 19:181–210

Velthof GL, Oenema O, Postma R, Van Beusichem ML (1997) Effects of type and amount of applied nitrogen fertilizer on nitrous oxide fluxes from intensively managed grassland. Nutr Cycl Agroecosyst 46:257–267. https://doi.org/10.1007/BF00420561

Vermoesen A, de Groot C-J, Nollet L, Boeckx P, van Cleemput O (1996) Effect of ammonium and nitrate application on the NO and N_2O emission out of different soils. Plant Soil 181:153–162. https://doi.org/10.1007/BF00011302

Viguria M, López DM, Arriaga H, Merino P (2015) ammonia and greenhouse gases emission from on-farm stored pig slurry. Water Air Soil Pollut 226:285. https://doi.org/10.1007/s11270-015-2548-6

Vlek PLG, Craswell ET (1979) Effect of nitrogen source and management on ammonia volatilization losses from flooded rice-soil systems. Soil Sci Soc Am J 43:352–358

Vlek PLG, Stumpe JM (1978) Effect of solution chemistry and environmental conditions on ammonia volatilization losses from aqueous systems. Soil Sci Soc Am J 42:416–421

Vos GJM, Bergevoet IMJ, Védy JC, Neyroud JA (1994) The fate of spring applied fertilizer N during the autumn-winter period: comparison between winter-fallow and green manure cropped soil. Plant Soil 160(2):201–213. https://doi.org/10.1007/BF00010146

Wang J, Pan X, Liu Y, Zhang X, Xiong Z (2012) Effects of biochar amendment in two soils on greenhouse gas emissions and crop production. Plant Soil 360(1):287–298. https://doi.org/10.1007/s11104-012-1250-3

Watkins N, Barraclough D (1996) Gross rates of N mineralization associated with the decomposition of plant residues. Soil Biol Biochem 28(2):169–175. https://doi.org/10.1016/0038-0717(95)00123-9

Watkins SH, Strand RF, Debell DS, Esch J Jr (1972) Factors influencing ammonia losses from urea applied to northwestern forest soils. Soil Sci Soc Am Proc 36:354–357

Watson CJ, Poland P, Miller H, Allen MBD, Garrett MK, Christianson CB (1994) Agronomic assessment and ^{15}N recovery of urea amended with the urease inhibitor nBTPT (N-(n-butyl) thiophosphoric triamide) for temperate grassland. Plant Soil 161:167–177. https://doi.org/10.1007/BF00046388

Webb J, Thorman RE, Fernanda-Aller M, Jackson DR (2014) Emission factors for ammonia and nitrous oxide emissions following immediate manure incorporation on two contrasting soil types. Atmos Environ 82:280–287. https://doi.org/10.1016/j.atmosenv.2013.10.043

Welch DC, Colls JJ, Demmers TGM, Wathes CM (2005) A methodology for the measurement of distributed agricultural sources of ammonia outdoors – Part 1: validation in a controlled environment. Atmos Environ 39(4):663–672. https://doi.org/10.1016/j.atmosenv.2004.10.023

Wetselaar R, Farquhar GD (1980) Nitrogen loss from tops of plants. Adv Agron 33:263–302

Whitehead DC, Raistrick N (1990) Ammonia volatilization from five nitrogen compounds used as fertilizers following surface application to soils. J Soil Sci 41:387–394. https://doi.org/10.1111/j.1365-2389.1990.tb00074.x

Wichink Kruit RJ, van Pul WAJ, Otjes RP, Hofschreuder P, Jacobs AFG, Holtslag AAM (2007) Ammonia fluxes and derived canopy compensation points over non-fertilized agricultural grassland in The Netherlands using the new gradient ammonia-high accuracy-monitor (GRAHAM). Atmos Environ 41:1275–1287. https://doi.org/10.1016/j.atmosenv.2006.09.039

Xu P, Zhang Y, Gong W, Hou X, Kroeze C, Gao W, Luan S (2015) An inventory of the emission of ammonia from agricultural fertilizer application in China for 2010 and its high-resolution spatial distribution. Atmos Environ 115:141–148. https://doi.org/10.1016/j.atmosenv.2015.05.020

Zantua MI, Bremner JM (1975) Comparison of methods of assaying urease activity in soils. Soil Biol Biochem 7:291–295

Zbieranowski AL, Aherne J (2012) Spatial and temporal concentration of ambient atmospheric ammonia in southern Ontario, Canada. Atmos Environ 62:441–450. https://doi.org/10.1016/j.atmosenv.2012.08.041

Zhang A, Liu Y, Pan G, Hussain O, Li L, Zheng J, Zhang X (2012) Effect of biochar amendment on maize yield and greenhouse gas emissions from a soil organic carbon poor calcareous loamy soil from Central China Plain. Plant Soil 351:263–275. https://doi.org/10.1007/s11104-011-0957-x

Zhao M, Tian Y, Zhang M, Yao Y, Ao Y, Yin B, Zhu Z (2015) Nonlinear response of nitric oxide emissions to a nitrogen application gradient: A case study during the wheat season in a Chinese rice-wheat rotation system. Atmos Environ 102:200–208. https://doi.org/10.1016/j.atmosenv.2014.11.052

Chapter 2
Impact of Tillage Methods on Environment, Energy and Economy

Egidijus Šarauskis, Zita Kriaučiūnienė, Kęstutis Romaneckas, and Sidona Buragienė

Abstract Soil tillage involves the mechanical manipulation of soils used for crop production. Tillage is done to prepare an optimal seedbed, to loosen compacted soil layers, to control weeds, to increase aeration, to incorporate plant residues into the soil, to facilitate water infiltration and soil moisture storage, and to control soil temperature. Nonetheless, soil tillage is one of the highest energy-consuming, environment-polluting and expensive technological processes in agriculture. Conventional tillage with ploughing is the most widely used practice. Conventional tillage has low efficiency, requires high-powered tractors with high fuel consumption and greenhouse gases emissions. Moreover, the cost of conventional tillage is high, and the influence on the soil structure, degradation, leaching of nutrients and the most fertile soil is negative. Here we review the impact of tillage methods on soil quality, environment and economy.

Due to the disadvantages of conventional tillage, sustainable tillage area increases each year by 4–6 million ha worldwide. Under sustainable tillage such as minimal or no-tillage, the total soil surface modified by the wheels of agricultural machinery is 20–40% lower than for conventional tillage. Sustainable tillage preserves better soil physical properties and biological processes. A comparison of tillage methods show that no-tillage has the highest energy efficiency ratio of 14.0, versus 12.4 for deep ploughing. The most expensive tillage operation is deep ploughing. The use of agricultural machinery under sustainable tillage conditions and preparation of soils without using a plough can reduce costs from 25% to 41%, compared with conventional tillage.

E. Šarauskis (✉) · S. Buragienė
Institute of Agricultural Engineering and Safety, Aleksandras Stulginskis University, Kaunas, Lithuania
e-mail: egidijus.sarauskis@asu.lt

Z. Kriaučiūnienė · K. Romaneckas
Institute of Agroecosystems and Soil Science, Aleksandras Stulginskis University, Kaunas, Lithuania

E. Lichtfouse (ed.), *Sustainable Agriculture Reviews 33*, Sustainable Agriculture Reviews 33, https://doi.org/10.1007/978-3-319-99076-7_2

Keywords Tillage · Soil Properties · CO_2 emissions · Fuel Consumption · Energy · Costs · Environment

2.1 Introduction

Environmentally friendly and energy saving agricultural technologies are integrated for agricultural production as such technologies deliver the most innovative, economical, energetic and environmental benefits (Morris et al. 2010; Šarauskis et al. 2012; Reicosky 2015; Mitchell et al. 2016). The essence of these technologies has several aims, such as to limit intensive mechanical and chemical impacts on soil and vegetation; provide soil productivity renovation; protect the environment; rationally use material, energetic and labour resources; meet strict environmental regulations; produce wholesome food; and guarantee economic effectiveness in the manufacturing of agricultural produce.

Data published by the European Conservation Agriculture Federation (ECAF) reveals that implementation of conservation agriculture techniques enables many environmental and economic benefits, such as (www.ecaf.org):

- An improvement in soil properties
- An increase in biodiversity
- A reduction in erosion
- A reduction in the contamination of downstream water
- A reduction in the number of floods and landslides
- A reduction of CO_2 emissions
- An increase in labour and fuel saving benefits
- An increase in cost-saving benefits

However, it is impossible to introduce new and sustainable agricultural technologies without the use of modern agricultural machinery that also meets augmented land protection and environmental requirements, the most important of which are as follows: soil must not be depleted, humus reduction and soil degradation should not occur, leaching of nutrients and the most fertile soil particles should be minimized, the soil should not be eroded and the soil structure should not be broken, the natural biological processes in the soil should be enhanced (Reicosky 2015).

Soil tillage involves the physical-mechanical manipulation or disturbance of soil for the purpose of crop production (Köller 2003; Reicosky and Allmaras 2003; Mitchell et al. 2016) and the main aims of tillage are: to prepare an optimal seedbed, to loosen compacted soil layers, to control pests, to increase aeration, to incorporate crop and weed residues into the soil, to inject or incorporate fertilizers and pesticides, to facilitate water infiltration and soil moisture storage, to stimulate net nitrogen mineralization, to control soil temperature and salinity, to mix soil layers and to increase rooting (Reicosky and Allmaras 2003; Mitchell et al. 2016). Conventional soil tillage and drilling machinery cannot fully meet the requirements of sustainable soil tillage, which involves the use of more complex soil tillage and seed introduction technological processes and specifies requirements for adhering to construction parameters and operation technological conditions when undertaking soil tillage and using drilling machinery. In this respect, scientific research previously conducted (Kushwaha et al. 1986; Linke 1998; Karayel 2009; Morris et al. 2010; Buragiene et al. 2015; Šarauskis et al. 2017) has confirmed the need for urgent studies to achieve sustainable practice.

Studies have shown that no-tilled soils under the impact of various environmental factors undergo a change in physical and mechanical properties when the soil surface is covered by residues from the previous yield (Buragiene et al. 2015); therefore, an interaction between the operating parts of the soil tillage machinery and the soil or crop residues invariably affects such changes.

2.2 Sustainable Tillage Technologies

Costly energy resources, climate change, loss of fertile soil layers, moisture retention, the need to save labour time and other factors are among the most important motives forcing the development of agricultural systems (Šimanskaite 2007; Morris et al. 2010; Soane et al. 2012; Šarauskis et al. 2014a, b; Buragiene et al. 2015) and their appropriate selection according to regionality. Therefore, the development of environment- and energy-sustaining tillage technologies is one of the most important goals required for modern agricultural progress. The use of sustainable agricultural technologies results in improved soil structure and phyto-sanitary conditions. When employing such technologies, plant residues are incorporated efficiently for fertilisation and soil protection, precision tillage machinery is used (thereby reducing the negative effect of technological processes on the environment), and labour, energy and agricultural production costs are reduced (Linke 1998; Tebrügge 2001; Hazarika et al. 2009; Lithourgidis et al. 2009; Soane et al. 2012; Derpsch et al. 2016).

2.2.1 Conservation Agriculture Worldwide

There has been a fast growth in the use of environmentally sustainable tillage and drilling technologies over the past 30 years, and their application has been promoted by the need for lower energy and labour costs. No-tillage or zero tillage technology, where drilling is performed directly into undisturbed soil, is particularly beneficial for protecting the soil and enabling economic and environmental advantages. In English, this method of drilling is known as '*Direct Drilling*', in German '*Direktsaat*' and in Russian '*Прямой посев*'. In certain cases, when drilling is performed into stubble, it is known as 'stubble drilling'. However, stubble drilling is only one of the technologies attributed to direct drilling, as drilling can be

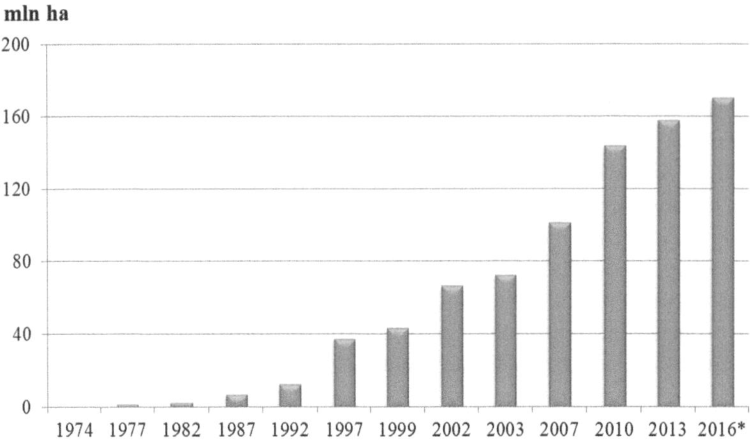

Fig. 2.1 Global uptake of conservation agriculture in million ha of arable cropland. *expected. (Prepared according to Kassam et al. 2015)

performed not only after growth of cereals, but also after grasses, root crops and other plants. Direct drilling is a no-tillage technology, in which the upper soil layer is minimally disturbed by the driller's coulters so that it only incorporates plant seeds.

Complex soil-crop-nutrient-water-landscape system management practices are known as Conservation Agriculture (Kassam et al. 2015). Conservation agriculture saves energy and mineral nitrogen used in farming, thereby reducing GHG emissions and enhancing biological activity in soils. However, no-tillage is not always sufficient for promoting sustainable and productive agriculture. Conservation agriculture also involves soil cover and cropping system diversification, and needs to be complemented with other techniques, such as integrated pest management, plant nutrient management, weed and water management. Total global areas under conservation agriculture systems are increasing (Derpsch and Friedrich 2009; López et al. 2012; Kassam et al. 2015) each year by 4–6 million hectares, as shown in Figure 2.1.

Conservation agriculture is most popular in the USA (35.6 million ha), Brazil (31.8 million ha), Argentina (29.2 million ha), Canada (18.3 million ha), Australia (17.7 million ha) and other countries where soil erosion is a major problem. The global extent of conservation agriculture cropland over the five-year period (2008–2013) increased by 47%, from 106 million ha (7.5% of global cropland) to approximately 157 million ha (11% of global cropland) (Kassam et al. 2009; Kassam et al. 2015). The area of agricultural land within the European Union (EU) accounts for approximately 161.6 million ha (Derpsch and Friedrich 2009); areas under conservation agriculture within the EU span approximately 2.0 million ha and are located mainly in Spain, Italy, Finland, France, Germany and the United Kingdom. Since 2008, the area of annual crops under conservation agriculture has changed by some 30%, from 1.6 million ha to 2.0 million ha in 2013, corresponding to 2.8% of arable cropland (Derpsch and Friedrich 2009; Kassam et al. 2015).

Ploughing is the main tillage technique used in most European countries (Arvidsson et al. 2013), although a growing switch from conventional tillage to different sustainable tillage technologies has occurred over the past 20 years. Such technologies do not use conventional tillage; instead they employ minimum tillage and implement different sustainable tillage technologies (Alvarez and Steinbach 2009; Morris et al. 2010; Buragiene et al. 2015), which are not suitable for all European agroecosystems. However, they conserve water and soil, thereby ensuring long-term ecological, social and economic sustainability (Lahmar 2010). When farming methods are properly selected and controlled, many of the adverse effects caused by agricultural machinery and agricultural systems can be eliminated.

It is evident that new farming systems and technologies should be implemented in many countries (Chamen et al. 2003), and it is recommended that direct drilling and other reduced tillage technologies are used on farms adapted to conservation agriculture (Celik et al. 2012). However, the economic success of using sustainable tillage technologies depends largely on the farm's locality and climatic factors (Uri 2000) and the spread of sustainable tillage systems is often prevented by a cool wet climate (Cannell and Hawes 1994), increased problems of weed spreading and the complicated incorporation of plant residues into soil.

The main technological aims of conservation agriculture are as follows (Linke 1998; Holland 2004; Lahmar 2010; Morris et al. 2010; Sommer et al. 2014; Derpsch et al. 2016):

- To reduce the negative mechanical impact on soil by reducing the intensity of topsoil loosening;
- To gradually reduce the use of tillage, apply zero tillage, minimum tillage or direct drilling where applicable;
- To reduce the depth of straw and other plant residue incorporation and mulching of soils, thus stimulating biological process on the soil surface;
- To increase field fertilisation with compost and green manure, improving the incorporation uniformity of liquid manure, slurry and other organic fertilisers;
- To apply correct doses of fertilizers according to plant needs and soil agrochemical properties, and to locally incorporate mineral fertilisers;
- To make use of mechanical weed control opportunities;
- To apply local pesticide spraying in microdoses according to phytosanitary conditions, amount of weeds and pest infestation.

2.2.2 Soil Protection Aspects of Tillage

Soil is the main environmental parameter involved in the technological process of tillage. The upper layer of soil is vitally important both for plant growth and for maintaining the environment. It performs significant ecological and economical functions as a living environment, water regulator, genetic resource, food producer and a source of raw materials and biomass. Healthy soil is required for plants to grow, as it supplies plants with air, water, nutrients and physically sustains growth. Fertile soil has a neutral or near neutral pH, is abundant with substances appropriate for plants to assimilate and contains sufficient humus and microorganisms; it is thus the most important agricultural resource and has long been acknowledged as such (Pekrun and Claupein 1998; Ogle et al. 2005; Alvarez 2005; Paz-Ferreiro et al. 2009).

The most important objective of tillage is to provide favourable conditions for plant growth. Selection of tillage technology depends on various soil properties and agrotechnological conditions, and the greatest influences on such a selection are: the plant species cultivated, meteorological conditions, relief, level of soil cultivation, content of organic matter, amount of weeds, moisture conditions, soil structure, available agricultural machinery and technological expertise.

Each tillage system has advantages and disadvantages. Application of conventional tillage by ploughing provides an opportunity to achieve a higher yield; however, due to the low performance of tillage technological operations and the need for high-powered tractors, the cost of such tillage is high. In addition, conventional tillage has a negative impact on the environment, soil and biodiversity (Linke 1998; Holland 2004; Šarauskis 2009; Arvidsson 2010; Morris et al. 2010; Šarauskis et al. 2014a, b), therefore it is unwise to make an irrational selection of tillage machinery or its technological modes. Furthermore, the use of increased loads of machinery causes higher amounts of toxic oxides to be emitted into the environment, strongly effecting the natural environmental ecosystem (Janulevičius et al. 2013).

In the search for alternatives to reduce intensive full inversion tillage, environmentally sustainable minimum tillage or direct drilling technologies are becoming increasingly popular. Such technologies reduce the impact of tillage machinery on soil degradation and soil properties, enabling reduced soil layer compaction; thereby ensuring natural water filtration and penetration of plant roots into diverse soil layers (Lamandé et al. 2007; Fritton 2008; Cavalieri et al. 2008).

The main objective of sustainable agriculture is to limit intensive technological (mechanical, chemical and biological) impacts on soil and plants and reduce their negative consequences, thus safeguarding continual regeneration of soil productivity, sustaining the biosphere and maintaining economically efficient production. However, sustainable agriculture requires adherence to certain rules, such as the more rational use of materials, energy and labour resources, compliance with strict environmental requirements and the production of healthy and cheap agricultural products. Nevertheless, perhaps the most important objective of conservation agriculture is to protect the soil and prevent soil impoverishment, stop deterioration and degradation of humus, reduce leaching of nutrients, protect the soil from erosion and structure disruption, stimulate natural biological processes, readjust the balance of organic matter conversion in the soil and improve aeration and the moisture content of topsoil (Holland, 2004).

Soil is an essential part of the natural environment and is actively influenced by human activities. Agriculture, forestry and natural ecosystems are continually affected by soil degradation processes. One of the main reasons for soil degradation is soil erosion, i.e. wearing away of the fertile upper layer of soil, and the main processes involved are water and wind erosion, compaction of soil layers, soil salinization and subsequent nutrient decline. Soil degradation can reduce yields, promote climate change, decrease biodiversity and cause a shortage of good quality water resources (De Paz et al. 2006). When soil is degraded, the most fertile layer of the soil is lost and nutrients required by plants are removed. Therefore, plant products need to apply larger quantities of fertilisers to make the cultivation of plants profitable. The problems of soil degradation problems in developing countries are even bigger due to prevailing climatic conditions, insufficient financing of agriculture and fast growing populations.

Soil erosion occurs naturally, however, prior to farming it did not disrupt terrestrial ecosystems as the soil surface was rarely left without a certain amount of vegetation cover. The consequences of human commercial activities are much more severe. Massive amounts of soil are lost, particularly in cultivated fields located far from forests and in hilly areas, as the natural consequences of heavy rain and wind are especially detrimental in such areas (Fig. 2.2). When soil remains uncultivated following harvest, a larger amount of the upper-most fertile soil layer is blown or washed away (Račinskas 1992; Pocienė and Kinčius 2008). Results of previous research (Bakker et al. 2004; Wilkinson 2005) have acknowledged that human activities speed up erosion, which means that the resources present in soil and the sustainability of natural ecosystems are at risk. In addition, soil erosion is accelerated by pollution of water bodies, siltation processes, areas lying as wasteland

Fig. 2.2 Wind and water soil erosion. (Photos by Romaneckas and Šarauskis)

and causes an increased loss of organic matter and a subsequent reduced water-storage capacity (Pimentel et al. 1995; Bakker et al. 2004; Boardman and Poesen 2006).

The protection of land resources is considered one of the most important objectives of environmental policies, which require a proper assessment of the magnitude of erosion and its geographical distribution. Such an assessment is also important for research into the impact of soil erosion on global cycles (Van Oost et al. 2007; Quinton et al. 2010). Researchers from different countries (Cerdan et al. 2010) have conducted erosion calculation research and established average and total amounts of erosion in European countries (Table 2.1), where average soil erosion ranges from 0.2 to 3.2 t ha^{-1} per year (the greatest amount of erosion occurred in Slovakia and the least in Finland).

On the basis of data obtained from multiannual field experiments, the annual soil erosion, E, or runoff forecast can be calculated using the following formula (Račinskas 1992; Pocienė and Kinčius 2008),

$$E = (1.61l + 0.09i + 0.18e - 14.46)B_m D_m\left[1 - \left(C_{met}^{ter} - 0.4\right)\right]0.6K \tag{2.1}$$

where:

l: slope length, m;
i: slope angle, in degrees;
B_m: humus effect on soil erosion rate per year;
K: climatic factor depending on the thickness of snow cover;
C_{met}^{ter}: anti-erosion capacity index of agricultural plants in spring during snow melting.

The intensity of soil erosion largely depends on the relief and soil granulometric composition, and has been established as the cause of intense erosion in hilly areas (Jankauskas and Jankauskienė 2005), where soil erosion largely depends on slope steepness, soil tillage technology used and the direction of tillage (direction of tillage perpendicular to the slope gradient results in much lower soil erosion compared to tillage parallel to the slope).

The tillage system selected has a large effect on variations in the hydraulic and mechanical properties of soil (Król et al. 2013) and inappropriate systems cause an increase in soil layer compaction, which is one of the most important factors stimulating soil physical degradation (Pagliai et al. 2003). About 33 million ha of soils in Europe are degraded due to soil compaction (Van den Akker and Canarache 2001); this is a huge concern for researchers and soil conservation services. Increased soil compaction results in a reduction in soil porosity and alters pore shapes and sizes (Pagliai et al. 2003, 2004). A decrease in pore size from soil compaction negatively affects soil microorganisms and fauna, thereby diminishing positive soil properties and ultimately resulting in lower yields. Furthermore, aeration (Czyż et al. 2001) and water infiltration can be negatively affected, which causes

Table 2.1 Mean and total soil erosion rates per European country (Cerdan et al. 2010)

Country	Mean erosion [t ha^{-1} $year^{-1}$]	Total erosion [10^5 t $year^{-1}$]	Area [10^3 km^2]	Total European erosion [%]
Slovakia	3.2	156	49.0	2.8
Denmark	2.6	109	42.1	2.0
Czech Republic	2.6	202	78.9	3.7
Italy	2.3	691	299.5	12.5
Bulgaria	1.9	211	110.8	3.8
Germany	1.9	674	356.7	12.2
Romania	1.8	423	237.8	7.6
Austria	1.6	135	84.0	2.4
Poland	1.5	480	311.7	8.7
France	1.5	805	547.4	14.5
Belgium	1.4	42	30.6	0.8
Portugal	1.2	109	88.4	2.0
Slovenia	1.2	23	20.3	0.4
Hungary	1.0	96	93.1	1.7
Spain	1.0	503	497.2	9.1
Lithuania	1.0	62	64.9	1.1
United Kingdom	0.9	222	241.7	4.0
Greece	0.8	98	129.5	1.8
Latvia	0.5	35	64.5	0.6
Sweden	0.5	225	446.3	4.1
Ireland	0.5	37	68.7	0.7
Estonia	0.4	20	44.5	0.4
Croatia	0.4	24	55.7	0.4
Netherlands	0.4	12	34.5	0.2
Finland	0.2	82	334.8	1.5

an increasing amount of surface water, thereby contributing to favourable conditions for soil erosion (Horn et al. 1995; Fleige and Horn 2000).

When deeper soil layers are compacted by the wheels of heavy tractors, combines and other agricultural machinery, the quality of surface and ground water and the value of soil resources (Soane and Van Ouwerkerk 1995) are diminished. The mass of agricultural machinery has been increasing rapidly over the last few decades and there is a tendency for further increase. In addition, structural changes in European (especially Eastern European) agriculture have caused a transition to larger-scale farms, which are more efficient, competitive and ensure cheaper agricultural produce; therefore, more efficient and heavier agricultural machinery will be used (Kutzbach 2000; Lamandé and Schjønning 2011; Zink et al. 2010). It is thus important for researchers to determine how to progress while ensuring both economic prosperity and the conservation of soil.

2.3 Impact of Tillage on Physical-Mechanical Properties of Soil

2.3.1 Soil Hardness

Soil hardness is one of the main characteristics of the physical properties of soil, and determines the ability of soil to resist penetration of a solid body. An understanding of soil hardness is especially important when determining tillage machinery working processes and also for designing, constructing and manufacturing tillage machinery and associated working parts. Soil hardness is also a very important indicator as it describes conditions for crop germination; at the beginning of growth plants need to overcome soil resistance using accumulated energy. Therefore, plants demonstrate poorer germination and growth in hard soil and their root systems are weaker, which could lead to a decline in yield.

Soil hardness varies depending on depth. Different technological tillage operations have differing effects on change in the soil hardness of the upper soil layer (topsoil) and the deep soil layer (subsoil). For example, to change only the soil hardness of the upper layer, conventional tillage via ploughing, disc-harrowing and loosening with shallow tillage cultivators is effective. However, this machinery is unable to reduce soil hardness in the deeper layers, and the application of deep ploughing or disc-harrowing can have the effect of promoting hardening of the deeper soil layers, forming a so-called plough or disc-harrow 'pan' (Batey 2009; Arvidsson 2010).

The more soil is tilled, cut, turned over, crumbled and mixed, the lower the soil hardness is within the arable layer. However, intensive tillage is very energy-consuming and requires a large amount of labour and energy resources. Therefore, the choice of tillage technology is an extremely important task. In addition, intense soil mixing and loosening does not always yield expected positive results, and tillage quality depends on soil texture, structure, density, moisture content and other properties (Guerif 1994; Rasmussen 1999; Romaneckas et al. 2015). Furthermore, such soil properties are inter-related and if one is changed the others are often altered either directly or indirectly.

Researchers investigating the influence of direct drilling technology on soil physical properties have found that soil resistance under direct drilling is not proportionate to soil density and the water content within it, as observed with other tillage technologies. One study proposed that the application of direct drilling ensures stable soil pores formed under the influence of plant root channels and soil organisms, and that these pores facilitate the creation of a stable soil structure (Unger and Jones 1998). Another study (Lapen et al. 2004) estimated the influence of soil water content on soil hardness; results indicated a reverse linear relationship between water content and soil resistance to penetration of a solid body under conditions of direct drilling. A parallel investigation also revealed that the dependence between water content and soil hardness also exists in conventional tillage technologies, however, this depends highly on the type of plant and season. Other scientists

(Ley et al. 1993; Kılıç et al. 2004; Šarauskis et al. 2014a) have determined that soil water content has an unequal impact on soil hardness when different tillage technologies are applied, and Shafiq et al. (1994) suggested that soil hardness is mainly increased by soil compaction. When the water content is high in soil, the compaction is also higher. Other studies (Goodman and Ennos 1999; Chung et al. 2013) have established that soil hardness is higher at a high bulk density of soil and a lower water content.

Soil hardness, k, can be calculated as follows:

$$k = \frac{F}{S}, \tag{2.2}$$

where F is the force required to penetrate a cone into soil, N and S is the cone area, cm^2.

The most important properties influencing soil hardness are its mechanical composition, moisture content and porosity. Soil hardness can thus be expressed as a function depending on other conditions that characterise the physical-mechanical properties of soil (Kulen and Kuppers 1986) as,

$$k = f(wPD_mS_t), \tag{2.3}$$

where w is soil moisture content, P is soil porosity, D_m is soil mechanical composition, and S_t is soil structure.

When conducting soil experiments, if the assessment accuracy of a certain property is higher, the overall assessment of the other properties will also be more accurate. If experiments are conducted in soil with the same mechanical composition, the above dependence will be even simpler (Kulen and Kuppers 1986),

$$k = f(wP). \tag{2.4}$$

Soil hardness is a very significant indicator used to assess soil quality and crop yield. Research conducted in Lithuania (Cesevicius et al. 2005) showed that soil hardness has a direct influence on the yield of spring barley, whereas soil bulk density and air permeability have no direct effect on barley yield but it have influence in interaction with soil hardness. Studies have found that when applying deep ploughing, shallow ploughing and direct drilling technologies, maximum soil hardness is obtained by direct drilling. Soil hardness is similar in the top soil layer (0–10 cm) immediately after drilling when using shallow and deep ploughing, but with direct drilling soil hardness is approximately 49%–54% higher. Feiza et al. (2008) found the lowest soil hardness to be observed postharvest in deep ploughed soil, shallowly ploughed soil was 12%–30% harder and soil hardness of direct drilled soil was 52%–71% greater compared to deep ploughed soil. Lopez et al. (1996) conducted long-term experiments on the effects of conventional and sustainable tillage practices on soil hardness in North East Spain, and established that upper soil layer hardness in direct drilling plots was 4 MPa while that in conventional

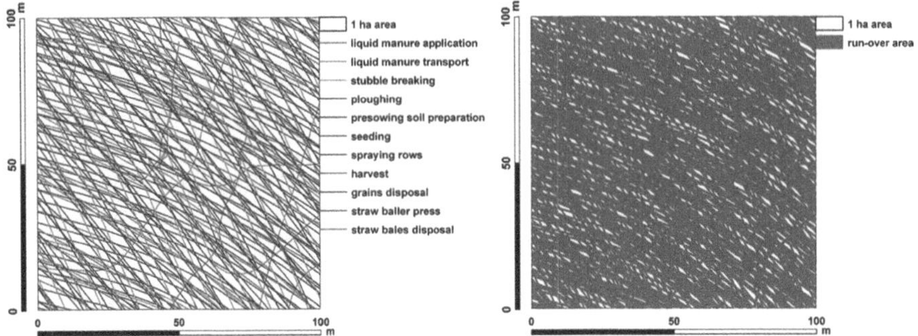

Fig. 2.3 Schematic representation of machinery passage using conventional soil tillage technology in a 1 ha area. Left: machine movement trajectories in field during one cropping season, Right: total run-over area. (Prepared according to Kroulík et al. 2009)

tillage plots was about 2 MPa. However, Horn (2004) suggested that compared with conventional tillage the long-term application of sustainable tillage technologies exerts a positive influence on physical soil properties, as soil becomes more resistant to physical impact and deformation. Similar opinions have been expressed by other researchers (Da Veiga et al. 2007; Singh and Malhi 2006), who found that no-tillage soil is more resistant to deformation compared to soil that is ploughed or tilled using other tillage machinery.

With respect to the parallel wheel tracks made by agricultural machinery and the size of soil particles, soil hardness measurements can be used to determine overall soil profile compaction and assess structural variance (Lowery and Morrison 2002). Conventional tillage technologies involve many technological operations and the soil is thus continually run-over by wheels. Each time the soil is run-over the surface is affected and soil compaction increases. Kroulík et al. (2009) performed experiments on field trafficking intensity using different soil tillage technologies and found that with conventional tillage technology used for cereals the entire soil surface was run-over by agricultural machinery at least once during processes of spreading liquid organic fertiliser prior to main tillage and when the straw after harvesting was pressed into bales and removed from the field. If we assume that the areas affected by the wheels of agricultural machinery is slightly wider than the wheel working width, 87.5% to 95.3% of the total soil surface is thus affected in the case of conventional tillage (Fig. 2.3).

When using minimum tillage technology (Fig. 2.4) the machinery wheels effect approximately 73% of the total soil surface, but in the case of direct drilling (Fig. 2.5) they affect approximately 56%. It is of note that some areas of the soil surface are run over two or more times, and it can thus be assumed that the application of different tillage technologies entails repeated running over of between 18.4% and 44.8% of the total soil surface area (Kroulík et al. 2009).

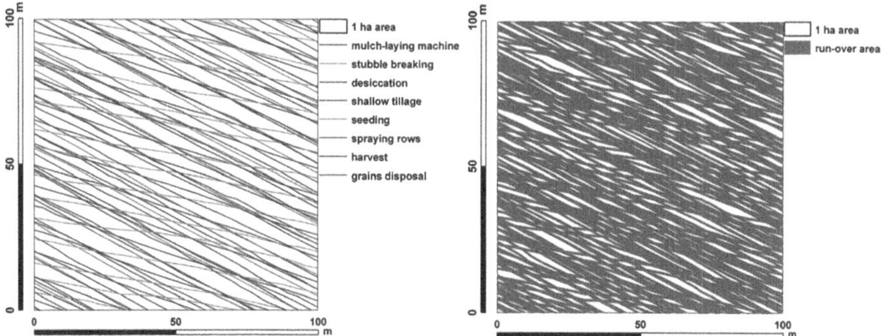

Fig. 2.4 Schematic representation of machinery passages for minimum tillage technology in 1 ha area. Left: machines movement trajectories in field during one cropping season, Right: total run-over area. (Prepared according to Kroulík et al. 2009)

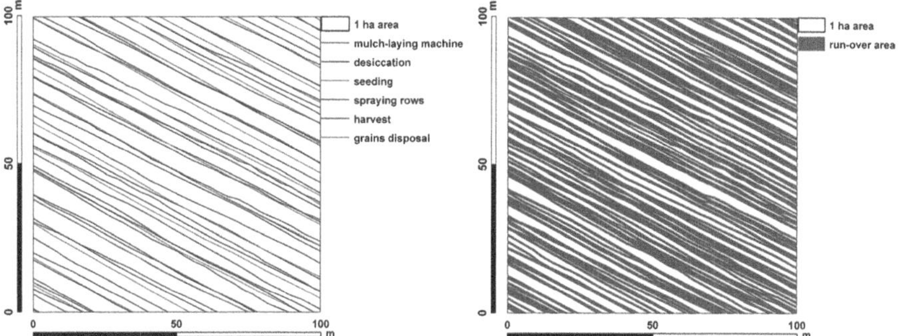

Fig. 2.5 Schematic representation of machinery passages for no-tillage technology in 1 ha area. Left: machines movement trajectories in field during one cropping season, Right: total run-over area. (Prepared according to Kroulík et al. 2009)

Although a number of studies have been conducted on soil hardness, they have usually been in association with other investigations and have aimed to characterise the possible influence on plant yield and its qualitative parameters. There is a lack of research assessing soil hardness in different soil layers in the long-term, where the same method of tillage is conducted in the same field for ten or more years.

2.3.2 *Soil Bulk Density*

Soil bulk density is measured using the weight of one cubic centimetre of dry undisturbed soil in grams (g cm^{-3}), although in literature Mg m^{-3} has been used (Reichert et al. 2004; Hazarika et al. 2009; Feiza et al. 2015; Kaczmarek et al. 2015).

The bulk density of loam (<1.0 g cm^{-3}) indicates that the soil is sufficiently crumbly or that it has a high humus content. When bulk density is between 1.1 and 1.2 g cm^{-3}, it is considered normal, and when it is 1.3–1.8 g cm^{-3} the soil is very compacted or has been run-over heavily. However, the bulk density of sandy loam or sandy soils may change only marginally by being loosened or run-over, and most plants adapt to grow within soil bulk density ranges of 0.9–1.3 g cm^{-3}. Optimum bulk density is distinguished when compacted soil regains its original condition and maintains a constant bulk density, and in fertile soils this is at about 1.2–1.4 g cm^{-3} (Reichert et al. 2004; Velykis and Satkus 2005). Soil bulk density changes in response to both environmental conditions and anthropogenic factors. The growing condition of cultivated plants depends on soil bulk density, as it effects the accumulation of air and moisture in soil.

Soil bulk density is mainly affected by tillage and becomes lower in line with more intensive tillage. This is mainly because small air spaces are formed between soil particles (clumps) of different structures during tillage. Alvarez and Steinbach (2009) investigated the impact of different tillage technologies on the physical properties of soil and found soil bulk density in direct drilling plots to be significantly higher than in ploughed plots. However, no significant differences in soil bulk density have been observed between the uses of various reduced tillage technologies. Researchers (López-Fando and Pardo 2012) conducted studies on the impact of tillage technologies on different soil properties in the central part of Spain, where three tillage technologies were applied: direct drilling, cultivation to a depth of 18–22 cm and ploughing to a depth of 25–30 cm. The results showed that bulk density differed significantly among the three tillage technologies. A comparison of soil bulk density at 0–5 cm and 5–10 cm depths revealed the highest density to be found in direct drilling research plots, and that in plots tilled by a plough or a cultivator was similar, although it was lower than with direct drilling. However, research results for deeper soil layers (10–20 cm and 20–30 cm) were conflicting; the highest bulk density was found in ploughed plots and the lowest in direct drilling plots.

A number of researchers (Logsdon et al. 1990; Hernanz et al. 2002, Šimanskaite 2007; Moret and Arrúe 2007; Roger-Estrade et al. 2009) have obtained similar research results and determined that soil bulk density depends not only on tillage but also on environmental conditions, in particular on the moisture content. A team of researchers (Coulouma et al. 2006) from the south of France investigated the influence of deep tillage on soil physical properties and found that there was a higher increase in the bulk density of soil under humid conditions of tillage, whereas under dry conditions during tillage no such increase was determined.

Roscoe and Buurman (2003) analysed the impact of different soil tillage practices on soil characteristics in Brazilian savannas and compared them with areas unaffected by agriculture. Interesting research results were obtained; they established that soil bulk density in a wild savanna was significantly lower than that tilled by agricultural implements, irrespective of the method of tillage employed. According to the authors of that article, the increase in bulk density could have been caused by soil compaction during tillage, as the heavy agricultural machinery exerted pressure

Fig. 2.6 Influence of regular wheels (left) and rubber track (right) on soil compaction and plant growth. (From: www.soucy-track.com)

on the soil and affected its properties. Aware that soil compaction can be a serious problem, a group of researchers from the USA (Logsdon and Karlen 2004) substituted direct drilling for tillage and found no negative impact on soil bulk density. Furthermore, the relationship between tillage technologies and soil bulk density was investigated in Lithuania. Bogužas et al. (2010) investigated the influence of tillage on soil bulk density and found no significant difference in soil bulk density when reduced autumn ploughing substituted conventional deep ploughing. In addition, research conducted by Feiza et al. (2006) on light loamy soil indicated no significant influence on soil bulk density using different tillage depths: soil bulk density in the 0–10 cm topsoil layer was 1.28–1.36 Mg m^{-3}.

Compaction of the soil surface and deeper layers depends very much on the weight of the agricultural machinery and the contact surface area between the wheels and soil. When the area is larger there is a lower influence from the weight of the machinery on soil compaction. Recently, use of a caterpillar (rubber track) undercarriage has been more common as its contact surface area is several times bigger than that of a wheeled chassis. If the soil is less compacted it will have a greater water and air supply, which will thus increase the harvest. Therefore, the use of rubber track is of significant benefit in the medium and long term. For example, a combine equipped with the most popular wheels will inflict a 1.76 kg cm^{-2} pressure on the soil, whereas the same combine equipped with a rubber track system reduces that pressure to 0.41 kg cm^{-2} (Fig. 2.6) (www.soucy-track.com).

Mechanisms causing vibration in agricultural machinery (tractors, combine harvesters and other self-propelled machinery) with internal combustion engines should also be considered. With every effort made by agricultural machinery manufacturers to minimalize vibration, it is still unavoidable. A small contact surface area further increases the influence of vibration and can compact much deeper soil layers.

Another very important technological aspect is that different soil layers are compacted differently by various tillage technologies. In conventional ploughing (depending on the number of furrows), the two tractor wheels frequently run along an already ploughed furrow (Würfel et al. 2002). As the furrow bottom is already compacted from the plough weight being transferred to the soil through plough shoes, the tractor wheels running along the ploughed furrow exert additional compaction on the deeper soil layers. Therefore, a significantly compacted layer of ploughed soil can be observed at 25–45 cm depths, and sometimes even deeper; this is known as a 'plough pan' (Batey 2009). The use of no-tillage technologies can help avoid compaction of the deep subsoil layers (Arvidsson 2010).

With the increasing spread of sustainable tillage technologies, it has also been noted that using the same disc tillage implements, such as disc harrows or disc stubble cultivators, on the same soil for several years can also cause a compacted soil layer, only at a shallower depth. This occurs from the weight of a tillage implement creating pressure on the soil, which is transferred through the concave edges of a disc. In addition, it must also be taken into account that during both ploughing and disc-harrowing, fine soil particles quickly settle on the furrow bottom, and the vibration caused by the tractor wheels presses them into the soil; if the subsoil layer contains sufficient moisture it becomes easily compacted (Šarauskis 2009).

Some researchers (Lampurlanes and Cantero-Martinez 2003) consider that the cultivation of crops requires optimum soil density, otherwise it can be harmful to plants. If the soil bulk density is too low there will be insufficient contact between the soil and plant roots, because air spaces that are too large appear and prevent formation of a capillary moisture regime that is favourable to plants. In contrast, when the bulk density is too high, the aeration properties of the soil deteriorate and soil hardness increases, resulting in a disrupted moisture regime in soil, altered opportunities of plant supply with nutrients, deteriorated root growth and development and declined plant fertility. Kushnarov (1986) proposed a mathematical relationship between soil bulk density and plant yield,

$$Q = 1 - a(\rho_0 - \rho_d)^2 + b(\rho_0 - \rho_d), \qquad (2.5)$$

where, $\rho_0$is optimum soil bulk density ensuring maximum yield, ρ_d is actual soil bulk density and a and b – empirical coefficients.

The relationship between soil bulk density and yield is presented in Fig. 2.7, where it shows that a higher soil density increases the probability of increasing the plant yield to a certain extent, but that the maximum harvest can be expected when the soil bulk density is about 1.2–1.4 g cm^{-3} and an increase in soil bulk density beyond this limit results in a decrease in yield.

Hamza and Anderson (2005) concluded in their research work that the application of ploughing and minimum tillage technologies produces a regular bulk density difference of up to 15%, regardless of which crops are grown. They explain this result by stating that the bulk density of undisturbed soil is always higher than that of ploughed soil. Multi-annual research data indicate that compacted soil by agricultural machinery results in a 5% reduction in plant yield during the first year and an 18% yield reduction after four years, regardless of plant type (Karapetyan 2005).

A literature analysis suggests that soil bulk density is a very important physical property of soil that influences both the soil and the growth of plants. Therefore, a considerable amount of global research attention has been paid to soil bulk density,

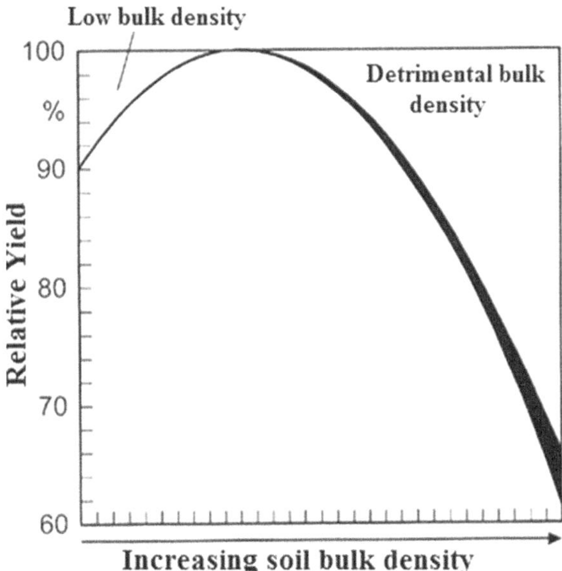

Fig. 2.7 Effect of soil bulk density on yield. (Prepared according to Würfel et al. 2002)

and most researchers agree that main factor resulting in soil bulk density alterations are related to soil tillage and associated machinery.

2.3.3 Soil Moisture Content

Soil moisture is an important part of the global atmospheric water cycle and is vitally important for the growth of agricultural plants. Most vegetation is more dependent on the water content at the root level than the amount of rainfall, and water deficiency disrupts plant development and the rate of growth. The elements affecting this system are summarised in the water cycle, which is adapted in Fig. 2.8 to centre around the soil moisture content as the stock.

After entering the soil, water moves gradually through the root zone. Each soil layer needs to be filled completely to saturation before water penetrates to a deeper layer. Water moves much faster through sandy soil than in clay or silty soils of fine texture and the soil gradually dries in relation to moisture evaporation from the soil surface, through plant leaves or in the process of plant root absorption.

Soil is composed of mineral particles, organic matter and pores filled with water and air. Pore quantity and size depend on the size of soil particles; the larger the particles the lower the pore quantity. The number of pores filled with water increases in relation to the moisture content of the soil.

Meteorological conditions cause unequal amounts of moisture to fall in various years (Chang and Lindwall 1990; Hsiao et al. 2007). Under drought conditions, direct drilling technology retains a higher moister content at 0–10 cm depth and is a

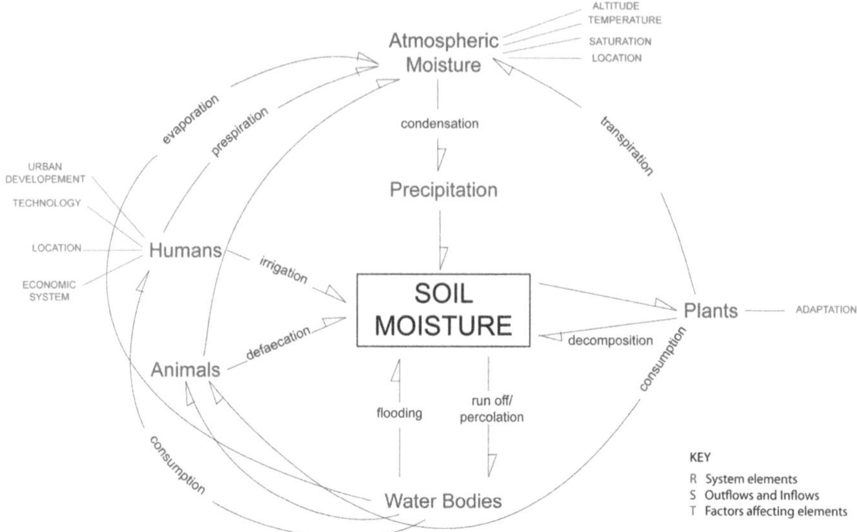

Fig. 2.8 Water cycle surrounding soil water content as the stock. (From: Of Complex Systems)

moisture sustaining environmental measure (Feiziene et al. 2009). Direct drilling technology is significantly more resilient to water erosion, and therefore its use is particularly relevant in areas with large amounts of rainfall (Auerswald et al. 1994). Although sustainable technological measures improve the availability of moisture for plants and increase soil moisture reserves, this method of tillage is not always superior to conventional soil tillage technology (Tessier et al. 1990).

Soil moisture content can be controlled using agro-technical measures, the most important of which is choice of appropriate tillage (Romaneckas et al. 2009). After drilling plants in untilled soil during a dry year, Šimanskaite et al. (2009) established that the soil moisture in the upper layer ranged from 16.0 to 22.8%, while in deeper layers it was 16.5% to 23.5%. Feiza et al. (2006) investigated the influence of tillage on soil properties and found that reduced tillage retained the moisture required for plants to germinate and grow for longer periods. The results of research into the tillage effects on soil properties conducted in southern Italy revealed that the soil moisture content was higher immediately after direct drilling compared to when using conventional tillage (De Vita et al. 2007); the higher water content in directly drilled soils indicates that there is lower water evaporation prior to drilling in relation to the lack of tillage being applied. However, after conducting experimental research on the effect of various tillage technologies (deep ploughing, shallow loosening and direct drilling) on soil moisture content Gruber et al. (2011) achieved somewhat different results; after spring tillage, the soil moisture content was marginally lower under direct drilling compared to when using other tillage technologies, but in autumn, there was no difference in the moisture content with respect to the differing tillage technology used.

Soil moisture content has a very important influence on tillage quality. For example, moist soil sticks to the working parts of tillage implements and does not crumble; water destroys soil clods, alters the structure and affects other soil properties. In addition, the soil moisture content strongly influences soil friction (the ability of soil to resist movement), which is characterised by the internal friction coefficient. With an increase in the soil moisture content, the friction coefficient begins to increase but later decrease (Hasankhani-Ghavam et al. 2015).

In addition, the soil moisture content has inevitable influence on the extent of soil compaction. Experiments have established that using the same size tyres at the same axle load with the same air pressure in the tyres resulted in different amounts of soil compaction in soils with differing moisture contents. With an increase in the moisture content, the compaction effect is transferred to the deeper layers of the soil (Würfel et al. 2002).

Another very important agrotechnological factor characterising soil tillage technologies and soil moister content is the residue of previous crops remaining on the soil surface, which occurs often during the use of sustainable tillage technologies and practically all the time during direct drilling. To a certain extent the residue serves as a 'sponge' that absorbs a greater amount of rainfall water, therefore even during severe precipitation there is a much lower risk that water will begin to run over the soil surface and contribute to soil erosion. The other important aspect is that in spring, when it is vital to retain soil moisture, sustainable tillage technologies leave plant residues on the soil surface, thereby reducing the amount of uncovered areas of soil and preventing the fast evaporation of moisture.

2.3.4 Soil Structure and Stability

Sustainable tillage and incorporation of the plant residue in the upper layer of soil is beneficial for soil structure and quality, both from an ecological perspective and in terms of environmental protection (Chivenge et al. 2007). Tillage without ploughing can be applied to soils with different properties under various climatic conditions. In addition, by applying this technology, good agricultural results can be expected even in heavy soils under dry climatic conditions (Ciuberkis et al. 2008).

Soil structure is the main characteristic in the functioning of soil, its ability to support plant life, and moderate environmental quality with particular emphasis on soil carbon sequestration and water quality (Bronick and Lal 2005). Soil is much more than a mere mixture of separate sand, silt and clay particles; only about 50% of soil consists of solid matter (sand, silt, clay, nutrients, minerals, organic matter and biological life forms) and the remaining part comprises pores that are filled with both water and air (Kiryushin 1996).

Soil structure changes naturally due to climatic conditions, such as wetting and drying or freezing and thawing (Rasmussen 1999). Healthy soil with a good structure has stable pores throughout all the soil layers into the profile, which provide an opportunity for water infiltration, root penetration and air circulation to occur (Pachepsky and Rawls 2003; Franzluebbers 2002). Tillage destroys soil structure, disrupts soil pores and reduces the amount of plant residue on the soil surface. If the soil has a poor structure (for example it is severely compacted), tillage can be beneficial because it crushes the compacted soil and enables the formation of pores. However, if the soil structure is stable and favourable for plant growth, tillage can disrupt the existing soil structure, reduce its stability and make it more vulnerable in relation to the weight of machinery. In soil with a poor structure, soil filtration is reduced, internal drainage is weakened, and problems occur with a water surplus and a lack of oxygen in the plant root zone (Jasa 2011).

Sustainable tillage and direct drilling are considered measures for improving damaged soil structure (Oyedele et al. 1999). Reduced tillage allows the formation of more stable soil particles, which are more resistant to the impact of environmental factors. Zhang et al. (2007) investigated the resistance of soil particles in differently tilled soils, and established that under reduced soil tillage technologies and direct drilling, the soil particles in the upper soil layer (0–5 cm) were significantly more resistant to water and only a small amount were disrupted by the effect of water. The advantages of direct drilling are that in the upper soil layer the amount of soil macroparticles (macroaggregates) (>0.25 mm) is significantly increased (approximately 8.1%) compared to conventional tillage. Such results are even more pronounced in warm and dry years (Fernández-Ugalde et al. 2009; Jin et al. 2011).

D'Haene et al. (2008) studied the influence of sustainable and conventional tillage on soil stability in different locations in Western Europe, where the stability index of soil particles was determined by dry and wet sieving methods. It was established that with the use of sustainable tillage technology the stability index of soil particles was up to 40% higher than when using conventional tillage technology.

As previously mentioned, heavy agricultural machinery (especially harvesters and transport vehicles) making multiple passes on the soil surface has a strong detrimental influence on its structure and stability (Mueller et al. 2009). Compaction of topsoil and subsoil layers and the worsening of soil structure under conventional tillage technology occurs in all soil types. Under conventional tillage techniques, crusts are more likely to form in the soil due to the loss of organic matter and the reduction of soil structural stability. However, with direct drilling, the pores are more evenly distributed throughout the soil surface. Therefore, soil structure is a very important parameter in choosing and assessing the applicability of tillage technologies (Roger-Estrade et al. 2009).

2.3.5 Soil Porosity and Air Movement in Soil

Soil air permeability is a very important physical property that influences the growth and development of plant roots, and is related to the presence of large soil pores, total soil porosity and the internal geometry of pores (Lindstrom 1990). Šimanskaite (2007) proposed that total porosity and aeration of soil are important soil property features essential for the water and air regimes within soil. According to the author, for all tillage technologies except direct drilling, the total porosity of soil decreases at the beginning of vegetative growth and increases at its end. Soil porosity depends directly on soil bulk density; an increase in density causes a decrease in porosity. Total soil porosity is calculated by estimating soil bulk density and particle density (Maikšteniene et al. 2007); the latter is commonly measured using a vacuum air pycnometer.

According to Kiryushin (1996), the most favourable soil aeration conditions are formed when total porosity is about 50%–60% of soil volume. Maikšteniene et al. (2008) suggested that the optimal soil air regime is when the aeration porosity of soil is 20%–25% of total porosity. Arvidsson (1998) indicated that a critical soil aeration porosity of 10% is required for plant growth. In addition, Feiziene et al. (2010) suggested that denser soil is less permeable to water and air, and with the increase in soil organic matter content the soil density decreases. Bogužas et al. (2010) reported that using no-tillage technology facilitates a rapid increase in the number of earthworms and a decrease in the size of soil particles (<0.25 mm) in soil; it can be assumed that there are large pore volumes existing between soil particles larger than 0.25 mm, and that these pores are filled with water or air.

Studies by Feiza et al. (2006) established that the total and aeration porosity of soil is close to optimal both under shallow and deep ploughing, and at the 0–10 cm and 10–20 cm soil layers this ranges within 46%–50% and 24%–30% respectively. Norwegian scientists (Ekeberg and Riley 1997) conducted research that compared deep autumn ploughing, deep loosening and direct drilling on morainic loam soil, and found that the lowest soil density and highest porosity were observed in soil under direct drilling. This was associated with a significantly increased number of soil macroparticles in untilled soil compared to within conventionally tilled soil.

Studies conducted at the Lithuanian Institute of Agriculture determined that air permeability in soil depends on field relief. In the upper (5–10 cm) layer of soil on a slope, the best soil permeability to air was observed under shallow ploughing; air permeability of the same soil layer after deep ploughing was lower. At the foot of the slope permeability to air was lower, both after shallow and deep ploughing, compared to that at the top of the slope (Feiza et al. 2008).

Soil aeration is a very dynamic component of soil quality that is greatly dependent on the amount of water present and the soil bulk density (Bhagat et al. 1996). Scientists (Czyz and Tomaszewska 1993; Dexter and Czyz 2000; Czyż 2004) have found that oxygen levels in soil are significantly reduced when it has been persistently compacted by heavy tillage equipment which can lead to extreme oxygen deficiency that is unfavourable for plant growth.

2.4 Environmental Aspects of Soil Tillage

The negative impact of tillage on soil also affects local ecosystems and the natural environment. The effect of soil compaction from heavy agricultural machinery or increased soil erosion contributes to an increased leaching of chemical substances to surface and ground waters, which subsequently pollutes water bodies. In addition, ditch banks and road sides are eroded, ditch beds choked up, and slope vegetation destroyed.

One of the most important goals of environmental sustainability is to reduce the effect of human activities on climate change, to which land, forests and water resources are very sensitive. During the last several decades, increasing amounts of CO_2, methane (NH_4), nitrous oxide (N_2O) and other gases have been emitted into the environment. These gases are known as greenhouse gases (Muñoz et al. 2010) as they capture infrared rays emitted from the earth, thus trapping heat in the atmosphere. The most significant of these gases is CO_2, the amount of which has increased in the atmosphere from 280 to 366 ppm compared to pre-industrial levels (Rastogi et al. 2002). Since the 1980s, the EU has been significantly aware of the international implications of contributing to climate change, and it played a main role in establishing the United Nations Framework Convention on Climate Change (UNFCCC) and the Kyoto Protocol. The associated documents set limits for the reduction of greenhouse gas emissions by industrialised countries; 15 countries from the EU committed to reducing overall amounts of greenhouse gas emissions by 8% compared to the 1990 level. In 2007 the EU proposed an increased commitment to reducing these emissions by at least 20% before 2020, and to achieve a reduction of 30% if other developed countries agreed to do the same. The Paris Agreement stated an intention to maintain 'the increase in the global average temperature to well below 2 °C above preindustrial levels and to pursue efforts to limit the temperature increase to 1.5 °C above pre-industrial levels, recognizing that this would significantly reduce the risks and impacts of climate change' (UNFCCC (2015) Draft Decision; Rogelj et al. 2016). In this respect, Boucher et al. (2016) identified research gaps and suggested new directions for research into a number of the facets of the Paris Agreement, including the 1.5 °C objective, articulation between near-term and long-term mitigation pathways, negative emissions, methods of verification, climate finance, non-Parties stakeholders and adaptation.

Agricultural activities promote the emission of greenhouse gases to the atmosphere, though to a lesser extent compared to other industries (Fig. 2.9). Agricultural land area covers more than a half of EU territory and contains huge supplies of carbon, which facilitates reduction of the amount of atmospheric CO_2. Nevertheless, agriculture faces a double challenge: the need to reduce emissions of greenhouse gases and to simultaneously adapt to new conditions for ameliorating climate change. Currently, greenhouse gas emissions from agriculture account for 9.4% (Duxbury 1995; Smithson 2008) to 12.5% (Kumar et al. 2012) of total emissions to the atmosphere.

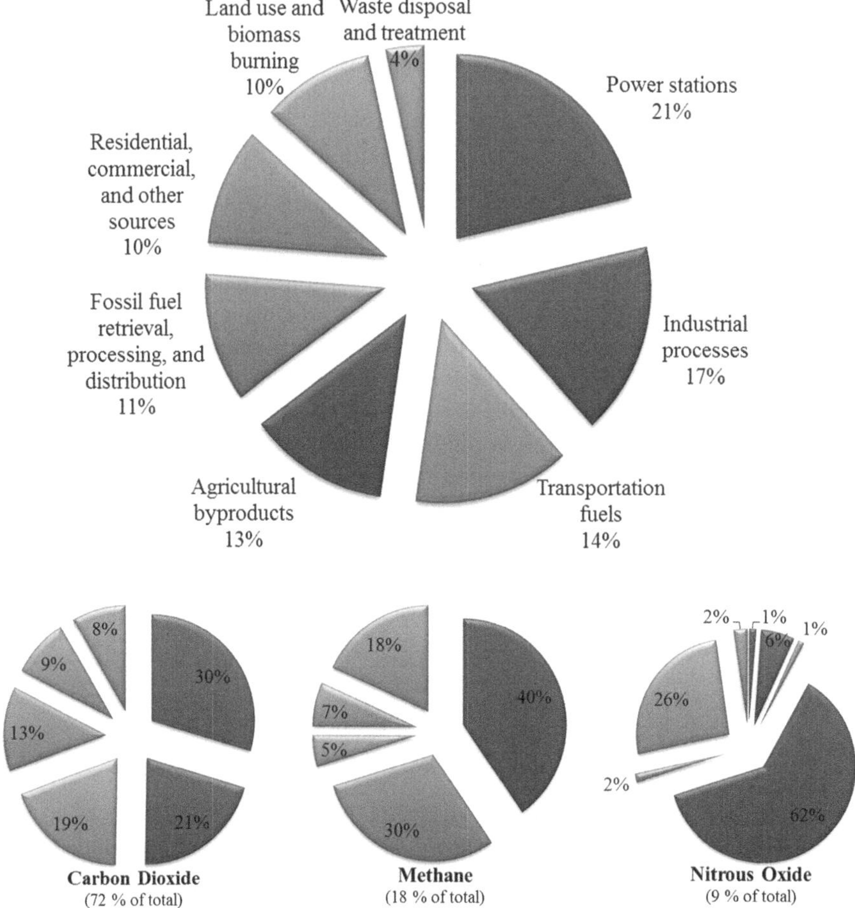

Fig. 2.9 Annual emission by various sectors. (Prepared according to Kumar et al. 2012)

Soil produces greenhouse gas. Tilled soil emits more CO_2 than untilled soil (Buragiene et al. 2015; Dossou-Yovo et al. 2016) due to the favourable conditions provided for microbiological activities after tillage, and the subsequent acceleration of the decomposition of residues from plants and those of animal origin. Soil tillage changes soil structure, promotes soil mixing with organic matter and alters water infiltration processes (Fleige and Horn 2000). However, it is acknowledged that emissions are not only related to tillage but also to other agricultural technological operations (particularly fertilisation) and fermentation processes (Buragiene et al. 2015; Dossou-Yovo et al. 2016).

Sustainable tillage is considered to one of the most important techniques that can be used to reduce CO_2 emissions from soil. Reduced intensity of tillage mitigates soil disruption and the activity of microorganisms, which thus alters CO_2 emissions

(Grant 1997). In addition, direct drilling and plant residues left on the soil surface continue the reduction in CO_2 emissions as they reduce the direct contact between the soil and environment; plant residues improve the isolation of carbon in soil (Schillinger and Young 2004). A comparison between direct drilling with plant residues left on the soil surface and direct drilling without residues shows that a greater amount of CO_2 is emitted when no plant residues remain (Lal and Kimble 1997). An increase in tillage intensity (i.e. a higher degree of soil disturbance, mixing and loosening) increases aeration and thus alters the amount of CO_2 emitted from the soil.

Research conducted in Germany established that consumption of 100 L of diesel fuel causes 376 kg CO_2 to be emitted into the environment (Tebrügge 2001). However, there is currently a lack of research determining exactly how each individual technological operation performed during tillage effects the amount of greenhouse gas emitted, but it is obvious that a reduction in the amount of tillage, drilling technological operations and traffic over the soil will substantially reduce fuel consumption and greenhouse gas emissions from tractors and agricultural machinery. Therefore, to make an accurate assessment of how these various tillage technological operations pollute the environment, it is necessary to know how much energy each of these technological operations require (Šarauskis et al. 2014b).

2.4.1 CO_2 Emissions from Soil Under Different Tillage Technologies

Ameliorating global climate warming caused by increased concentrations of greenhouse gases in the atmosphere is a challenging issue. As previously mentioned, one source of CO_2 emissions is agricultural activities (Duxbury 1994, 1995; Kumar et al. 2012). The soil is both a source of emissions and a reservoir for CO_2 accumulation when bound by plant biomass during photosynthesis, and when organic matter decomposes, CO_2 can be incorporated into soil as organic carbon. Actions disrupting the soil structure, such as tillage, can increase CO_2 emissions from soil with respect to an increase in soil aeration. In contrast, an absence of tillage and maintaining plant residue on the soil surface postharvest can increase the amount of carbon within the soil (Jastrow et al. 1996).

Respiration via plant roots, soil fauna and microflora also significantly contribute to CO_2 emissions from soil (Lal et al. 1995) and represent the fundamental soil mechanisms involved in soil carbon removal from soil (Curtin et al. 2000). The carbon content in soil is involved in the carbon cycle, which is a complicated system whereby carbon is cycled through the atmosphere, terrestrial biosphere and oceans. Plants absorb CO_2 from the atmosphere during photosynthesis, produce their tissues from carbon and return it back to the atmosphere during withering and decomposing. CO_2 can remain in the atmosphere for 50–200 years, depending on the mechanisms involved in repeatedly returning it to the land or oceans. CO_2 released from soil consists of three biological respiration processes: microbiological, plant root and soil fauna respiration (Macfadyen 1970). CO_2 accounts for about 60%–72% of total greenhouse gas emissions (Macfadyen 1970; Kumar et al. 2012), and to reduce greenhouse gas emissions it is necessary to sequestrate carbon in soil. The use of a range of tillage machinery systems, incorporating manure and crop residues, improving biodiversity and mulching can thus increase the amount of soil carbon.

Various studies suggest that factors such as soil structure, moisture, pH, carbon content, stable and unstable soil organic matter and the nitrogen content in soil affect CO_2 emissions from it. During plant root respiration, photosynthetic products passed to roots and various organic residues are carbon sources for microbiological soil respiration (Kowalenko and Ivarson 1978; Bronick and Lal 2005; Feiziene et al. 2010). The carbon present in plant residues is released in the form of CO_2, and the amount released is 30% higher during conventional tillage compared to direct drilling (La Scala et al. 2001, 2006). Although Parkin and Kaspar (2003) consider that a decrease in organic carbon within soil due to tillage is almost unnoticeable in the short term, Prior et al. (2000) state that tillage in the short-term increases CO_2 emissions from soil, due to the physical disruption of soil pores and the release of CO_2 held within them. A Spanish researcher, Álvaro-Fuentes et al. (2007) found that after disruption of the soil structure during tillage, the amount of CO_2 emissions from soil increased suddenly over a relatively short period lasting less than 3 h post tillage. In addition, the amount of CO_2 emitted after tillage was proportional to tillage intensity; although the maximum amount of CO_2 was emitted from ploughed soil, CO_2 emissions from soil under direct drilling were stably lower during the entire experiment. Danish scientists (Chatskikh and Olesen 2007) concluded that following conventional tillage, there was a greater increase in the amount of CO_2 emitted compared to during direct drilling and when using other sustainable tillage technologies. Studies of CO_2 emissions from soil under conventional tillage and direct drilling were conducted for 330 days in northern France, and results showed no difference in the amount of CO_2 emitted during 41% of days. However, emissions from soil under direct drilling were higher on 53% of the days but when using conventional tillage emissions were only higher on 6% of days (Oorts et al. 2007).

Tillage accelerates oxidation of soil organic carbon and thus large amounts of CO_2 are released into the atmosphere over a period of several weeks after tillage (Reicosky et al. 1997; Ellert and Janzen 1999; Rochette and Angers 1999; Prior et al. 2000; La Scala et al. 2006; Buragiene et al. 2011, 2015). A group of authors (La Scala et al. 2008, 2009) suggested a differential equation to represent the amount of CO_2 released from the soil as a direct relationship with C decay (mass) in soil depending on time as,

$$\frac{dC_{soil}(t)}{dt} = -kC_{soil}(t), \tag{2.6}$$

where C_{soil} is the amount of labile soil C of readily decomposable organic matter in g m^{-2}; k is the decay (mass) constant, $time^{-1}$; t–time after tillage, days. Having solved this equation, the same authors (La Scala et al. 2008) obtained the dependence from which it is possible to estimate the content of available labile soil C for decay at any period of time as,

$$C_{soil}(t) = C_0 e^{-kt}, \tag{2.7}$$

where C_0 is initial amount of available labile organic C in soil, g m^{-2}. It is important to note that the decay constant (k) is considered as a constant only for short-term field experiments (1 month). As a rule, k is presented in literature as having an exponential dependence on soil temperature and soil moisture content (Parton et al. 1994; La Scala et al. 2009).

CO_2 emissions from soil in relation to microbiological respiration can be estimated using equation 2.7, which is particularly relevant when soil is not covered with plant residue. However, not all labile organic carbon is easily transformed into CO_2; some part of carbon can be included into the biomass depending on the activity of microorganisms (Stevenson and Cole 1999; La Scala et al. 2009). It is possible to make a reasoned assumption that the emission of organic C and CO_2 is proportional to the negative rate of carbon breakdown and a higher breakdown rate increases CO_2 emissions from soil (La Scala et al. 2008, 2009),

$$CO_2(t) = -\frac{dC_{soil}(t)}{dt}. \tag{2.8}$$

$C_{soil}(t)$ for tillage technologies can be calculated as follows,

$$C_{soil}(t) = C_{NT}(t) + C_T(t), \tag{2.9}$$

where C_{NT} is the organic carbon content in soil prior to tillage (or direct drilling technology), g m^{-2}; C_T is the organic carbon content in soil after tillage, g m^{-2}.

Tillage provides more favourable conditions for the release of organic carbon present inside soil aggregates, and it becomes more easily available for microbiological processes within soil. According to equations 2.8 and 2.9, the influence of tillage on CO_2 emissions from soil can be estimated according to the following equation (La Scala et al. 2009):

$$CO_2(t) = -\frac{dC_{NT}}{dt} - \frac{dC_T}{dt}. \tag{2.10}$$

Experiments were conducted in Spain to investigate the impact of conventional and sustainable tillage technologies on CO_2 emissions from soil with observations of emission seasonality. CO_2 emissions were found to be higher to the environment during the plant vegetation period with the application of conventional tillage technology. However, when sustainable tillage technology was applied, CO_2 emissions from soil were approximately 24% lower (Sanchez et al. 2002). Lee et al. (2009) established the dependence of CO_2 emissions from soil on the type of plants cultivated and the growth stage, but observed no significant influence of the use of tillage technologies on CO_2 emissions. During the period from tillage to drilling, when the soil is bare, tillage can have a significant influence on CO_2 emissions from soil. For example, spring tillage techniques, which loosen and mix the soil more intensely, have a stronger influence on CO_2 emissions from soil during the first two-week period.

Environmental factors also have an influence on CO_2 emissions from the soil. Zhang et al. (2011) proposed that after tillage there is an increased influence from water and temperature on emissions from soil. Furthermore, they considered that the impact of tillage on CO_2 emissions from soil continues for about 35 days. Morell et al. (2010) found that CO_2 emissions increase when the soil is moist and after persistent rainfall, when emissions are substantial.

Some authors have obtained similar CO_2 emission results from the soil under direct drilling, sustainable and conventional tillage (Fortin et al. 1996), whereas others (Hendrix et al. 1988) have obtained higher emissions under direct drilling technology. Furthermore, other authors (Ball et al. 1999; Vinten et al. 2002) consider that CO_2 emissions from the soils under direct drilling are higher during a certain time period and at other times the emissions are lower. Finally, other scientists (Reicosky and Lindstrom 1993; Dao 1998; Kessavalou et al. 1998; Alvarez et al. 2001; Al-Kaisi and Yin 2005) consider that significantly lower CO_2 emissions are released from the soil for a short period after direct drilling compared to soils that are ploughed.

Buragiene et al. (2015) researched the effects of five autumn tillage systems with different intensities on CO_2 emissions from soils during the maize vegetation period. Deep conventional ploughing was performed at a depth of 23–25 cm, shallow ploughing at a depth of 12–15 cm, deep loosening at depth of 25–27 cm, shallow loosening at depth of 12–15 cm and in a fifth system no-tillage was used. Analysis of soil CO_2 emissions during the maize vegetative period indicated that they varied depending on the tillage system used (Table 2.2); the highest soil CO_2 emissions were observed for conventional deep ploughing tillage and the lowest for the no-tillage system.

The dynamics of soil CO_2 emissions during summer, which corresponds with the maize growth period, indicate that environmental factors (air temperature, soil temperature and precipitation) significantly ($P < 0.05$) influence emissions of CO_2 from the soil. However, influences from the tillage systems were also observed. After summarising all average measurements of experimental research for all three years during the vegetation period, it was established that the highest average soil CO_2 emissions occurred when using the DP tillage system (2.18 $\mu mol\ m^{-2}\ s^{-1}$), followed by the SP system (1.95 $\mu mol\ m^{-2}\ s^{-1}$), the DC system (1.96 $\mu mol\ m^{-2}\ s^{-1}$), the SC system (1.89 $\mu mol\ m^{-2}\ s^{-1}$) and the NT system (1.59 $\mu mol\ m^{-2}\ s^{-1}$) (Buragiene et al. 2015).

It is very evident that contradictory experimental results have been obtained with respect to CO_2 emissions from soil under all the various tillage systems and throughout the various time periods. It is not yet clear to what extent CO_2 emissions from the soil increase after tillage, why this occurs, and how long the occur for, or

Table 2.2 CO_2 emissions (μmol m^{-2} s^{-1}) from soil during maize vegetation period (Buragiene et al. 2015)

Tillage systems	June		July		August	
	Beginning[1]	End[2]	Beginning[1]	End[2]	Beginning[a]	End[b]
2009						
DP	2.09 ± 0.44	2.42 ± 0.51	3.00 ± 0.35	2.76 ± 0.64	3.46 ± 0.22	1.25 ± 0.30
SP	1.74 ± 0.53	1.71 ± 0.34	3.09 ± 0.36	2.69 ± 0.39	3.13 ± 0.37	0.89 ± 0.32
DC	1.83 ± 0.55	1.75 ± 0.52	3.18 ± 0.30	2.21 ± 0.60	3.50 ± 0.23	0.92 ± 0.21
SC	1.46 ± 0.41	1.54 ± 0.45	2.87 ± 0.39	2.01 ± 0.63	2.94 ± 0.38	0.84 ± 0.36
NT	1.35 ± 0.45	1.51 ± 0.37	2.58 ± 0.42	1.57 ± 0.63	2.50 ± 0.29	0.64 ± 0.15
2010						
DP	1.36 ± 0.41	4.01±0.05	0.59 ± 0.14	0.81 ± 0.30	3.29 ± 0.90	0.70 ± 0.24
SP	1.23 ± 0.41	3.94 ± 0.08	0.57 ± 0.12	0.88 ± 0.26	2.52±1.09	1.11 ± 0.32
DC	0.79 ± 0.31	3.78 ± 0.10	0.53 ± 0.05	0.97 ± 0.32	1.98 ± 1.17	1.05 ± 0.34
SC	1.30 ± 0.34	3.88 ± 0.12	0.56 ± 0.10	0.71 ± 0.29	2.70 ± 1.06	1.20 ± 0.35
NT	0.49 ± 0.05	3.74 ± 0.13	0.48 ± 0.04	0.90 ± 0.31	2.27 ± 1.17	1.16 ± 0.42
2011						
DP	1.86 ± 0.76	1.79 ± 0.63	3.18 ± 0.41	2.84 ± 0.67	3.16 ± 0.33	0.58 ± 0.18
SP	1.31 ± 0.74	1.25±0.59	3.14 ± 0.42	2.59 ± 0.63	2.77 ± 0.45	0.50 ± 0.14
DC	1.88 ± 0.79	1.66 ± 0.55	3.45 ± 0.33	2.22 ± 0.77	3.15 ± 0.42	0.51 ± 0.13
SC	1.80 ± 0.69	1.74 ± 0.57	2.84 ± 0.38	2.09 ± 0.82	3.03 ± 0.51	0.44 ± 0.05
NT	1.29 ± 0.57	1.18 ± 0.43	2.48 ± 0.57	1.70 ± 0.75	2.34 ± 0.34	0.42 ± 0.07

Note: *DP* deep ploughing, *SP* shallow ploughing, *DC* deep cultivation, *SC* shallow cultivation, *NT* no-tillage

[a]During first 10 days of month

[b]During last 10 days of month

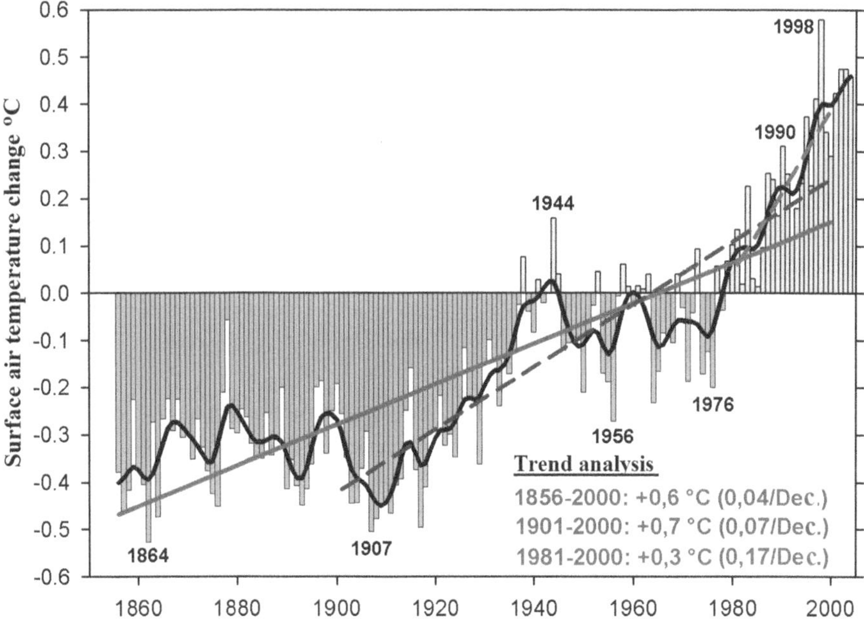

Fig. 2.10 Global annual surface air temperature changes and linear trends for the period 1856–2004. (By Jones et al. 1999; Schönwiese and Janoschitz 2005; Schaller and Weigel 2007)

what the impact is of soil physical-mechanical properties, soil temperature and climatic conditions. It is, however, evident, that extensive research is necessary to clarify these issues.

2.4.2 Characteristics of Variations in Climatic Conditions and Soil Temperature

Agriculture is one of the most sensitive sectors of the economy and is heavily dependent on meteorological and climatic conditions; therefore, various climatic anomalies are direct challenges to farmers. However, by reducing greenhouse gas emissions into the environment from agriculture, a significant contribution can be made to climate change mitigation.

Climate change observations began several hundreds of years ago, and can be used to make comparisons with modern values and observations. It is known that the average global air temperature increased by about 0.7 °C over the last century (Fig. 2.10); this increase was even higher in Europe (about 0.95 °C). It has also been noted that there are seasonal difference to these changes, and that changes are higher in winter than in summer. In addition, the global annual surface air temperature has increased (Schaller and Weigel 2007).

Air temperature has a direct impact on soil temperature, as it affects the various chemical, physical and biological processes within soil that affect the development of plants and subsequent crop yield. Higher temperatures speed up the germination of plants, accelerate microbiological activity, increase root activity and growth intensity and have an influence on root cell permeability to water. However, these positive effects begin to diminish when the soil temperatures are too high.

The soil temperature regime is dependent on energy absorption in the upper layer and heat transfer properties to deeper layers. Soil temperature, T, as a function of time, t, and soil depth, z, is described by the following equation (Gisi 1990):

$$T(z,t) = T_m + A_o e^{-z\sqrt{\pi f/D_H}} \sin\left[(2\pi t f) - z\sqrt{\pi f/D_H}\right], \tag{2.11}$$

where T_m is the average soil temperature during experimental day period, A_O is the maximum temperature amplitude on soil surface during experimental day period, z is soil depth, D_H is thermal diffusivity characterising the heat distribution speed within soil, D_H values are presented in the chart (Gisi 1990) and f is the frequency of temperature fluctuations during the experimental day period.

The velocity of thermal flow into soil, v_t, can be calculated from the following formula (Gisi 1990),

$$v_t = 2\sqrt{\pi D_H f}. \tag{2.12}$$

The thermal exchange process within soil depends on meteorological conditions, soil thermal conductivity, thermal capacity, water content in the soil and other soil properties. In addition, one of the main factors influencing the soil thermal process is tillage, as well as the effect of various plants and residues covering the soil surface.

Different tillage technologies have a different impact on the soil temperature (Curtin et al. 2000; Al-Kaisi and Yin 2005). For example, Dalmago et al. (2004) determined that during the initial plant growth period (up to 30 days from the start of germination) soil temperature was lower under direct drilling compared to conventional tillage technology, and that the temperature difference was particularly prominent at 2.5 cm depth. This is mainly attributed to the fact that the soil was covered with chopped plant residues when using direct drilling technology. In subsequent periods, the increase in leaf area of plants resulted in a decrease in the soil temperature difference, and thus differences were insignificant. Other scientists (Morote et al. 1990; Salton and Mielniczuk 1995; Curtin et al. 2000; Calderon and Jackson 2002; Al-Kaisi and Yin 2005) made similar temperature measurements and suggested that direct drilling prevented the soil from fast warming during the initial period of vegetation compared to using conventional tillage, due to crop residues on the soil surface. Plant vegetation also impacts variations in soil temperature in relation to the shadow made by the biomass (Amos et al. 2005). Bergamaschi et al. (2004) examined the dependence of plant foliage development on tillage systems and found that there was a lower plant leaf density under direct drilling technology than with conventional tillage technology; there were variations in the

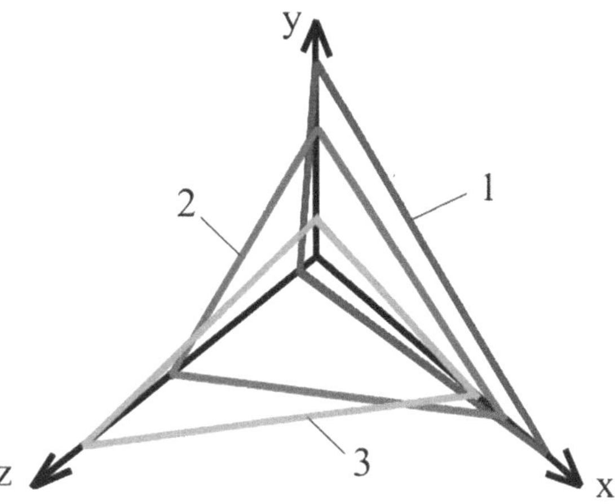

Fig. 2.11 Relation between performance indicators (x, y, z) and different tillage systems (1, 2, 3). x: crop yield and quality; y: tillage systems costs; z: soil, environmental and biodiversity benefits; *1*: conventional tillage; *2*: reduced tillage; *3*: no-tillage. (Prepared according to Morris et al. 2010)

amount of shadow from the plant leaves, resulting in different soil surface temperatures.

Soil temperature is also affected by soil bulk density. Temperatures of heavily compressed soils are 1–2 °C lower than those where soil bulk density is within an optimal range. There are changes in the processes of moisture capillary rise and evaporation with variations in soil bulk density, and large temperature variations affect plant growth and development (particularly in initial stages).

According to other studies, temperature variation in soil is dependent on the soil tillage intensity and environmental factors. In addition, thermal processes in soil have a very strong impact on crop growth and are very closely related to soil physical-mechanical properties and plant residues.

2.5 Energy and Economic Aspects of Tillage

In relation to tillage technologies, environmental sustainability is equally important as the energy-related and economic aspects. In this respect, each tillage technology has advantages and disadvantages. The application of conventional tillage technology makes it more possible to achieve higher and better quality yields; however, the cost of this technology is usually higher due to low productivity and the need to use high-power tractors. In contrast, although the use of direct drilling technologies is likely to result in lower yields and quality, the costs of tillage and drilling technological operations are lower. In addition, the impact of the latter technologies on the environment and biodiversity is also more positive compared to conventional ploughing. Morris et al. (2010) provides a visual presentation (Fig. 2.11) of the dependencies of tillage technologies and various activity indexes, using a trilinear coordinate system.

Hernanz et al. (1995) conducted long-term experiments on the growth of winter wheat, winter barley and other crops. Three different tillage systems were then compared in terms of energy consumption: conventional tillage (ploughing to a depth of 30 cm), minimum tillage (cultivation to a depth of 15 cm) and direct drilling. Results showed when considering machinery, fuel, seeds, fertilisers and other costs prior to harvesting, the use of minimum tillage and direct drilling technologies for cereals can save between 7% and 11% of energy costs compared to the use of conventional tillage.

Research conducted by Lithourgidis et al. (2009) showed that the use of sustainable tillage in Greece has many advantages compared to conventional tillage: the environment is protected and costs for machinery, repairs, maintenance and fuel are saved. In addition, the experiments of Baker et al. (2007) showed that elimination of tillage can save up to 80% of fuel costs and up to 60% of labour time spent on maintenance and repair of machinery.

In Croatia, the energy aspects of three different tillage and drilling technologies were analysed: conventional, reduced and direct drilling (Filipovic et al. 2006). It was found that fuel consumption when applying conventional tillage and drilling accounted for between 48 and 61 L ha^{-1}, but reduced tillage had a decreased fuel consumption of 1.5–2.0 times less than conventional technology. Furthermore, application of direct drilling technology enabled a reduction in fuel consumption of between 5 and 8 times less than when using conventional technology.

Turkish researchers found that direct drilling reduces the amount of fuel used by approximately 6-fold, and saves approximately four times the amount of labour time used with conventional tillage and drilling technology. A higher yield of approximately 400 kg ha^{-1} was obtained by drilling into conventionally tilled soil (Yalcin et al. 2005). Furthermore, the energy efficiency of three different intensity soil tillage technologies was investigated in Italy, where a comparison between deep soil ploughing (35 cm), minimum tillage (15 cm) and direct drilling revealed that the highest energy efficiency was attributed to deep ploughing (Bertocco et al. 2008). Safa and Tabatabaeefar (2008) analysed fuel consumption related to technological operations involved in wheat cultivation by agricultural machinery on different size farms: fuel consumption for soil tillage amounted to 75–121 L ha^{-1}, and for drilling it was only between 14.2 and 20.7 L ha^{-1}. In support of this result, Germanas (2008) proposed that fuel consumption accounts for about 8–10 L ha^{-1} under direct drilling.

Tebrügge and Böhrnsen (1995) found that conventional tillage technology (deep ploughing, pre-seeding tillage and drilling) resulted in approximately 2 h ha^{-1} labour time and a fuel consumption of approximately 35 L ha^{-1}. The labour time required for direct drilling technology was five-fold less than using conventional technology, and the fuel costs were 5–6 times lower. They also estimated that the traction force required for direct drilling involved about 8–10 kW per metre of working width, whereas that for conventional tillage technology required approximately 16–18 kW m^{-1}.

Abandoning certain tillage technological operations thus enables a reduction in diesel fuel consumption and a simultaneous reduction in the total production costs of plant production, while fostering sustainable agricultural development. One of the

Table 2.3 Energy inputs, outputs and efficiency ratios using different tillage systems for maize cultivation (From: Šarauskis et al. 2014b)

Energy inputs/outputs/ efficiency ratio	Unit	Tillage systems of maize cultivation				
		Deep ploughing	Shallow ploughing	Deep cultivation	Shallow cultivation	No-tillage
Human labour	MJ ha^{-1}	9.54	8.66	7.31	7.17	3.23
Diesel fuel	MJ ha^{-1}	2662.70	2274.62	2318.18	2116.22	1123.85
Agricultural machinery (including self-propelled machines)	MJ ha^{-1}	1739.56	1578.82	1332.36	1307.35	589.38
Maize seed	MJ ha^{-1}	459.00				
Herbicides	MJ ha^{-1}	634.25				1814.25
Nitrogen (N)	MJ ha^{-1}	10,532.00				
Phosphate (P_2O_5)	MJ ha^{-1}	1316.14				
Potassium (K_2O)	MJ ha^{-1}	774.69				
Total of energy inputs	MJ ha^{-1}	18,127.9	17,578.2	17,373.9	17,146.8	16,153.5
Energy output of maize grain and biomass	MJ ha^{-1}	224,179	236,033	196,931	205,751	226,184
Energy efficiency ratio	–	12.4	13.4	11.3	12.0	14.0

most important tasks for modern energy policies is the correctly balanced management of the energy consumption intensity; energy is a significant factor in the socio-economic development of any state (Tolón-Becerra et al. 2010; Omer 2008) and ensures energy security, economic competitiveness and environmental protection (Ang et al. 2010).

Bakasenas (2008) examined the modelling structure of energy consumption using pre-seeding tillage machinery. He established that the fuel energy consumption of the pre-seeding tillage machinery is 31%–43% that of the total energy consumption used by all machinery. An assessment of sustainable tillage and direct drilling systems for the cultivation of maize was conducted in Lithuania in relation to energy and economic aspects (Šarauskis et al. 2014b), and energy efficiency balance calculations reflected the ratio of the energy output with the maize yield using various tillage systems (in terms of energy indicators and energy inputs associated with the use of agricultural machinery, diesel fuel, seeds, fertiliser, herbicides and other materials (Table 2.3)).

The best energy efficiency ratio (14.0) was obtained using the no-tillage system; that for shallow ploughing was 13.4, and for conventional maize cultivation using deep ploughing the energy efficiency ratio was 12.4. An economic assessment of mechanised technological operations used in maize cultivation showed that the highest costs were related to tillage. The most expensive tillage operation evaluated was that of deep ploughing, and assessment of the cost of various technologies indicated that sustainable tillage systems saves between 8.0 and 23 EUR ha^{-1} compared with use of the conventional deep ploughing system (Šarauskis et al. 2014b). In addition, Sørensen and Nielsen (2005) investigated the cost of agricultural machinery use under reduced-tillage conditions and proposed that preparation of soils without using a plough can reduce costs from 25% to 41%, compared with conventional tillage.

Other very important factors in the cost of tillage operations relate to the dimensions of the field cultivated. For example, a larger, flatter field with fewer obstacles and stones facilitates greater energy efficiency and lower tillage costs. However, for fields that are stony or that have an irregular shape or relief, an appropriate value for the total correction factor should be estimated (Šarauskis et al. 2012, 2014b).

2.6 Conclusion

The main aim of agriculture is to produce the largest amount of food with the minimum environmental impact. It is thus very important to choose appropriate crop and soil management systems that enable minimal soil disturbance and soil degradation processes and which improve the physical, mechanical, biological and chemical properties of the soil (Hobbs et al. 2008). Sustainable tillage such as no-tillage, strip tillage, minimal tillage and reduced tillage technologies play an important role in agriculture and are used to: protect soil from erosion and runoff; reduce leaching of chemical substances into water bodies; improve soil moisture storage due to optimised soil porosity, encourage a greater number of earthworms and more extensive plant rooting; save labour time, fuel consumption and the cost of tillage operations; minimize CO_2 emissions and reduce the negative impacts of tillage on the environment. Crop residues incorporated into the soil or left on its surface also have important functions in terms of improving soil structure and stability and increasing soil organic matter (Morris et al. 2010; Lahmar 2010; Sommer et al. 2014). The use of sustainable tillage has energy-saving, economic advantages, delivers sufficiently high yields, thereby delivering a greater output with a reduced input. Other advantages also include climate change mitigation through reduced emissions by providing a lower fuel use of 60%–70%, and a 50% reduction in the amount of machinery and labour required (Basch et al. 2012).

References

Al-Kaisi MM, Yin X (2005) Tillage and crop residue effects on soil carbon and carbon dioxide emission in corn–soybean rotations. J Environ Qual 34(2):437–445. https://doi.org/10.2134/jeq2005.0437

Alvarez R (2005) A review of nitrogen fertilizer and conservation tillage effects on soil organic carbon storage. Soil Use Manag 21(1):38–52. https://doi.org/10.1111/j.1475-2743.2005.tb00105.x

Alvarez R, Steinbach HS (2009) A review of the effects of tillage systems on some soil physical properties, water content, nitrate availability and crops yield in the Argentine Pampas. Soil Tillage Res 104:1–15. https://doi.org/10.1016/j.still.2009.02.005

Alvarez R, Alvarez CR, Lorenzo G (2001) Carbon dioxide fluxes following tillage from a mollisol in the Argentine Rolling Pampa. Eur J Soil Biol 37(3):161–166. https://doi.org/10.1016/S1164-5563(01)01085-8

Álvaro-Fuentes J, Cantero-Martínez C, López MV, Arrúe JL (2007) Soil carbon dioxide fluxes following tillage in semiarid Mediterranean agroecosystems. Soil Tillage Res 96(1):331–341. https://doi.org/10.1016/j.still.2007.08.003

Amos B, Arkebauer TJ, Doran JW (2005) Soil surface fluxes of greenhouse gases in an irrigated maize-based agroecosystem. Soil Sci Soc Am J 69(2):387–395. https://doi.org/10.2136/sssaj2005.0387

Ang BW, Mu AR, Zhou P (2010) Accounting frameworks for tracking energy efficiency trends. Energy Econ 32(5):1209–1219. https://doi.org/10.1016/j.eneco.2010.03.011

Arvidsson J (1998) Influence of soil texture and organic matter content on bulk density, air content, compression index and crop yield in field and laboratory compression experiments. Soil Tillage Res 49(1):159–170. https://doi.org/10.1016/S0167-1987(98)00164-0

Arvidsson J (2010) Energy use efficiency in different tillage systems for winter wheat on a clay and silt loam in Sweden. Eur J Agron 33(3):250–256. https://doi.org/10.1016/j.eja.2010.06.003

Arvidsson J, Westlin A, Sörensson F (2013) Working depth in non-inversion tillage – effects on soil physical properties and crop yield in Swedish field experiments. Soil Tillage Res 126:259–266. https://doi.org/10.1016/j.still.2012.08.010

Auerswald K, Mutchler CK, McGregor KC (1994) The influence of tillage-induced differences in surface moisture content on soil erosion. Soil Tillage Res 32(1):41–50. https://doi.org/10.1016/0167-1987(94)90031-0

Bakasenas A (2008) Comparative analysis of energy input in presowing soil tillage and crop sowing. Agric Eng 40(3–4):5–15

Baker CJ, Saxton KE, Ritchie WR, Chamen WCT, Reicosky DC, Ribeiro F, Justice SE, Hobbs PR (2007) No-tillage seeding in conservation agriculture. CABI, FAO, Rome 326 p

Bakker MM, Govers G, Rounsevell MD (2004) The crop productivity–erosion relationship: an analysis based on experimental work. Catena 57(1):55–76. https://doi.org/10.1016/j.catena.2003.07.002

Ball BC, Scott A, Parker JP (1999) Field N_2O, CO_2 and CH_4 fluxes in relation to tillage, compaction and soil quality in Scotland. Soil Tillage Res 53(1):29–39. https://doi.org/10.1016/S0167-1987(99)00074-4

Basch G, Kassam A, González-Sánchez E, Streit S (2012) Making Sustainable Agriculture real in CAP 2020: the role of conservation agriculture. ECAF, Brussels, 43 p

Batey T (2009) Soil compaction and soil management – a review. Soil Use Manag 25(4):335–345. https://doi.org/10.1111/j.1475-2743.2009.00236.x

Bergamaschi H, Dalmago GA, Bergonci JI, Bianchi CAM, Heckler BMM, Comiran F (2004) Solar radiation intercepted by maize crops as function of soil tillage systems and water availabilities. In: 13th international soil conservation organisation conference, Brisbane, Australia CD-ROM

Bertocco M, Basso B, Sartori L, Martin EC (2008) Evaluating energy efficiency of site-specific tillage in maize in NE Italy. Bioresour Technol 99(15):6957–6965. https://doi.org/10.1016/j.biortech.2008.01.027

Bhagat RM, Bhuiyan SI, Moody K (1996) Water, tillage and weed interactions in lowland tropical rice: a review. Agric Water Manag 31(3):165–184. https://doi.org/10.1016/0378-3774(96)01242-5

Boardman J, Poesen J (2006) Soil erosion in Europe. Wiley, Chichester, 855 p. https://doi.org/10.1002/0470859202.fmatter

Bogužas V, Kairyte A, Jodaugiene D (2010) Soil physical properties and earthworms as affected by soil tillage systems, straw and green manure management. Zemdirbyste-Agriculture 97(3):3–14

Boucher O, Bellassen V, Benveniste H, Ciais P, Criqui P, Guivarch C, Le Treut H, Mathy S, Séférian R (2016) Opinion: in the wake of Paris Agreement, scientists must embrace new directions for climate change research. Proc Natl Acad Sci 113(27):7287–7290. https://doi.org/10.1073/pnas.1607739113

Bronick CJ, Lal R (2005) Soil structure and management: a review. Geoderma 124(1):3–22. https://doi.org/10.1016/j.geoderma.2004.03.005

Buragiene S, Šarauskis E, Romaneckas K, Sakalauskas A, Užupis A, Katkevičius E (2011) Soil temperature and gas (CO_2 and O_2) emissions from soils under different tillage machinery systems. J Food Agric Environ 9(2):480–485

Buragiene S, Šarauskis E, Romaneckas K, Sasnauskiene J, Masilionyte L, Kriauciuniene Z (2015) Experimental analysis of CO_2 emissions from agricultural soils subjected to five different tillage systems in Lithuania. Sci Total Environ 514:1–9. https://doi.org/10.1016/j.scitotenv.2015.01.090

Calderon FJ, Jackson LE (2002) Rototillage, disking, and subsequent irrigation. J Environ Qual 31(3):752–758. https://doi.org/10.2134/jeq2002.7520

Cannell RQ, Hawes JD (1994) Trends in tillage practices in relation to sustainable crop production with special reference to temperate climates. Soil Tillage Res 30(2-4):245–282. https://doi.org/10.1016/0167-1987(94)90007-8

Cavalieri KMV, Arvidsson J, da Silva AP, Keller T (2008) Determination of precompression stress from uniaxial compression tests. Soil Tillage Res 98(1):17–26 https://doi.org/10.1016/j.still.2007.09.020

Celik I, Turgut MM, Acir N (2012) Crop rotation and tillage effects on selected soil physical properties of a Typic Haploxerert in an irrigated semi-arid Mediterranean region. Int J Plant Prod 6(4):457–480

Cerdan O, Govers G, Le Bissonnais Y, Van Oost K, Poesen J, Saby N, Gobin A, Vacca A, Quinton J, Auerswald K, Klik A, Kwaad FJPM, Raclot D, Ionita I, Rejman J, Rousseva S, Muxart T, Roxo MJ, Dostal T (2010) Rates and spatial variations of soil erosion in Europe: a study based on erosion plot data. Geomorphology 122(1):167–177. https://doi.org/10.1016/j.geomorph.2010.06.011

Cesevicius G, Feiza V, Feiziene D (2005) Tausojančių žemės dirbimo būdų ir augalinių liekanų įtaka dirvožemio fizikinėms savybėms ir vasarinių miežių derliui. Vagos 69(22):7–18 (in Lithuanian)

Chamen T, Alakukku L, Pires S, Sommer C, Spoor G, Tijink F, Weisskopf P (2003) Prevention strategies for field traffic-induced subsoil comp action: a review: part 2. Equipment and field practices. Soil Tillage Res 73(1):161–174. https://doi.org/10.1016/S0167-1987(03)00108-9

Chang C, Lindwall CW (1990) Comparison of the effect of long term tillage and crop rotation on physical properties of a soil. Can Agric Eng 32(1):53–55

Chatskikh D, Olesen JE (2007) Soil tillage enhanced CO_2 and N_2O emissions from loamy sand soil under spring barley. Soil Tillage Res 97(1):5–18. https://doi.org/10.1016/j.still.2007.08.004

Chivenge PP, Murwira HK, Giller KE, Mapfumo P, Six J (2007) Long-term impact of reduced tillage and residue management on soil carbon stabilization: implications for conservation agriculture on contrasting soils. Soil Tillage Res 94(2):328–337. https://doi.org/10.1016/j.still.2006.08.006

Chung SO, Sudduth KA, Motavalli PP, Kitchen NR (2013) Relating mobile sensor soil strength to penetrometer cone index. Soil Tillage Res 129:9–18. https://doi.org/10.1016/j.still.2012.12.004

Ciuberkis S, Ozeraitiene D, Bernotas S, Ambrazaitiene D (2008) Changes in the soil properties as affected by conventional and minimal soil tillage systems. Zemdirbyste-Agriculture 95 (2):16–28

Coulouma G, Boizard H, Trotoux G, Lagacherie P, Richard G (2006) Effect of deep tillage for vineyard establishment on soil structure: a case study in Southern France. Soil Tillage Res 88 (1):132–143. https://doi.org/10.1016/j.still.2005.05.002

Curtin D, Wang H, Selles F, McConkey BG, Campbell CA (2000) Tillage effects on carbon fluxes in continuous wheat and fallow–wheat rotations. Soil Sci Soc Am J 64(6):2080–2086

Czyż EA (2004) Effects of traffic on soil aeration, bulk density and growth of spring barley. Soil Tillage Res 79(2):153–166. https://doi.org/10.1016/j.still.2004.07.004

Czyz EA, Tomaszewska J (1993) Changes of aeration conditions and the yield of sugar beet on sandy soil of different density. Pol J Soil Sci 26(1):1–9

Czyż EA, Tomaszewska J, Dexter AR (2001) Response of spring barley to changes of compaction and aeration of sandy soil under model conditions. Int Agrophys 15(1):9–12

D'Haene K, Vermang J, Cornelis WM, Leroy BL, Schiettecatte W, De Neve S, Gabriels D, Hofman G (2008) Reduced tillage effects on physical properties of silt loam soils growing root crops. Soil Tillage Res 99(2):279–290. https://doi.org/10.1016/j.still.2008.03.003

Da Veiga M, Horn R, Reinert DJ, Reichert JM (2007) Soil compressibility and penetrability of an Oxisol from southern Brazil, as affected by long-term tillage systems. Soil Tillage Res 92 (1):104–113. https://doi.org/10.1016/j.still.2006.01.008

Dalmago GA, Bergamaschi H, Comiran F, Bianchi CAM, Bergonci JI, Heckler BMM (2004) Soil temperature in maize crops as function of soil tillage systems. In: Proceedings of the 13th international soil conservation organisation conference, Brisbane, Australia, pp 4–8

Dao TH (1998) Tillage and crop residue effects on carbon dioxide evolution and carbon storage in a Paleustoll. Soil Sci Soc Am J 62(1):250–256. https://doi.org/10.2136/sssaj1998.03615995006200010032x

De Paz JM, Sánchez J, Visconti F (2006) Combined use of GIS and environmental indicators for assessment of chemical, physical and biological soil degradation in a Spanish Mediterranean region. J Environ Manag 79(2):150–162. https://doi.org/10.1016/j.jenvman.2005.06.002

De Vita P, Di Paolo E, Fecondo G, Di Fonzo N, Pisante M (2007) No-tillage and conventional tillage effects on durum wheat yield, grain quality and soil moisture content in southern Italy. Soil Tillage Res 92(1):69–78. https://doi.org/10.1016/j.still.2006.01.012

Derpsch R, Friedrich T (2009) Development and current status of no-till adoption in the world. In: Proceedings on CD, 18th triennial conference of the International Soil Tillage Research Organization (ISTRO), June 15–19, 2009, Izmir, Turkey

Derpsch R, Lange D, Birbaumer G, Moriya K (2016) Why do medium-and large-scale farmers succeed practicing CA and small-scale farmers often do not? – Experiences from Paraguay. Int J Agric Sustain 14(3):269–281. https://doi.org/10.1080/14735903.2015.1095974

Dexter AR, Czyz EA (2000) Effects of soil management on the dispersibility of clay in a sandy soil. Int Agrophys 14(3):269–272

Dossou-Yovo ER, Brüggemann N, Jesse N, Huat J, Ago EE, Agbossou EK (2016) Reducing soil CO_2 emission and improving upland rice yield with no-tillage, straw mulch and nitrogen fertilization in northern Benin. Soil Tillage Res 156:44–53. https://doi.org/10.1016/j.still.2015.10.001

Duxbury JM (1994) The significance of agricultural sources of greenhouse gases. Fertil Res 38 (2):151–163. https://doi.org/10.1007/BF00748775

Duxbury JM (1995) The significance of greenhouse gas emissions from soils of tropical agroecosystems. Soil management and greenhouse effect. CRC Press, Boca Raton, pp 279–292

Ekeberg E, Riley HCF (1997) Tillage intensity effects on soil properties and crop yields in a long-term trial on morainic loam soil in southeast Norway. Soil Tillage Res 42(4):277–293. https://doi.org/10.1016/S0167-1987(97)00008-1

Ellert BH, Janzen HH (1999) Short-term influence of tillage on CO_2 fluxes from a semi-arid soil on the Canadian Prairies. Soil Tillage Res 50(1):21–32. https://doi.org/10.1016/S0167-1987(98)00188-3

Feiza V, Feiziene D, Deveikyte I (2006) Reduced tillage in spring: 1. Influence on soil physical properties. Zemdirbyste-Agriculture 93(3):35–55

Feiza V, Feiziene D, Kadziene G (2008) Agro-physical properties of Endocalcari-pihypogleyic Cambisol arable layer in long-term soil management systems. Zemes ukio mokslai 15:13–23

Feiza V, Feiziene D, Sinkeviciene A, Boguzas V, Putramentaite A, Lazauskas S, Deveikyte I, Seibutis V, Steponaviciene V, Pranaitiene S (2015) Soil water capacity, pore-size distribution and CO_2 e-flux in different soils after long-term no-till management. Zemdirbyste-Agriculture 102(1):3–14. https://doi.org/10.13080/z-a.2015.102.001

Feiziene D, Feiza V, Kadžiene G (2009) The influence of meteorological conditions on soil water vapour exchange rate and CO_2 emission under different tillage systems. Zemdirbyste-Agriculture 96(2):3–22

Feiziene D, Feiza V, Vaideliene A, Povilaitis V, Antanaitis S (2010) Soil surface carbon dioxide exchange rate as affected by soil texture, different long-term tillage application and weather. Zemdirbyste-Agriculture 97(3):25–42

Fernández-Ugalde O, Virto I, Bescansa P, Imaz MJ, Enrique A, Karlen DL (2009) No-tillage improvement of soil physical quality in calcareous, degradation-prone, semiarid soils. Soil Tillage Res 106(1):29–35. https://doi.org/10.1016/j.still.2009.09.012

Filipovic D, Kosutic S, Gospodaric Z, Zimmer R, Banaj D (2006) The possibilities of fuel savings and the reduction of CO_2 emissions in the soil tillage in Croatia. Agric Ecosyst Environ 115 (1):290–294. https://doi.org/10.1016/j.agee.2005.12.013

Fleige H, Horn R (2000) Field experiments on the effect of soil compaction on soil properties, runoff, interflow and erosion. Adv Geoecol 32:258–268

Fortin MC, Rochette P, Pattey E (1996) Soil carbon dioxide fluxes from conventional and no-tillage small-grain cropping systems. Soil Sci Soc Am J 60(5):1541–1547. https://doi.org/10.2136/sssaj1996.03615995006000050036x

Franzluebbers AJ (2002) Water infiltration and soil structure related to organic matter and its stratification with depth. Soil Tillage Res 66(2):197–205. https://doi.org/10.1016/S0167-1987(02)00027-2

Fritton DD (2008) Evaluation of pedotransfer and measurement approaches to avoid soil compaction. Soil Tillage Res 99(2):268–278. https://doi.org/10.1016/j.still.2008.03.004

Germanas L (2008) Evaluation indexes of sowing by disc coulters. Agric Eng 40(3-4):16–25

Gisi U (1990) Bodenökologie. Georg Thience Verlag, Stuttgart 304 p. (in German)

Goodman AM, Ennos AR (1999) The effects of soil bulk density on the morphology and anchorage mechanics of the root systems of sunflower and maize. Ann Bot 83(3):293–302. https://doi.org/10.1006/anbo.1998.0822

Grant RF (1997) Changes in soil organic matter under different tillage and rotation: mathematical modeling in ecosys. Soil Sci Soc Am J 61(4):1159–1175. https://doi.org/10.2136/sssaj1997.03615995006100040023x

Gruber S, Möhring J, Claupein W (2011) On the way towards conservation tillage-soil moisture and mineral nitrogen in a long-term field experiment in Germany. Soil Tillage Res 115:80–87. https://doi.org/10.1016/j.still.2011.07.001

Guerif J (1994) Effects of compaction on soil strength parameters. In: Soane BD, Ouwerkerk CV (eds) Soil compaction crop production. Elsevier, Amsterdam, pp 191–214

Hamza MA, Anderson WK (2005) Soil compaction in cropping systems: a review of the nature, causes and possible solutions. Soil Tillage Res 82(2):121–145. https://doi.org/10.1016/j.still.2004.08.009

Hasankhani-Ghavam F, Abbaspour-Gilandeh Y, Shahgoli G, Rahmanzadeh-Bahram H (2015) Design, manufacture and evaluation of the new instrument to measure the friction coefficient of soil. Agric Eng Int CIGR J 17(1):101–109

Hazarika S, Parkinson R, Bol R, Dixon L, Russell P, Donovan S, Allen D (2009) Effect of tillage system and straw management on organic matter dynamics. Agron Sustain Dev 29(4):525–533. https://doi.org/10.1051/agro/2009024

Hendrix PF, Han CR, Groffman PM (1988) Soil respiration in conventional and no-tillage agroecosystems under different winter cover crop rotations. Soil Tillage Res 12(2):135–148. https://doi.org/10.1016/0167-1987(88)90037-2

Hernanz JL, Giron VS, Cerisola C (1995) Long-term energy use and economic evaluation of three tillage systems for cereal and legume production in central Spain. Soil Tillage Res 35 (4):183–198. https://doi.org/10.1016/0167-1987(95)00490-4

Hernanz JL, López R, Navarrete L, Sanchez-Giron V (2002) Long-term effects of tillage systems and rotations on soil structural stability and organic carbon stratification in semiarid central Spain. Soil Tillage Res 66(2):129–141. https://doi.org/10.1016/S0167-1987(02)00021-1

Hobbs PR, Sayre K, Gupta R (2008) The role of conservation agriculture in sustainable agriculture. Philos Trans R Soc B 363(1491):543–555

Holland JM (2004) The environmental consequences of adopting conservation tillage in Europe: reviewing the evidence. Agric Ecosyst Environ 103(1):1–25. https://doi.org/10.1016/j.agee.2003.12.018

Horn R (2004) Time dependence of soil mechanical properties and pore functions for arable soils. Soil Sci Soc Am J 68(4):1131–1137. https://doi.org/10.2136/sssaj2004.1131

Horn R, Domżżał H, Słowińska-Jurkiewicz A, Van Ouwerkerk C (1995) Soil compaction processes and their effects on the structure of arable soils and the environment. Soil Tillage Res 35 (1):23–36. https://doi.org/10.1016/0167-1987(95)00479-C

Hsiao TC, Steduto P, Fereres E (2007) A systematic and quantitative approach to improve water use efficiency in agriculture. Irrig Sci 25(3):209–231. https://doi.org/10.1007/s00271-007-0063-2

Jankauskas B, Jankauskienė G (2005) Erosion-preventive agricultural practices and adjustment of farm specialisation and alternative human activities to these practices. Zemdirbyste-Agriculture 91(3):27–29

Janulevičius A, Juostas A, Pupinis G (2013) Engine performance during tractor operational period. Energy Convers Manag 68:11–19. https://doi.org/10.1016/j.enconman.2013.01.001

Jasa P (2011) No-till soil and water issues. In: agronomy conference, South Dakota agri-business association, pp 14–15. Available: http://www.sdaba.org/agronomy-conference-2011.htm Accessed 12 May 2016

Jastrow JD, Miller RM, Boutton TW (1996) Carbon dynamics of aggregate-associated organic matter estimated by carbon-13 natural abundance. Soil Sci Soc Am J 60(3):801–807. https://doi.org/10.2136/sssaj1996.03615995006000030017x

Jin H, Hongwen L, Rasaily RG, Qingjie W, Guohua C, Yanbo S, Xiaodong Q, Lijin L (2011) Soil properties and crop yields after 11 years of no tillage farming in wheat–maize cropping system in North China Plain. Soil Tillage Res 113(1):48–54. https://doi.org/10.1016/j.still.2011.01.005

Jones PD, New M, Parker DE, Martin S, Rigor IG (1999) Surface air temperature and its changes over the past 150 years. Rev Geophys 37:173–199. https://doi.org/10.1029/1999RG900002

Kaczmarek Z, Gajewski P, Mocek A, Owczarzak W, Glina B (2015) Physical and water properties of selected Polish heavy soils of various origins. Soil Sci Annu 66(4):191–197. https://doi.org/10.1515/ssa-2015-0036

Karapetyan MA (2005) The basic concept of the environmental compatibility of the system "Machine – Tractor – Technology – Soil". Mekhanizatsiya i elektrifikatsiya selskogo khozyaystva (Mechanization and Electrification of Agriculture) 9:30–32 (in Russian)

Karayel D (2009) Performance of a modified precision vacuum seeder for no-till sowing of maize and soybean. Soil Tillage Res 104:121–125. https://doi.org/10.1016/j.still.2009.02.001

Kassam AH, Friedrich T, Shaxson F, Pretty J (2009) The spread of conservation agriculture: justification, sustainability and uptake. Int J Agric Sustain 7(4):1–29

Kassam A, Friedrich T, Derpsch R, Kienzle J (2015) Overview of the worldwide spread of conservation agriculture. Field actions science reports, vol. 8, Available: http://factsreports.revues.org/3966. Accessed 20 Sept 2016

Kessavalou A, Doran JW, Mosier AR, Drijber RA (1998) Greenhouse gas fluxes following tillage and wetting in a wheat-fallow cropping system. J Environ Qual 27(5):1105–1116. https://doi.org/10.2134/jeq1998.00472425002700050016x

Kılıç K, Özgöz E, Akbaş F (2004) Assessment of spatial variability in penetration resistance as related to some soil physical properties of two fluvents in Turkey. Soil Tillage Res 76(1):1–11. https://doi.org/10.1016/j.still.2003.08.009

Kiryushin VI (1996) Ekologicheskie osnovy zemledeliya (Ecological Bases of Agriculture). Moscow 367 p. (in Russian)

Köller K (2003) Techniques of soil tillage. In: Titi AE (ed) Soil tillage in agroecosystems. CRC Press, New York, 25 p

Kowalenko CG, Ivarson KC (1978) Effect of moisture content, temperature and nitrogen fertilization on carbon dioxide evolution from field soils. Soil Biol Biochem 10(5):417–423. https://doi.org/10.1016/0038-0717(78)90068-8

Król A, Lipiec J, Turski M, Kuś J (2013) Effects of organic and conventional management on physical properties of soil aggregates. Int Agrophys 27(1):15–21. https://doi.org/10.2478/v10247-012-0063-1

Kroulík M, Kumhála F, Hůla J, Honzík I (2009) The evaluation of agricultural machines field trafficking intensity for different soil tillage technologies. Soil Tillage Res 105(1):171–175. https://doi.org/10.1016/j.still.2009.07.004

Kulen A, Kuppers X (1986) Modern agricultural mechanics. Agropromizdat:231–234 (in Russian)

Kumar S, Himanshu SK, Gupta KK (2012) Effect of global warming on mankind – a review. Int Res J of Environ Sci 1(4):56–59

Kushnarov AS (1986) Umensheniye vrednogo vozdeystviya na pochvu rabochikh organov i khodovykh sistem mashinnykh agregatov pri vnedrenii industrialnykh tekhnologiy vozdelyvaniya s.-kh. kultur. Moscow, 56 p. (in Russian)

Kushwaha RL, Vaishnav AS, Zoerb GC (1986) Soil bin evaluation of disc coulters under no-till crop residue conditions. Trans ASAE 29(1):40–44

Kutzbach HD (2000) Trends in power and machinery. J Agric Eng Res 76(3):237–247. https://doi.org/10.1006/jaer.2000.0574

La Scala N, Lopes A, Marques J, Pereira GT (2001) Carbon dioxide emissions after application of tillage systems for a dark red latosol in southern Brazil. Soil Tillage Res 62(3):163–166. https://doi.org/10.1016/S0167-1987(01)00212-4

La Scala N, Bolonhezi D, Pereira GT (2006) Short-term soil CO_2 emission after conventional and reduced tillage of a no-till sugar cane area in southern Brazil. Soil Tillage Res 91(1):244–248. https://doi.org/10.1016/j.still.2005.11.012

La Scala N, Lopes A, Spokas K, Bolonhezi D, Archer DW, Reicosky DC (2008) Short-term temporal changes of soil carbon losses after tillage described by a first-order decay model. Soil Tillage Res 99(1):108–118. https://doi.org/10.1016/j.still.2008.01.006

La Scala N, Lopes A, Spokas K, Archer DW, Reicosky DC (2009) Short-term temporal changes of bare soil CO_2 fluxes after tillage described by first-order decay models. Eur J Soil Sci 60 (2):258–264. https://doi.org/10.1111/j.1365-2389.2008.01102.x

Lahmar R (2010) Adoption of conservation agriculture in Europe: lessons of the KASSA project. Land Use Policy 27(1):4–10. https://doi.org/10.1016/j.landusepol.2008.02.001

Lal R, Kimble JM (1997) Conservation tillage for carbon sequestration. Nutr Cycl Agroecosyst 49 (1-3):243–253. https://doi.org/10.1023/A:1009794514742

Lal R, Kimble JM, Stewart BA (1995) World soils as a source or sink for radiatively-active gases. Soil management and greenhouse effect. Advances in soil science. CRC Press, Boca Raton, pp 1–8

Lamandé M, Schjønning P (2011) Transmission of vertical stress in a real soil profile. Part I: site description, evaluation of the Söhne model, and the effect of topsoil tillage. Soil Tillage Res 114 (2):57–70. https://doi.org/10.1016/j.still.2011.05.004

Lamandé M, Schjønning P, Tøgersen FA (2007) Mechanical behaviour of an undisturbed soil subjected to loadings: effects of load and contact area. Soil Tillage Res 97(1):91–106. https://doi.org/10.1016/j.still.2007.09.002

Lampurlanes J, Cantero-Martinez C (2003) Soil bulk density and penetration resistance under different tillage and crop management systems and their relationship with barley root growth. Agron J 95(3):526–536. https://doi.org/10.2134/agronj2003.5260

Lapen DR, Topp GC, Edwards ME, Gregorich EG, Curnoe WE (2004) Combination cone penetration resistance/water content instrumentation to evaluate cone penetration–water content relationships in tillage research. Soil Tillage Res 79(1):51–62. https://doi.org/10.1016/j.still.2004.03.023

Lee J, Hopmans JW, van Kessel C, King AP, Evatt KJ, Louie D, Rolston DE, Six J (2009) Tillage and seasonal emissions of CO_2, N_2O and NO across a seed bed and at the field scale in a Mediterranean climate. Agric Ecosyst Environ 129(4):378–390. https://doi.org/10.1016/j.agee.2008.10.012

Ley GJ, Mullins CE, Lal R (1993) Effects of soil properties on the strength of weakly structured tropical soils. Soil Tillage Res 28(1):1–13. https://doi.org/10.1016/0167-1987(93)90051-P

Lindstrom J (1990) Methods for measurement of soil aeration. Swedish University of Agricultural Sciences. Rep Dissertations 5(1):1–19

Linke C (1998) Direktsaat – eine Bestandsaufnahme unter besonderer Berücksichtigung technischer, agronomischer und ökonomischer Aspekte. Dissertation, University of Hohenheim, Stuttgart, 482 p. (in German)

Lithourgidis AS, Damalas CA, Eleftherohorinos IG (2009) Conservation tillage: a promising perspective for sustainable agriculture in Greece. J Sustain Agric 33(1):85–95. https://doi.org/10.1080/10440040802587280

Logsdon SD, Karlen DL (2004) Bulk density as a soil quality indicator during conversion to no-tillage. Soil Tillage Res 78(2):143–149. https://doi.org/10.1016/j.still.2004.02.003

Logsdon S, Allmaras RR, Wu L, Swan JB, Randall GW (1990) Macroporosity and its relation to saturated hydraulic conductivity under different tillage practices. Soil Sci Soc Am J 54 (4):1096–1101. https://doi.org/10.2136/sssaj1990.03615995005400040029x

Lopez MV, Arrúe JL, Sánchez-Girón V (1996) A comparison between seasonal changes in soil water storage and penetration resistance under conventional and conservation tillage systems in Aragon. Soil Tillage Res 37(4):251–271. https://doi.org/10.1016/0167-1987(96)01011-2

López MV, Blanco-Moure N, Limón MÁ, Gracia R (2012) No tillage in rainfed Aragon (NE Spain): effect on organic carbon in the soil surface horizon. Soil Tillage Res 118:61–65. https://doi.org/10.1016/j.still.2011.10.012

López-Fando C, Pardo MT (2012) Use of a partial-width tillage system maintains benefits of no-tillage in increasing total soil nitrogen. Soil Tillage Res 118:32–39. https://doi.org/10.1016/j.still.2011.10.010

Lowery B, Morrison JE (2002) Soil penetrometers and penetrability. In: Dane JH, Topp GC (eds) Methods of soil analysis, Part 4 physical methods. Soil science Society of America Book Series no. 5, pp 363–388

Macfadyen A (1970) Soil metabolism in relation to ecosystem energy flow and to primary and secondary production. In: Phillipson J (ed) "Methods of Study in Soil Ecology", Paris Symposium, pp 167–172

Maikšteniene S, Šlepetiene A, Masilionyte L (2007) The effect of mouldboard nouldbo and ploughless primary soil tillage on the properties of Endocalcari – Endohypogleyic Cambisol and on energetic efficiency of agrosystems. Zemdirbyste-Agriculture 94(1):3–23

Maikšteniene S, Krištaponyte I, Masilionyte L (2008) The effects of long-term fertilisation systems on the variation of major productivity parameters of Gleyic Cambisols. Zemdirbyste-Agriculture 95(1):22–39

Mitchell JP, Carter LM, Reicosky DC, Shrestha A, Pettygrove GS, Klonsky KM, Marcum DB, Chessman D, Roy R, Hogan P, Dunning L (2016) A history of tillage in California's Central Valley. Soil Tillage Res 157:52–64. https://doi.org/10.1016/j.still.2015.10.015

Morell FJ, Álvaro-Fuentes J, Lampurlanés J, Cantero-Martínez C (2010) Soil CO_2 fluxes following tillage and rainfall events in a semiarid Mediterranean agroecosystem: effects of tillage systems and nitrogen fertilization. Agric Ecosyst Environ 139(1):167–173. https://doi.org/10.1016/j.agee.2010.07.015

Moret D, Arrúe JL (2007) Dynamics of soil hydraulic properties during fallow as affected by tillage. Soil Tillage Res 96(1):103–113. https://doi.org/10.1016/j.still.2007.04.003

Morote CGB, Vidor C, Mendes NG (1990) Soil temperature as affected by mulching and irrigation. Revista Brasileira de Ciência do Solo 14(1):81–84

Morris NL, Miller PCH, Orson JH, Froud-Williams RJ (2010) The adoption of non-inversion tillage systems in the United Kingdom and the agronomic impact on soil, crops and the environment – a review. Soil Tillage Res 108(1):1–15. https://doi.org/10.1016/j.still.2010.03.004

Mueller L, Kay BD, Deen B, Hu C, Zhang Y, Wolff M, Eulenstein F, Schindler U (2009) Visual assessment of soil structure: part II. Implications of tillage, rotation and traffic on sites in Canada, China and Germany. Soil Tillage Res 103(1):188–196. https://doi.org/10.1016/j.still.2008.09.010

Muñoz C, Paulino L, Monreal C, Zagal E (2010) Greenhouse gas (CO_2 and N_2O) emissions from soils: a review. Chilean J Agric Res 70(3):485–497

Of Complex Systems Available: http://systemexplorations.wordpress.com/2011/09/05/of-complex-systems/. Accessed 27 Sept 2016

Ogle SM, Breidt FJ, Paustian K (2005) Agricultural management impacts on soil organic carbon storage under moist and dry climatic conditions of temperate and tropical regions. Biogeochemistry 72(1):87–121. https://doi.org/10.1007/s10533-004-0360-2

Omer AM (2008) Energy, environment and sustainable development. Renew Sust Energ Rev 12 (9):2265–2300. https://doi.org/10.1016/j.rser.2007.05.001

Oorts K, Merckx R, Gréhan E, Labreuche J, Nicolardot B (2007) Determinants of annual fluxes of CO_2 and N_2O in long-term no-tillage and conventional tillage systems in northern France. Soil Tillage Res 95(1):133–148. https://doi.org/10.1016/j.still.2006.12.002

Oyedele DJ, Schjønning P, Sibbesen E, Debosz K (1999) Aggregation and organic matter fractions of three Nigerian soils as affected by soil disturbance and incorporation of plant material. Soil Tillage Res 50(2):105–114. https://doi.org/10.1016/S0167-1987(98)00200-1

Pachepsky YA, Rawls WJ (2003) Soil structure and pedotransfer functions. Eur J Soil Sci 54 (3):443–452. https://doi.org/10.1046/j.1365-2389.2003.00485.x

Pagliai M, Marsili A, Servadio P, Vignozzi N, Pellegrini S (2003) Changes in some physical properties of a clay soil in Central Italy following the passage of rubber tracked and wheeled tractors of medium power. Soil Tillage Res 73(1):119–129. https://doi.org/10.1016/S0167-1987(03)00105-3

Pagliai M, Vignozzi N, Pellegrini S (2004) Soil structure and the effect of management practices. Soil Tillage Res 79(2):131–143. https://doi.org/10.1016/j.still.2004.07.002

Parkin TB, Kaspar TC (2003) Temperature controls on diurnal carbon dioxide flux. Soil Sci Soc Am J 67(6):1763–1772. https://doi.org/10.2136/sssaj2003.1763

Parton WJ, Woomer PL, Martin A, Swift MJ (1994) Modelling soil organic matter dynamics and plant productivity in tropical ecosystems. In: Woomer PL, Swift MJ (eds) The biological management of tropical soil fertility, pp 171–188

Paz-Ferreiro J, Trasar-Cepeda C, Leirós MC, Seoane S, Gil-Sotres F (2009) Biochemical properties in managed grassland soils in a temperate humid zone: modifications of soil quality as a consequence of intensive grassland use. Biol Fertil Soils 45(7):711–722. https://doi.org/10.1007/s00374-009-0382-y

Pekrun C, Claupein W (1998) Forschung zur reduzierten Bodenbearbeitung in Mitteleuropa: eine Literaturübersicht. Pflanzenbauwissenschaften 2(4):160–175

Pimentel D, Harvey C, Resosudarmo P, Sinclair K, Kurz D, McNair M, Crist S, Shpritz L, Fitton L, Saffouri R, Blair R (1995) Environmental and economic costs of soil erosion and conservation benefits. Science-AAAS-Weekly Paper Edition 267(5201):1117–1122

Pocienė A, Kinčius L (2008) Dirvožemio erozija ir jos prevencija (Soil erosion and its prevention). Kaunas, Arvida 79 p. (in Lithuanian)

Prior SA, Reicosky DC, Reeves DW, Runion GB, Raper RL (2000) Residue and tillage effects on planting implement-induced short-term CO_2 and water loss from a loamy sand soil in Alabama. Soil Tillage Res 54(3):197–199. https://doi.org/10.1016/S0167-1987(99)00092-6

Quinton JN, Govers G, Van Oost K, Bardgett RD (2010) The impact of agricultural soil erosion on biogeochemical cycling. Nat Geosci 3(5):311–314. https://doi.org/10.1038/ngeo838

Račinskas A (1992) Dirvožemio erozija (Soil erosion). Mokslas, Vilnius 135 p. (in Lithuanian)

Rasmussen KJ (1999) Impact of ploughless soil tillage on yield and soil quality: a Scandinavian review. Soil Tillage Res 53(1):3–14. https://doi.org/10.1016/S0167-1987(99)00072-0

Rastogi M, Singh S, Pathak H (2002) Emission of carbon dioxide from soil. Curr Sci 82(5):510–517

Reichert JM, da Silva VR, Reinert DJ (2004) Soil moisture, penetration resistance, and least limiting water range for three soil management systems and black beans yield. In: Conserving soil and water for society: sharing solutions. 13th international soil conservation organisation conference – Brisbane paper 721:1–4

Reicosky DC (2015) Conservation tillage is not conservation agriculture. J Soil Water Conserv 70 (5):103A–108A. https://doi.org/10.2489/jswc.70.5.103A

Reicosky DC, Allmaras RR (2003) Advances in tillage research in North American cropping systems. J Crop Prod 8(1-2):75–125. https://doi.org/10.1300/J144v08n01_05

Reicosky DC, Lindstrom MJ (1993) Fall tillage method: effect on short-term carbon dioxide flux from soil. Agron J 85(6):1237–1243. https://doi.org/10.2134/agronj1993.00021962008500060027x

Reicosky DC, Dugas WA, Torbert HA (1997) Tillage-induced soil carbon dioxide loss from different cropping systems. Soil Tillage Res 41(1):105–118. https://doi.org/10.1016/S0167-1987(96)01080-X

Rochette P, Angers DA (1999) Soil surface carbon dioxide fluxes induced by spring, summer, and fall moldboard plowing in a sandy loam. Soil Sci Soc Am J 63(3):621–628. https://doi.org/10.2136/sssaj1999.03615995006300030027x

Rogelj J, Den Elzen M, Höhne N, Fransen T, Fekete H, Winkler H, Schaffer R, Sha F, Riahi K, Meinshausen M (2016) Paris Agreement climate proposals need a boost to keep warming well below 2 °C. Nature 534(7609):631–639. https://doi.org/10.1038/nature18307

Roger-Estrade J, Richard G, Dexter AR, Boizard H, De Tourdonnet S, Bertrand M, Caneill J (2009) Integration of soil structure variations with time and space into models for crop management: a review. Sustain Agric:813–822. https://doi.org/10.1007/978-90-481-2666-8_49

Romaneckas K, Romaneckiene R, Šarauskis E, Pilipavičius V, Sakalauskas A (2009) The effect of conservation primary and zero tillage on soil bulk density, water content, sugar beet growth and weed infestation. Agron Res 7(1):73–86

Romaneckas K, Šarauskis E, Avižienyte D, Buragiene S, Arney D (2015) The main physical properties of planosol in maize (Zea mays L.) cultivation under different long-term reduced tillage practices in the Baltic region. J Integr Agric 14(7):1309–1320. https://doi.org/10.1016/S2095-3119(14)60962-X

Roscoe R, Buurman P (2003) Tillage effects on soil organic matter in density fractions of a Cerrado Oxisol. Soil Tillage Res 70(2):107–119. https://doi.org/10.1016/S0167-1987(02)00160-5

Safa M, Tabatabaeefar A (2008) Fuel consumption in wheat production in irrigated and dry land farming. World J Agric Sci 4(1):86–90

Salton JC, Mielniczuk J (1995) Relations between tillage systems, temperature and humidity of a dark red podzolic Eldorado South. Revista Brasileira de Ciência do Solo 19(2):313–319

Sanchez ML, Ozores MI, Colle R, López MJ, De Torre B, Garcıa MA, Pérez I (2002) Soil CO_2 fluxes in cereal land use of the Spanish plateau: influence of conventional and reduced tillage practices. Chemosphere 47(8):837–844. https://doi.org/10.1016/S0045-6535(02)00071-1

Šarauskis E (2009) Technological processes of operation of environmentally-friendly soil tillage and sowing machines. Review of scientific works presented for habilitation procedure 38 p

Šarauskis E, Buragiene S, Romaneckas K, Sakalauskas A, Jasinskas A, Vaiciukevicius E, Karayel D (2012) Working time, fuel consumption and economic analysis of different tillage and sowing systems in Lithuania. Eng Rural Develop 11:1–5

Šarauskis E, Buragiene S, Romaneckas K, Masilionyte L, Kriauciuniene Z, Sakalauskas A, Jasinskas A, Karayel D (2014a) Deep, shallow and no-tillage effects on soil compaction parameters. Eng Rural Develop 13:31–36

Šarauskis E, Buragiene S, Masilionytė L, Romaneckas K, Avizienyte D, Sakalauskas A (2014b) Energy balance, costs and CO_2 analysis of tillage technologies in maize cultivation. Energy 69:227–235. https://doi.org/10.1016/j.energy.2014.02.090

Šarauskis E, Vaitauskiene K, Romaneckas K, Jasinskas A, Butkus V, Kriauciuniene Z (2017) Fuel consumption and CO emission analysis in different strip tillage scenarios. Energy 118:957–968. https://doi.org/10.1016/j.energy.2016.10.121

Schaller M, Weigel HJ (2007) Analyse des Sachstands zu Auswirkungen von Klimaveränderungen auf die deutsche Landwirtschaft und Maßnahmen zur Anpassung. Landbauforschung Völkenrode. Sonderheft 316, 247 p

Schillinger WF, Young DL (2004) Cropping systems research in the world's driest rainfed wheat region. Agron J 96(4):1182–1187. https://doi.org/10.2134/agronj2004.1182

Schönwiese CD, Janoschitz R (2005) Klima-Trendatlas Deutschland 1901-2000. Berichte des Instituts für Atmosphäre und Umwelt der Universität Frankfurt/Main. Available: https://www.uni-frankfurt.de/45447808/Inst_Ber_4_21.pdf. Accessed 15 Nov 2016

Shafiq M, Hassan A, Ahmad S (1994) Soil physical properties as influenced by induced compaction under laboratory and field conditions. Soil Tillage Res 29(1):13–22. https://doi.org/10.1016/0167-1987(94)90098-1

Šimanskaite D (2007) The effect of ploughing and ploughless soil tillage on soil physical properties and crop productivity. Žemes ukio mokslai 1:9–19

Šimanskaite D, Feiza V, Lazauskas S, Feiziene D, Kadžiene G (2009) Soil tillage systems impact on hydrophysical properties of Gleyic Cambisol. Zemdirbyste-Agriculture 96(1):23–38

Singh B, Malhi SS (2006) Response of soil physical properties to tillage and residue management on two soils in a cool temperate environment. Soil Tillage Res 85(1):143–153. https://doi.org/10.1016/j.still.2004.12.005

Smithson R (2008) Evolution pressed. New Sci 197(2640):24

Soane BD, Van Ouwerkerk C (1995) Implications of soil compaction in crop production for the quality of the environment. Soil Tillage Res 35(1):5–22. https://doi.org/10.1016/0167-1987(95)00475-8

Soane BD, Ball BC, Arvidsson J, Basch G, Moreno F, Roger-Estrade J (2012) No-till in northern, western and south-western Europe: a review of problems and opportunities for crop production and the environment. Soil Tillage Res 118:66–87. https://doi.org/10.1016/j.still.2011.10.015

Sommer R, Thierfelder C, Tittonell P, Hove L, Mureithi J, Mkomwa S (2014) Fertilizer use should not be a fourth principle to define conservation agriculture: response to the opinion paper of Vanlauwe et al. (2014)'A fourth principle is required to define conservation agriculture in sub-Saharan Africa: the appropriate use of fertilizer to enhance crop productivity'. Field Crop Res 169:145–148. https://doi.org/10.1016/j.fcr.2014.05.012

Sørensen CG, Nielsen V (2005) Operational analyses and model comparison of machinery systems for reduced tillage. Biosyst Eng 92(2):143–155. https://doi.org/10.1016/j.biosystemseng.2005.06.014

Soucy Track Available: http://www.soucy-track.com/en-CA/avantages/agricultural-sector/agronomic-performance Accessed 20 Nov 2016

Stevenson FJ, Cole MA (1999) Cycles of soils: carbon, nitrogen, phosphorus, sulfur, micronutrients. Wiley, New York 429 p

Tebrügge F (2001) No-tillage visions – protection of soil, water and climate and influence on management and farm income. In: Garcia-Torres L, Benites J, Martınez-Vilela A (eds) Conservation agriculture – a worldwide challenge. World Congress on Conservation Agriculture 1:303–316. https://doi.org/10.1007/978-94-017-1143-2_39

Tebrügge F, Böhrnsen A (1995) Direktsaat–Auswirkungen auf bodenökologische Faktoren und Ökonomie. Landtechnik–Agric Eng 50(1):6–7 (in German)

Tessier S, Peru M, Dyck FB, Zentner FP, Campbell CA (1990) Conservation tillage for spring wheat production in semi-arid Saskatchewan. Soil Tillage Res 18(1):73–89. https://doi.org/10.1016/0167-1987(90)90094-T

The European Conservation Agriculture Federation (ECAF) Available: http://www.ecaf.org/. Accessed 25 Nov 2016

Tolón-Becerra A, Lastra-Bravo X, Botta GF (2010) Methodological proposal for territorial distribution of the percentage reduction in gross inland energy consumption according to the EU energy policy strategic goal. Energy Policy 38(11):7093–7105. https://doi.org/10.1016/j.enpol.2010.07.028

UNFCCC (2015) Draft decision: CP.21, FCCC/CP/2015/L.9/Rev.1. Available: unfccc.int/resource/docs/2015/cop21/eng/l09r01.pdf. Accessed 12 Nov 2016

Unger PW, Jones OR (1998) Long-term tillage and cropping systems affect bulk density and penetration resistance of soil cropped to dryland wheat and grain sorghum. Soil Tillage Res 45(1):39–57. https://doi.org/10.1016/S0167-1987(97)00068-8

Uri ND (2000) Perceptions on the use of no-till farming in production agriculture in the United States: an analysis of survey results. Agric Ecosyst Environ 77(3):263–266. https://doi.org/10.1016/S0167-8809(99)00085-7

Van den Akker JJH, Canarache A (2001) Two European concerted actions on subsoil compaction. Landnutzung und Landentwicklung 42:15–22

Van Oost K, Quine TA, Govers G, De Gryze S, Six J, Harden JW, Ritchie JC, McCarty GW, Heckrath G, Kosmas C, Giraldez JV, Marques da Silva JR, Merckx R (2007) The impact of agricultural soil erosion on the global carbon cycle. Science 318(5850):626–629. https://doi.org/10.1126/science.1145724

Velykis A, Satkus A (2005) Effect of winter crop and minimized tillage on soil physical properties. Zemes ukio mokslai 3:8–17

Vinten AJA, Ball BC, O'sullivan MF, Henshall JK (2002) The effects of cultivation method, fertilizer input and previous sward type on organic C and N storage and gaseous losses under spring and winter barley following long-term leys. J Agric Sci 139(03):231–243. https://doi.org/10.1017/S0021859602002496

Wilkinson BH (2005) Humans as geologic agents: a deep-time perspective. Geology 33(3):161–164. https://doi.org/10.1130/G21108.1

Würfel T, Vetter R, Unterseher E, Elsäßer M (2002) Merkblätter für die Umweltgerechte Landbewirtschaftung 25:1–8 (in German)

Yalcin H, Cakir E, Aykas E (2005) Tillage parameters and economic analysis of direct seeding, minimum and conventional tillage in wheat. J Agron 4(4):329–332

Zhang GS, Chan KY, Oates A, Heenan DP, Huang GB (2007) Relationship between soil structure and runoff/soil loss after 24 years of conservation tillage. Soil Tillage Res 92(1):122–128. https://doi.org/10.1016/j.still.2006.01.006

Zhang H, Wang X, Feng Z, Pang J, Lu F, Ouyang Z, Liu W, Hui D (2011) Soil temperature and moisture sensitivities of soil CO_2 efflux before and after tillage in a wheat field of Loess Plateau, China. J Environ Sci 23(1):79–86. https://doi.org/10.1016/S1001-0742(10)60376-2

Zink A, Fleige H, Horn R (2010) Load risks of subsoil compaction and depths of stress propagation in arable luvisols. Soil Sci Soc Am J 74(5):1733–1742. https://doi.org/10.2136/sssaj2009.0336

Chapter 3
Coffee Production and Climate Change in Ethiopia

Birhanu Tsegaye Sisay

Abstract Ethiopia is the center of origin and diversity of arabica coffee. Arabica coffee is the most widely consumed, over 70% in volume of production and over 90% of traded value globally. 157,437 thousands of 60 kg bags coffee were produced in 2016 in the world, including 101,552 thousand bags of arabica coffee. Ethiopia is the leading coffee producer in Africa, and the 5th in the world. Ethiopian coffee is known for its unique characteristics, aroma and flavor. Coffee production in Ethiopia recorded an average annual growth rate of 2.6% during the last 50 years, increasing to 3.6% since 1990. Ethiopia has also a strong domestic coffee consumption culture, which frequently accounts for over half of the production. Coffee is produced in Ethiopia under different production systems, i.e. forest, semi-forest, plantation and garden. The area of plantation and home garden coffee are increasing despite the decrease in forest and semi-forest coffee.

Ethiopian coffee forest area is shrinking from time to time, largely due to increasing population, land use conflict, high deforestation, expansion of large-scale coffee and tea farms, and other agricultural practices. Ethiopia has a wide range of coffee genetic diversity. Around 11,691 arabica coffee germplasm accessions from different coffee growing areas throughout Ethiopia were collected and conserved ex-situ in field gene banks. The major challenges facing the coffee sector is the threat of coffee genetic erosion and various production constraints like disease and pest prevalence, replacement of coffee by other crops, coffee market price fluctuation. Concerning climate change, data from weather stations of Ethiopia showed that the mean annual temperature has increased by 1.3 °C between 1960 and 2006, at an average rate of 0.28 °C per decade, and by 0.3 °C per decade in the south western region. In addition, spring and summer rains have declined by 15–20% since the mid-1970s and late 2000s, in southern, south-western and south-eastern Ethiopia. The mean annual temperature of Ethiopia is projected to increase by 1.1–3.1 °C by the 2060s, and 1.5–5.1 °C by the 2090s.

B. T. Sisay (✉)
Wolkite University, College of Agriculture and Natural Resource, Wolkite, Ethiopia

E. Lichtfouse (ed.), *Sustainable Agriculture Reviews 33*, Sustainable Agriculture Reviews 33, https://doi.org/10.1007/978-3-319-99076-7_3

Keywords Coffee · Genetic diversity · Coffee forest ecology · Sustainability · Climate change

3.1 Introduction

The total coffee produced in 2016 on the world was 157,437 thousand of 60 kg bags, from which 101,552 thousand 60 kg bags was accounted to Arabica alone and the remaining 55, 885 was for Robusta coffee. In 2016, from the total amount of coffee produced in the world, 17, 208 thousand 60 kg bags was produced in Africa, 45, 083 thousand 60 kg bags in Asia and Oceania, 20, 269 thousand 60 kg bags in Mexico and central America, and 74, 877 thousand 60 kg bags was produced in south America (ICO 2017a, b, c). In crop year 2012/13, world coffee production reached 145.1 million bags, the largest on record. With the exception of Africa, all coffee growing regions recorded a steady growth in their production over the time period (ICC 2014). Africa's share in world production has hence decreased from 25% to an average of 14%. Since 1990, production levels have generally stagnated, registering less than 20 million bags every year. The subsequent decline in production was initially attributable to structural factors given low yields and ageing coffee trees as well as the economic liberalization programmes implemented in the 1990s. Other factors were related to the regional conflicts affecting certain countries. Production in crop year 2012/13 is estimated at 16.7 million bags (ICC 2014).

The most dynamic growth in African production was observed in Ethiopia, which recorded an average annual growth rate of 2.6% during the last 50 years, increasing to 3.6% since 1990. The country's production trend is generally upward despite some downward interruptions, reaching 6.4 million bags in 2012/13. Ethiopia is also unique in Africa by having a strong domestic coffee consumption culture, which frequently accounts for over half of production (ICC 2014). According to the ICO (International Coffee Organization) report in 2017, Ethiopia is the leading coffee producer in Africa by producing 7100 thousand 60 kg bags, followed by Uganda (4900 thousand 60 kg bags) and Côte d'Ivoire (1500 thousand 60 kg bags) in 2016 cropping year. Ethiopia has also recorded 5.7% change of coffee production in 2016 compared to the previous cropping period of 2015 (ICO 2017a, b, c). This shows remarkable potential of the country to produce fine specialty coffee (Table 3.1).

Despite the increase in the world coffee production from 2014 to 2016, coffee consumption has also grown from time to time. ICO (2017a, b, c) report indicated that, 155,469 thousand of 60 kg bags of coffee was consumed globally in the year 2015/16. The world coffee consumption has increased from 2012/13 to 2015/16 by 1.9%, in Africa by 1.0% and in Ethiopia by 2.9%. Generally, an increased coffee consumption of 2.1% was recorded by exporting countries as compared to importing countries, which was 1.8% increase in the cropping years of 2012/13 to 2015/16 (ICO 2017a, b, c). Nearly 45–50% of Ethiopia's coffee production is locally consumed. Most of this coffee is considered lower quality and may have been originally destined for export, but was rejected since it did not meet Ethiopia

Table 3.1 Total coffee production by some exporting countries (in thousands of 60 kg bags)

Country		Crop year 2010/11	2011/12	2012/13	2013/14	2014/15	2015/16	2016/17
Brazil	(A/R)	53,428	50,592	55,420	54,698	52,299	50,388	55,000
Vietnam	(R/A)	20,000	26,500	23,402	27,610	26,500	28,737	25,500
Colombia	(A)	8523	7652	9927	12,163	13,339	14,009	14,500
Indonesia	(R/A)	9129	10,644	11,519	11,265	11,418	12,317	11,491
Ethiopia	(A)	7500	6798	6233	6527	6625	6714	6600
Peru	(A)	4069	5373	4453	4338	2883	3304	4222
Uganda	(R/A)	3267	3115	3914	3633	3744	3650	3800
Côte d'Ivoire	(R)	982	1966	2072	2107	1750	1893	2000
Kenya	(A)	641	757	875	838	765	789	783
Burundi	(A)	353	204	406	163	248	274	258
Rwanda	(A)	323	251	259	258	238	278	251
Malawi	(A)	17	26	23	28	25	21	18

Source: ICO (2016)
A Arabica, *R* Robusta

Table 3.2 Domestic coffee consumption by some exporting countries in thousand 60 kg bags

Country	Crop year	2010/11	2011/12	2012/13	2013/14	2014/15	2015/16	2016/17
Brazil	(A/R)	19,132	19,720	20,330	20,085	20,333	20,500	20,500
Indonesia	(R/A)	3333	3667	3900	4167	4333	4500	4600
Madagascar	(R)	467	450	430	410	390	370	360
Philippines	(R/A)	2125	2175	2325	2550	2800	3000	3000
Colombia	(A)	1308	1439	1441	1469	1505	1672	1700
Costa Rica	(A)	407	381	423	335	381	436	359
Côte d'Ivoire	(R)	317	317	317	317	317	317	317
Ethiopia	(A)	3383	3383	3400	3650	3675	3700	3700
Honduras	(A)	345	345	345	345	345	345	345
India	(R/A)	1800	1917	2000	2100	2200	2250	2250
Mexico	(A)	2354	2354	2354	2354	2354	2354	2354
Thailand	(R/A)	775	1100	1130	1200	1250	1300	1300
Uganda	(R/A)	204	210	216	221	229	234	240
Venezuela	(A)	1650	1650	1650	1650	1650	1650	1650
Vietnam	(R/A)	1583	1650	1825	2000	2200	2300	2300

Source: ICO (2016)

commodity exchange quality standards. Interestingly, even though it may be a lower quality than what is exported, the price of coffee in the local market place is sometimes higher than the international price (Tefera 2015) (Table 3.2).

There are over 120 species of coffee in the genus *Coffea*. However, the only two species of economic importance are Arabica coffee (*Coffea arabica)* and Robusta coffee (*Coffea canephora*). Many scholars agreed that Ethiopia is the center of origin and diversity of Arabica coffee. Arabica coffee is the most widely consumed, dominating over 70% in volume of production and over 90% of traded value globally. More than 80 developing countries mainly earn their foreign currency from coffee (Gole 2015). Exports by exporting countries during the period from 1990/91 to 2011/12 show varying performances. Of the 20 leading exporting countries, Vietnam recorded a high growth rate in its exports (13.9%), followed by Peru (6.2%), Nicaragua (7.3%), Honduras (5.2%), India (5.7%) and Ethiopia (6.2%). In terms of volume, Vietnam exported 20 million bags in 2012/13 compared to 1.1 million bags in 1990/91. Brazil exported 31.2 million bags in 2012/13, while Colombia recorded a negative growth rate in its annual exports (−1.5%) as a result of the prolonged falls in its production during recent years. Decreases in exports were also recorded in Cameroon (−7.6%), Côte d'Ivoire (−4.1%), Kenya (−3.1%) and El Salvador (−2.5%) (ICC 2014). For Ethiopia, coffee is the most important export commodity, with a share of 20–25% of the total foreign exchange earnings. Ethiopia has exported 210 thousand 60 kg bags in October 2016 and this has grown to 277 thousand 60 kg bags in October 2017 and this shows 31.8% change (ICO 2017a, b, c) (Table 3.3).

Ethiopia produces and exports premium quality coffee to the world market. ICO (2016) report indicate that Ethiopia is the leading producer in Africa, and the 5th in the world, following Brazil, Vietnam, Colombia and Indonesia. If we consider Arabica alone, Ethiopia is the 3rd largest producer after Brazil and Colombia (ICO

Table 3.3 Exports of all forms of coffee by some exporting countries in thousand of 60 kg bags

	Years						
Country	2010	2011	2012	2013	2014	2015	2016
Brazil	33,167	33,806	28,549	31,662	36,429	37,018	34,267
Vietnam	14,229	17,717	22,920	19,718	26,097	20,655	27,422
Colombia	7822	7734	7170	9670	10,954	12,716	12,831
Indonesia	5489	6159	10,722	10,882	6175	8379	6545
India	4647	5414	5044	5033	5131	5262	6086
Honduras	3349	3947	5508	4185	4252	5030	5306
Peru	3817	4697	4310	3971	2720	2790	3960
Uganda	2657	3142	2685	3672	3442	3596	3543
Guatemala	3468	3697	3750	3575	3043	2961	3072
Ethiopia	3324	2675	3203	2870	3117	2985	3001
Mexico	2498	2907	3556	3132	2402	2519	2384
Nicaragua	1712	1468	1987	1661	1901	1753	1961
Côte d'Ivoire	1912	772	1712	1962	1600	1418	1495
Papua New Guinea	929	1225	925	811	807	711	1133
Costa Rica	1200	1243	1374	1344	1209	1128	1007

Source: ICO (2016)

2015). Ethiopia has the largest highland area suitable for Arabica production and, hence has the potential to be a leading producer in both quality and quantity. The country has favorable ecological factors such as suitable altitude, ample rainfall and optimum temperatures, appropriate planting materials, and fertile soil. Coffee grows in Ethiopia in several places at various altitudes ranging from 550 to 2750 m above sea level (ECEA 2008). Ethiopian coffee is known for its unique characteristics, aroma and flavor. This shows that there is very high coffee production and export potential in Ethiopia and to earn premium prices. The country has also become of particular interest to the world for its inherent quality and coffee production potential.

The Ethiopian coffee production has great potential of resilience to the effects of climate change, due to (i) the diverse and adaptive management practices, (ii) presence of high genetic diversity of both wild and cultivated coffee, and (iii) vast highland plateau suitable for coffee. The management practices for coffee production involves growing coffee under deep, medium and no-shade (ECFF 2017). The fragmented Ethiopian coffee forest area is shrinking from time to time, largely due to increasing population, land use conflict, high deforestation, expansion of large-scale coffee and tea farms, other agricultural practices and fluctuating international coffee prices (Gole 2003). Most of the shade grown coffee system resembles natural forest in structural complexity and diversity. Shade modifies the micro-climate, and can moderate extreme temperature by at least 5 °C. As a result, the known coffee types and brands are either replaced by other more profitable crops or their cultivation is expanding into less suitable areas (ECFF 2017).

3.2 Coffee Production in Ethiopia

Ethiopia, which is considered as coffee's birthplace, produces high quality Arabica coffee for both the domestic and international markets. Coffee plays a major role in Ethiopia's economy and is deeply intertwined with cultural traditions and day-to-day living. There are an estimated 15 million people, or approximately 15% of the country's total population, who derive their livelihoods from coffee (Tefera 2015).

Different scholars explained out that coffee is produced in Ethiopia under different production systems, namely: forest coffee, semi forest coffee, garden coffee and plantation coffee (ECEA 2008; Gole 2015). The coffee production systems are mostly forest based; and the differences between the systems are manifested by the level of forest management intensities. Accordingly, the level of forest management ranges from little or none in the forest coffee to intensive management in the home-garden and plantation systems (Gole 2015). On the other hand, forest (shade) coffee and sun coffee can be considered as the two main coffee production systems in Ethiopia (ECFF 2017). The traditional production systems account for 90–95% of the production, while plantation may range 5–10%. Even if recent survey data is not available for accurate figure but generally, the areas of plantations and home gardens are increasing (Gole 2015). However, inconsistent data was reported by different authors and organizations in the share of this production system from the total coffee produced in the country.

It is estimated that 525,000 hectares (5250 km^2) of land are planted by coffee in Ethiopia (Tefera 2015), although the actual area is probably in excess of 20,000 km^2. Coffee provides Ethiopia with its most important agricultural commodity, contributing around one quarter of its total export earnings (Minten, et al. 2014). In 2014/15 Ethiopia exported around 180,000 metric tons of coffee (ICO 2015) at a value estimated to be in excess of 800 million USD. Coffee farming alone provides a livelihood income for around 15 million Ethiopians (16% of the population), based on four million smallholder farms (Tefera 2015; Minten, et al. 2014; Tefera and Tefera 2014). For many of these farmers, coffee is their single most important source of income.

The main coffee growing areas are found within Oromia Region and Southern Nations, Nationalities, and Peoples' Region (SNNPR), with modest production in Amhara Region and minor output in Benishangul-Gumuz Region (ECFF 2017). Most coffee is grown in areas of humid (moist) evergreen forest. This type of rainforest is found at 650–2600 m above sea level, with coffee mainly confined to altitudes of 1200–2100 m. Almost 95% of Ethiopian coffee is produced in the administrative zones of Keffa, Sidamo, Ilubabor, Wellega, Gedeo and Harerghe (Hailu 2011). The following map demonstrate coffee producing areas in Ethiopia (Fig. 3.1).

The quality standards of Ethiopian coffee are classified according to their origin of production (ECEA 2008). Among the best-known coffee varieties in Ethiopia; Harar, Wellega, Limu, Sidama and Yirgacheffee take the priority (ECEA 2008). But, there are numerous lesser-known coffee regions that have equally distinct flavor profiles. The range of flavor profiles adds a unique element to Ethiopian coffee, and makes it especially well-suited to development within the specialty coffee market (ECFF 2017). The following table summarizes the quality standards of Ethiopian coffee classified according to their origin of production (Table 3.4).

3.3 Coffee Genetic Diversity and Conservation Strategies

In Ethiopia, there are immense genetic potentials among and within the wild types, local landraces and released arabica coffee varieties with great diversity for any desirable traits such as yield, quality, disease resistance, drought tolerance, low caffeine content, etc. and wider ecological conditions (Kufa and Burkhardt 2013). The best hope for crop improvement lies in the progenitors or wild relatives of the cultivated plants that harbor rich genetic resources for tolerance against abiotic and biotic stresses (Nevo 1998; Schoen and Brown 1993).

The high genetic diversity of Ethiopian coffee is considered to be of great value both nationally and internationally, as it represents a pool to develop improved breeds of coffee. There are currently hundred types of coffee under cultivation,

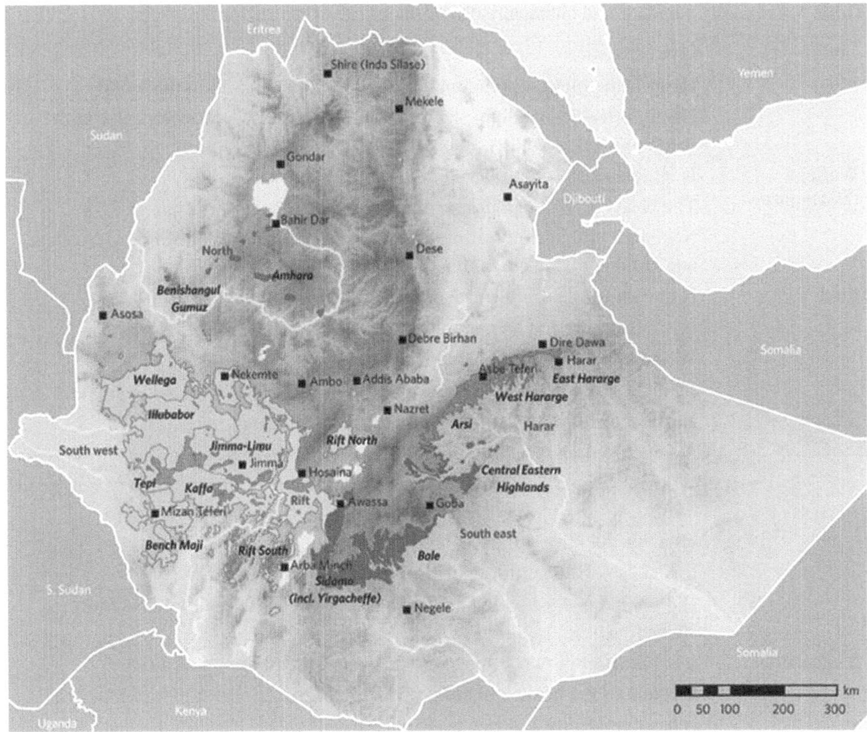

Fig. 3.1 The main coffee growing zones and areas of Ethiopia. (Source: Justin et al. 2017)
The coffee zones represented by coloured polygons: red/pink, North Zone (coffee areas: Amhara and Benishangul Gumuz); light blue, South West Zone (coffee areas: Wellega, Illubabor, Jimma-Limu, Kaffa, Tepi and Bench Maji); light green, Rift Zone (coffee areas: Rift North and Rift South); dark blue, South East Zone (coffee areas: Sidamo, Yirgacheffe, Bale and Central Eastern Highlands); dark green, Harar Zone (coffee areas: Arsi, West Hararge and East Hararge) (Justin et al. 2017)

each with varying aromas, tastes, colors especially in the Southwestern part of the country. Ethiopian coffee ranks highly in intrinsic quality of the bean due to the diverse agro-ecological zones and immense genetic diversity aforementioned. It is also de-facto wild and organic due to predominant subsistence and traditional production systems (Kufa 2006; FAO 2014). The use of chemical inputs, such as pesticides, fungicides and artificial fertilizers is rarely practiced, and although certification is not common (Tefera, and Tefera 2014), Ethiopian coffee can often be considered as organic by default, and may indeed exceed the standards set for organic certification (ECFF 2017).

The wild populations of *C. arabica* in the mountain rainforests are the most important genetic pool of the crop (Gole 2015). It is naturally restricted to two isolated mountain forests on the western and eastern sides of the Great Rift Valley in

Table 3.4 Quality standard and characteristics of major Ethiopian coffee by origin

Coffee type	Characteristics
Harar	Harar is the highest premium coffee in Ethiopia as well in the world
	Harar coffee has medium size bean, with a greenish-yellowish color with medium level of acidity and a distinctive mocha flavor
Wollega (Nekempt)	Is Highland grown produced in western Ethiopia
	The beans of Wollega has medium-to-bold bean with fruity taste
	Nekempti coffee export designations include: Kelem Wollega, East Wollega and Gimbi coffees are coffee mostly sun-dried
Limu	Is known for its spicy and wine flavor, and good acidity
	It is most preferred and popular in Europe and the U.S.
	Washed Limu is one of Ethiopia's premium coffees
	Medium sized bean and greenish bluish in color mostly round in shape
Sidamo	Has greenish-grayish color and medium-sized beans
	Sidamo accounts for 30% of all Ethiopian coffee production
	Washed Sidamo, called sweet coffee is known for its balanced taste and good flavor
	It has fine acidity and good body
	It is always blended for gourmet or specialty coffee
	High grade unwashed Sidama coffees are known for their intense fruity characteristics
Yirgacheffee	It is highland grown coffee and has intense flavor known as flora
	It is one of the best highland-grown coffees
	It has fine acidity and rich body
	Many roasters are attracted to its fine and fruit flavor and are willing to pay a premium price for it
	Internationally known and recognized as Yirgachaffe Brand Name

Source: ECEA (2008), Willem (2011) and Gole (2015)

Southern Ethiopia (Wintgens 2004). Tesfaye (2006) reported high genetic variability within and between different wild populations in Ethiopia. He further noted that wild coffee plants are genetically distinct and more diverse when compared to the cultivated varieties grown in Ethiopia and around the world. The mountain rainforest areas in Ethiopia are at different field gene banks in Ethiopia (Kufa 2010).

Efforts are under way to conserve coffee germplasms in Ethiopia by Jimma Agricultural Research Center (JARC) and Ethiopian Biodiversity Institute (EBI). Surendra (2008) indicated that Ethiopia alone possesses around 99.8% of the world total arabica's genetic diversity. In efforts to collect and document the use of coffee genes in breeding programs, researchers have collected a total of around 11,691 Arabica coffee germplasm accessions from different coffee growing areas throughout Ethiopia. The collections are conserved *ex situ* in field gene banks at Jimma Agricultural Research Center and its sub-centers (5960 accessions) and at Choche (5731 accessions) in Jimma zone of Oromia state, Ethiopia (Gole 2015). The collection at Choche is mainly for conservation and managed by the Ethiopian

Biodiversity Institute (Gole 2015). Despite *ex-situ* field gene banks, conserving coffee germplasm *in-situ* is also important.

However, currently one of the major challenge facing the coffee sector is the threat of coffee genetic erosion and production constraints because of the degradation of major ecosystems that have been supporting the coffee farming system for several generations. Coffee genetic resources are under severe threat from genetic erosion in the centers of origin and diversity. These coupled with the increasing patterns of climate change are threatening the natural coffee gene pools (Kufa 2010), requiring urgent measures for preserving environmental sustainability and coffee biodiversity at their country of origin and genetic diversity in Africa. Kufa (2010) also emphasized on the severity of human disturbance threats and underlined the need for urgent characterization and conservation measures of coffee genetic resources for future development of the coffee sector worldwide. Kufa and Burkhardt (2013) pointed out that the extent of variations in root growth varied across geographical areas, demonstrating the strong link between coffee genetic and climatic factors.

The diversity in coffee genes, species and ecosystems, traditional farming practices and technological innovations such as mitigation and adaptation strategies to climate change need to be exploited in the African continent to produce superior quality coffee types and remain competitive in the world market (Kufa 2015).

3.4 Climate Change

In the centers of origin and diversity of *Coffea arabica*, coffee species are, however, under a severe threat of genetic erosion and irreversible loss largely due to increasing population, expansion of agricultural farms, coffee crop replacement by other crops, coffee market price fluctuation and climate change impact are the most important one.

The impact of climate change on degradation of farm lands and natural resources can contribute to reduced agricultural productivity and food insecurity in developing countries. Intensive coffee production, on the other hand, can hamper efforts to protect, maintain, and enhance habitats and species. Hence, unless appropriate global initiatives are urgently realized, the present green Africa can be easily converted into deserts with profound and damaging consequences for natural resource bases, biodiversity, economies and livelihoods. In most African countries, natural forest ecosystems with high levels of biodiversity are under serious threat, largely due to increasing population pressures and subsequent deforestation and land degradation. The destruction of natural coffee habitats coupled with changes in weather patterns can adversely affect coffee genetic resources and the livelihoods of millions of people in Africa and elsewhere (Kufa 2015).

Historical data from weather stations for Ethiopia provides specific details of the general warming trend. Data shows that the mean annual temperature has increased by 1.3 °C between 1960 and 2006, at an average rate of 0.28 °C per decade

(McSweeney et al. 2010), and by 0.3 °C per decade in the south western region (Jury and Funk 2013). significant trend observed in mean rainfall in any season for which climate data is available (1960–2006). However, based on data from quality-controlled climate station observations, it has been shown that spring (Belg) and summer (Kiremt) rains have declined by 15–20% since the mid-1970s and late 2000s, in southern, south-western and south-eastern Ethiopia (Funk et al. 2008; Funk et al. 2005).

The mean annual temperature of Ethiopia is projected to increase by 1.1–3.1 °C by the 2060s, and 1.5–5.1 °C by the 2090s, with the scale of the projections depending on the emission scenario (McSweeney et al. 2010). Projections from different General Circulation Models (GCMs) are broadly consistent in indicating increases in annual rainfall in Ethiopia (IPCC 2013; Niang et al. 2014), but these increases are largely due to increasing rainfall in the October–December period in southern Ethiopia (McSweeney et al. 2010). Projections of change in the rainy seasons of April–June and July–September, which affect larger areas of Ethiopia, are much more uncertain but tend towards slight increases in the south west and deceases in the north east (McSweeney et al. 2010). In the Ethiopian Highlands, a region of high and complex topography, projections from the GCMs also indicate likely increases in rainfall, and extreme rainfall, by the end of the twenty-first century (Niang et al. 2014). However, these future trends are not consistent with observed decreasing rainfall for many places in Ethiopia (from the 1950s to the present day).

Human-induced land-use change is one of the major sources of greenhouse gas (GHG) emission, resulting in climate changes. Though Ethiopia is contributing less than 2 tons of CO_2 per capita, it is one of the country's most affected by the consequences of climate change. Hence, as part of the global response to climate change, Ethiopia has developed its Climate Resilient Green Economy (CRGE) Strategy in 2011 (Gole 2015). The vision of the country is to reach middle income country status by 2025, while following the green economy development path, with net-zero GHG emission, taking 2010 as the baseline (FDRE 2011).

The climate resilient green economy report documents that Ethiopia is already exposed to elevated climate variability and a series of extreme weather events. Consequently, climate change will have a profound impact on the coffee sector. Coffee is highly sensitive to fluctuations in rainfall. Undesirable weather events have had an acute impact on output. Unlike robusta, arabica coffee is an extremely climate-sensitive crop. There is increasing evidence that rising temperatures in coffee landscapes may be one of the most threatening issues facing Ethiopia's coffee subsector (UNDP 2012). Preliminary assessments suggest that wild coffee population in Ethiopia will be vanished by year 2018 due to impact of climate change (UNDP 2012).

Climate change will negatively impact much of the current coffee farming landscape of Ethiopia, however, substantial areas that were previously unsuitable for coffee will become suitable as the century progresses. This is due to the upslope shift of coffee growing suitability (the niche) as higher altitude areas (e.g. above 2000 m) improve and lower altitude areas worsen, as the climate changes. Relocation of coffee farms to higher altitudes will require positive action with critical attention to feasibility and planning. Before decisions are made concerning the establishment of new coffee farms and areas, suitability has to be determined in terms of climate, agronomy (e.g. soil, shade, slope), the potential for land-use contradictions (e.g. land tenure, other existing land-use activities), market value, and logistics. A certain amount of shift could occur quite naturally, as farmers see the potential and financial incentive to start growing coffee on their land in areas previously unsuitable (ECFF 2017).

Assisted migration to 'new' areas will be a key component for ensuring resilience in the Ethiopian coffee economy due to the replacement of areas lost at lower altitudes (below 1400–1500 m) as the climate there becomes unsuitable for coffee growing. The new areas are mostly in the South West coffee zone, but North, Rift (Rift South coffee area), and South East (Sidamo coffee area) coffee zones may also provide opportunities (ECFF 2017). In many cases afforestation and reforestation will be necessary to provide the right growing environment for coffee, but this would also bring additional benefits in terms of increasing forest cover, preserving biodiversity and providing improvements in ecosystem services. Prevention of forest loss is also a key issue, especially in the South West and South East coffee zones where recent deforestation rates are high and could have a more significant impact than climate change, at least in the short to medium term (ECFF 2017).

3.5 Mitigation Strategies

To mainstream the sustainable production, the strategy places emphasis on scaling up and disseminating best practices and new technologies to boost productivity. Special attention is given to practices that allow better adaptation to climate variability and the expansion of shade management systems. Strengthening conservation and management of natural resources will be another focal area. However, the strategy clearly notes that such goals will only be realized if the appropriate institutions are strengthened to provide the necessary support services to improve productivity (UNDP 2012).

One approach to mitigate warming climates at the landscape level is to encourage farmers to readopt traditional forest shade production systems. However, despite the benefits offered by traditional shade systems, many smallholders continue to cut down forest trees to increase the amount of solar irradiation to coffee canopies (UNDP 2012).

This calls for special attention to systematic investigations in order to tailor site-specific conservation options. Thus, characterization of the diverse coffee types would also allow sustainable, wise exploitation and maintenance of the wealth of wild coffee genetic resources in their original habitats. Immediate measures are required to identify and design ways of implementing relevant conservation strategies against the possible threats from climate change to coffee ecology and production at country of origin. Implementation of global coffee genetic resources conservation initiatives and other relevant projects could be among the top priority actions to maintain quality environments, conserve and benefit from the unique coffee germplasm base in Africa. However, detail works on the identification and implementation of sustainable ecological, economical and social conservation of wild coffee diversity and its natural forest ecosystems are crucial. Moreover, investigations on the levels of forest management, identification of suitable coffee genotypes with desirable traits, including low caffeine, disease resistant, drought tolerant and urgent implementation of sound incentive mechanisms (like forest product branding and certification) are of paramount importance to benefit from the potential forest genetic resources and ecosystem services. This need, strong coffee partnerships in both coffee producing and importing countries to coordinate and facilitate sustainability initiatives for the future development of the coffee sector.

Air, soil temperature and moisture can be altered by specific farming interventions, and provide a buffer against inadequate growing conditions and extreme weather events. The most obvious interventions are irrigation, shade management and mulching, but terracing has clear benefits, and pruning is also advocated in some cases. Observations of on-farm adaptation and improvements in coffee farming practices across Ethiopia indicate the potential to provide various levels of resilience for many but not all farmers. Careful assessment of each farming site is required to see which interventions, or combination of measures, would be most suitable (ECFF 2017).

Improved cultivars and selections of indigenous arabica coffee may provide some potential, especially for disease resilience and improved productivity, but significant climate resilience is unlikely based on field observations of field trials, and available genetic diversity information.

References

Ango TG, Börjeson L, Senbeta F, Hylander K (2014) Balancing ecosystem services and disservices: smallholder farmers' use and management of forest and trees in an agricultural landscape in Southwestern Ethiopia. Ecol Soc 19(1):30. https://doi.org/10.5751/ES-06279-190130

Belete Y, Belachew B, Fininsa C (2014) Performance evaluation of indigenous Arabica coffee genotypes across different environments. J Plant Breed Crop Sci 6(11): 171–178. Available online at: https://doi.org/10.5897/JPBCS2014.0447., http://www.academicjournals.org/journal/JPBCS/article-abstract/65D03BC48010

Buechley E, Sekercioglu AA, Gebremichael G, Ndung'u JK, Mahamued BA, Beyene T, Mekonnen T, Lens L (2015) Importance of Ethiopian shade coffee farms for forest bird conservation. Biol Conserv 188:50–60. https://doi.org/10.1016/j.biocon.2015.01.011 Available at http://www.sciencedirect.com/science/article/pii/S0006320715000324?via%3Dihub)

ECEA (Ethiopian Commodity Exchange Authority) (2008) Analysis of coffee supply, production, utilization and marketing issues and challenges in Ethiopia. Policy analysis and economic research team. Addis Ababa, Ethiopia, vol 1, pp IV–V. File:///C:/Users/HP%20PRO%20BOOK/Downloads/Downloads/Coffee/Coffee%20Report%20(edited).pdf

ECFF (Environment and Coffee Forest Forum) and Kew garden (2017) Coffee farming and climate change in Ethiopia; Summary report on impacts, Forecasts, resilience and opportunities. https://www.kew.org/sites/default/files/Coffee%20Farming%20and%20Climate%20Change%20in%20Ethiopia.pdf

FAO (Food and Agriculture Organization) (2014) Analysis of price incentives for coffee in Ethiopia for the time period 2005–2012. Technical notes series, MAFAP, by Kuma Worako T, MasAparisi A. and Lanos B, Rome. https://agriknowledge.org/downloads/g158bh310

FDRE (Federal Democratic Republic of Ethiopia) (2011) Ethiopia's climate resilient green economy: the green economy strategy. Federal Democratic Republic of Ethiopia, Addis Ababa. https://www.uncclearn.org/sites/default/files/ethiopia_crge.pdf

Funk C, Senay G, Asfaw A, Verdin J, Rowland J, Korecha D, Eilerts G, Michaelsen J, Amer S, Choularton R (2005) Recent drought tendencies in Ethiopia and equatorial-subtropical Eastern Africa. U.S. Agency for International Development, Washington, DC. http://chg.geog.ucsb.edu/pub/pubs/FEWSNET_2005.pdf

Funk C, Dettinger MD, Michaelsen JC, Verdin JP, Brown ME, Barlow M, Hoell A (2008) Warming of the Indian Ocean threatens eastern and Southern African food security but could be mitigated by agricultural development. Proc Natl Acad Sci U S A 105:11081–11086. https://doi.org/10.1073/pnas.0708196105 Available online at https://www.ncbi.nlm.nih.gov/pmc/articles/PMC2497460/

Gole TW (2003) Vegetation Ecology of the Yayu forest in South West Ethiopia: impacts of human use and implications for *in situ* conservation of wild *Coffea arabica* L. populations. Ecology and Development Series, No. 10. https://cuvillier.de/de/shop/publications/3387

Gole TW (2015) Environment and coffee forest forum. Coffee: Ethiopia's gift to the world, the traditional production systems as living examples of crop domestication, and sustainable production and an assessment of different certification schemes. Addis Ababa, Ethiopia. https://www.naturskyddsforeningen.se/sites/default/files/dokument-media/coffee_-_ethiopias_gift_to_the_world_ecff_2015.pdf

Gray Q, Tefera A, Tefera T (2013) Ethiopia: Coffee Annual Report. GAIN Report No. ET 1302. Available online at: https://gain.fas.usda.gov/Recent%20GAIN%20Publications/Coffee%20Annual_Addis%20Ababa_Ethiopia_6-4-2013.pdf

Hailu H (2011) Growth and physiological response of two *Coffea arabica* L. Populations under high and low irradiance. Thesis submitted to Addis Ababa University. http://www.intracen.org/WorkArea/DownloadAsset.aspx?id=51214

ICC (International Coffee Council) (2014) World coffee trade (1963–2013): a review of the markets, challenges and opportunities facing the sector. 112th Session, 3–7 March 2014, London, United Kingdom. Available online at: http://www.ico.org/news/icc-111-5-r1e-world-coffee-outlook.pdf

ICO (International Coffee Organization) (2015) Historical Data on the Global Coffee Trade. http://www.ico.org/new_historical.asp

ICO (International Coffee Organization) (2016) Historical Data on the Global Coffee Trade. http://www.ico.org/new_historical.asp

ICO (International Coffee Organization) (2017a) Total production by all exporting countries. http://www.ico.org/prices/po-production.pdf

ICO (International Coffee Organization) (2017b) World coffee consumption. http://www.ico.org/prices/new-consumption-table.pdf

ICO (International Coffee Organization) (2017c) Monthly export statistics – October 2017. http://www.ico.org/prices/m1-exports.pdf

IPCC (Intergovernmental Panel on Climate Change) (2013) Summary for policymakers. In: Stocker TF, Qin D, Plattner G-K, Tignor M, Allen SK, Boschung J, Nauels A, Xia Y, Bex V, Midgley PM (eds) Climate change 2013: the physical science basis. Contribution of working Group I to the fifth assessment report of the intergovernmental panel on climate change. Cambridge University Press, Cambridge/New York https://www.ipcc.ch/pdf/assessment-report/ar5/wg1/WGIAR5_SPM_brochure_en.pdf

Jean-Pierre L, Surendra K (2008) Preserving diversity for specialty coffees. A focus on production systems and genetic resources of arabica coffee in Ethiopia. In: SCAA 20th Annual Conference, Minneapolis, 2–5 May 2008, pp 1. https://agritrop.cirad.fr/568110/1/document_568110.pdf

Jury MR, Funk C (2013) Climatic trends over Ethiopia: regional signals and drivers. Int J Climatol 33: 1924–1935. http://onlinelibrary.wiley.com/doi/10.1002/joc.3560/pdf

Justin M, Jenny W, Susana B, Timothy W, Gole TW, Challa ZK, Sebsebe D, Davis AP (2017) Resilience potential of the Ethiopian coffee sector under climate change. Nat Res J Nat Plant 3: 17081. Available online at: https://doi.org/10.1038/nplants.2017.81

Kufa T (2006) Eco-physiological diversity of wild Arabica coffee populations in Ethiopia: growth, water relation and hydraulic characteristics along a climatic gradient. Doctoral thesis, Center for Development Research (ZEF), University of Bonn, Germany, 305 pp. https://www.zef.de/fileadmin/webfiles/downloads/zefc_ecology_development/ecol_dev_46_text.pdf

Kufa T (2010) Environmental sustainability and coffee diversity in Africa. In: Paper presented in the international coffee organization world coffee conference, 26–28 February 2010, Guatemala City. http://dev.ico.org/event_pdfs/wcc2010/presentations//wcc2010-kufa.pdf

Kufa T (2015) Environmental sustainability and coffee diversity in Africa. https://www.researchgate.net/publication/228672549_Environmental_Sustainability_and_Coffee_Diversity_in_Africa

Kufa T, Burkhardt J (2013) Studies on root growth of *coffea arabica* populations and its implication for sustainable management of natural forests. J Agric Crop Res 1(1):1–9 http://sciencewebpublishing.net/jacr/archive/2013/July/pdf/Kufa%20and%20Burkhardt.pdf

McSweeney C, New M, Lizcano G (2010) UNDP climate change country profiles: Ethiopia, pp 27. http://www.un-gsp.org/sites/default/files/documents/ethiopia.oxford.report.pdf

Minten B, Tamru S, Kuma T, Nyarko Y (2014) Structure and performance of Ethiopia's coffee export sector. In: ESSP Working Paper No. 66. IFPRI and EDRI, Addis Ababa. https://ageconsearch.umn.edu/bitstream/197157/2/Session%202%20-%20Minten%20Tamru%20Kuma%20Nyarko.pdf

Nevo E (1998) Genetic diversity in wild cereals: regional and local studies and their bearing on conservation ex-situ and in-situ. Genet Resour Crop Evol 45:355–370. https://doi.org/10.1023/A:1008689304103

Niang I, Ruppel OC, Abdrabo MA, Essel A, Lennard C, Padgham J, Urquhart P (2014) Africa In: Climate change 2014: impacts, adaptation, and vulnerability. https://www.ipcc.ch/pdf/assessment-report/ar5/wg2/WGIIAR5-Chap22_FINAL.pdf

Schoen DJ, Brown AHD (1993) Conservation of allelic richness in wild crop relatives is aided by assessment of genetic markers. Proc Natl Acad Sci U S A 90(22):10623–10627 https://www.ncbi.nlm.nih.gov/pmc/articles/PMC47829/?page=1

Senbeta WF (2006) Biodiversity and ecology of afromontane rainforests with wild *Coffea arabica* L. populations in Ethiopia. Ecology and Development Series No. 38, Center for Development Research. University of Bonn. https://cuvillier.de/uploads/preview/public_file/4642/3865378072.pdf

Senbeta F, Gole TW, Kelbessa E, Denich M (2013) Diversity of useful plants in coffee forest of Ethiopia. J Ethnobotany Res Appl 11:049–069

Surendra K (2008) Arabicas from the garden of Eden-*Coffea ethiopica*. http://www.scae.com/news/214/arabicas, 14 pp

Tefera A (2015) Ethiopia: coffee annual report. GAIN Report (number ET1514–26/5/2015). USDA Foreign Agricultural Service, pp 6. https://gain.fas.usda.gov/Recent%20GAIN%20Publications/Coffee%20Annual_Addis%20Ababa_Ethiopia_5-26-2015.pdf

Tefera A, Tefera T (2014) *w*. GAIN Report (number ET1402–13/5/2014). USDA Foreign Agricultural Service, pp 11. https://gain.fas.usda.gov/Recent%20GAIN%20Publications/Coffee%20Annual_Addis%20Ababa_Ethiopia_6-4-2013.pdf

Tesfaye GK (2006) Genetic Diversity of wild Coffea arabica populations in Ethiopia as a contribution for conservation and use planning. Doctoral thesis at Faculty of Mathematics & Natural Science, University of Bonn, ZEF Ecology and Development Series, 44. https://www.zef.de/index.php?id=2232&tx_zefportal_staff[ref]=2252&tx_zefportal_staff[uid]=553&no_cache=1

UNDP (United Nation Development Program) (2012) Draft proposal for coffee plat form in Ethiopia. November 2012, Addis Ababa, pp 7–8. http://www.undp.org/content/dam/undp/

documents/projects/ETH/Ministry%20of%20Trade%20-Ethiopia%20Coffee%20Platform%20proposal%20_%20ab2.docx

Willem JB (2011) Ethiopian coffee buying manual: practical guide for buying and importing ethiopian specialty coffee beans. USAID. https://www.scribd.com/document/332240269/Ethiopian-Coffee-Buying-Guide-pdf

Wintgens JN (2004) Coffee: growing, processing, sustainable production. A guide for growers, traders, and researchers. WILEYVCH Verlag GmbH and Co. KGaA, Weinheim, Germany. http://www.scirp.org/(S(351jmbntvnsjt1aadkposzje))/reference/ReferencesPapers.aspx?ReferenceID=582686

Chapter 4
Impact of Climate Change on Coastal Agro-Ecosystems

Saon Banerjee, Suman Samanta, and Pramiti Kumar Chakraborti

Abstract Climate change is a major threat for ecosystems, food security, forests and other natural resources. Proper steps must be taken to reduce the vulnerability of the farming communities living in coastal areas, especially in the developing countries. This chapter reviews the impact of climate change on the coastal agro-ecosystem, and practices to improve sustainability. We found that 27 countries are the most vulnerable due to accelerated sea level rise. In some coastal areas, up to 40% biodiversity loss has already been observed. About 70% income is generated from crop cultivation and the rest is from fisheries and other animal husbandry activities. Hence, climate resilient agriculture can secure the rural livelihood. Adaptation measures may include agro-forestry practices, establishment of orchards, nutrient recycling, salinity management and rational use of water. Techniques of climate resilient agriculture vary with techniques available, needs of the farming community, resources and infrastructure.

Keywords Coastal ecosystem · Climate resilient agriculture · Climate change · Sea level rise · Cyclone

4.1 Introduction

Agriculture can be viewed as one form of ecological engineering for manipulation of populations, communities and ecosystem for human purposes. This concept considers the agricultural operations as ecosystem manipulation instead of only

S. Banerjee (✉)
AICRP on Agrometeorology, Directorate of Research, Bidhan Chandra KrishiViswavidyalaya, Kalyani, West Bengal, India

S. Samanta
Department of Environmental Studies, Visva-Bharati, Santiniketan, West Bengal, India

P. K. Chakraborti
AICRPAM-NICRA Project, Directorate of Research, Bidhan Chandra KrishiViswavidyalaya, Kalyani, West Bengal, India

E. Lichtfouse (ed.), *Sustainable Agriculture Reviews 33*, Sustainable Agriculture Reviews 33, https://doi.org/10.1007/978-3-319-99076-7_4

production process (Weiner 2003). A variety of agro-ecosystems exist side by side in a particular region, they are associated with different methods of farming, such as organic, integrated or conventional (Wezel et al. 2009). In present days, the different ecosystems are under threat mainly due to global warming and climate change. However, climate change and variability are not new. Many societies have coped with and adapted to climate variability and many other stressors during the past centuries (Mertz et al. 2009). Climate change is considered to be one of the major threats to sustainable development because of its effects on ecosystem, health, infrastructure, food security, forests, etc. (IPCC 2007). Global food security and imbalance in agro-ecosystems, threatened by climate change, are the most important challenges in the twenty-first century to supply sufficient food for the increasing global population (Magadza 2000; Lal 2005).

In recent years, more and more attention has been paid to the risks associated with climate change, which will increase uncertainty with respect to food production. The coastal ecosystems are more vulnerable to climate change (Dickson et al. 2007). Although some aquatic plants have their ability to adapt themselves in the diversified coastal environment due to climate change, the problems of soil salinity, sea level rise, etc., pose a threat to coastal agriculture (Cronk and Fennessy 2001). Moreover, the coastal zones of the world are mostly populated and any change would affect a considerable number of population. The effects of global warming would be first felt on these coastal zones. The rate of sea level rise increased from the nineteenth to twentieth century. The total twentieth century rise is estimated to be 0.17 m (Jena and Mishra 2011). Proper steps must be taken to reduce the vulnerability of farming community living in coastal areas, especially in the developing countries (Tompkins and Adger 2005). The present article intends to enumerate the application of climate resilient agriculture towards sustainability of coastal agro-ecosystem.

4.2 Coastal Agro-Ecosystems

The science of living organisms and their interaction with the natural environment is ecology. Ecology's root extends to the origins of humanity. Agricultural practices are too dates back to the time of civilization. So both agriculture and ecosystem are walking hand in hand and sometimes they become complementary to each other.

Ecology can be grouped according to the types of organisms or habitats being studied. Examples of habitats include marine and coastal ecosystems, rain-forests, deserts, etc. Coastal ecosystems may also be defined as a broad interaction between land and sea influencing each other in true sense. Coastal areas include marine and terrestrial ecosystems ranging from coastal lowlands to coral reefs with their unique characteristics and coastal ecology is the study of coastal ecosystems (Hoorweg and Muthiga 2009).

Coastal ecosystems are most productive but highly threatened ecosystems in the world. Coastal ecology includes both marine and terrestrial ecosystems. A large

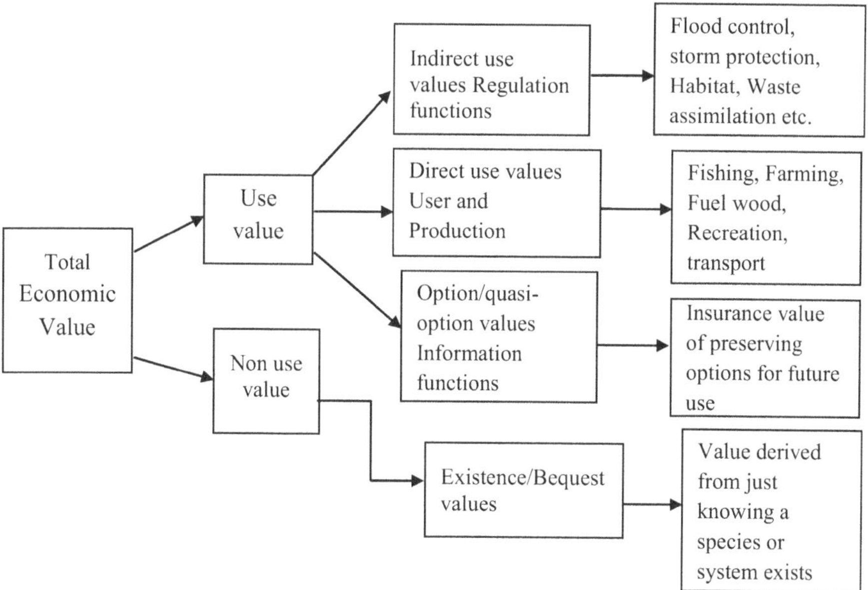

Fig. 4.1 Values of coastal ecosystems. (Adapted from Costanza et al. 1997)

range of functions and values are associated with coastal ecosystem (Fig. 4.1). Among them, regulation and supporting services such as shoreline stabilization, flood control, detoxification of polluted waters and waste disposal are indirect use value. The direct use values includes the utility which are derived directly from the use of many living and nonliving resources. The fishery sector is one of the major sources of income and livelihood for millions of people around the world. About 540 million or nearly 8% of the world's population are directly or indirectly involved with fishery production system (FAO 2010a, b).

Coastal ecosystems serve as breeding and nursery grounds for fish and other aquatic organisms as well as seasonal migration grounds for marine mammals and birds, and also includes plants such as mangroves, seaweed and sea grasses that require brackish to salty water to grow (Hoorweg and Muthiga 2009). World's coastal regions possess vast range environments which are shown in Table 4.1 (Burke et al. 2000). Mangroves and mudflat ecosystems play key roles for sheltering, feeding and spawning grounds for finfish and shellfish (Vidthayanon and Premcharoen 2002). One of the important sources of income for coastal communities in developing countries are the ornamental marine species like corals, invertebrates and fish having high economic value (Lem 2001).

Coastal ecosystems are frequently exposed to aberrant weather conditions and nowadays it becomes more vulnerable due to global warming. Since the 1950s, the quantity and magnitude of natural disasters increased significantly in the coastal

Table 4.1 Characteristics feature of the coastal environment

Zones	Features
Near shore terrestrial	Dunes, cliff, rocky and sandy shores; coastal xeromorphic habitats, urban, industrial and agricultural landscape
Intertidal	Estuaries, deltas, lagoons, mangrove forests, mud flats, salt marshes, salt pans, other coastal wet lands, ports and marinas, aqua cultural beds
Benthic	Kelp forest, sea grasses, coral reefs and soft bottom environment above the continental shelf, artificial reefs and structures
Pelagic	Open waters above continental shelf, freestanding fish farms e.g. plankton blooms, neustone zone, sea ice herring schools

Adapted from Burke et al. (2000)

areas, which may be due to climate change (Jena and Mishra 2011; Sovacool 2014). The increase in urbanization near the coast has serious environmental impacts. Worldwide, more than 600 million people are living in coastal zones which are less than 10 m in elevation; of which 360 million reside in urban areas (McGranahan et al. 2007). Let us take an example of India, a developing country, sharing more than 8000 Km long coastline with Arabian Sea, Bay of Bengal and Indian Ocean. More or less one third of the country's population lives in coastal zones with highly increasing population density. The major activities found along the Indian coastal zone are traditional activities like, fishing, tourism, agricultural activities, oil exploration, commercial and residential development. Fishing activities not only provides important source of food but also provides employment, income and foreign exchange for India (Senapati and Gupta 2014).

Within India we found mangrove ecosystem along the coastal areas of West Bengal which are distributed over North and South 24 Parganas, East Midnapur and southern parts of Howrah. The coastal region of West Bengal lies between $87^0 25'$ E and 89^0 E latitude and $21^0 30'$ N and $23^0 15'$ longitude, covering an large area along the Bay of Bengal coast. The major part of the coastal area in West Bengal falls within the boundary of the districts of South and North 24 Parganas. The coastal region of the districts of South and North 24 Parganas is popularly known as Sundarbans, due to the fact that the area is under mangrove forest dominated by the noble mangrove tree named Sundari *(Heritierafomes).* The forest has both regional and global importance for its diverse ecological resources (Islam 2003). In the said zone, main livelihood and major land use pattern in this coastal zone includes crop cultivation. Nearly 80–90% of the cultivated land is used for agricultural crop production. The next important land use is associated with fish cultivation in ponds and ditches. Almost 70% income is generated from crop cultivation and rest is from fisheries and other animal husbandry activities. Apart from rice, the other crops grown are chili, cucurbits, tomato, beet and bettlevine (Bandyopadhyay et al. 2003). A large number of natural vegetation is observed in Sunderban. In this zone salinity plays a key role in regulating the density of and diversity of phytoplankton (Raha et al. 2012).

4.3 Climate Change Impact on Coastal Ecology

With increasing rate of greenhouse gases in atmosphere due to urbanization and industrialization, climate change has emerged as the most prominent environmental issue. Most of the third world countries, mainly developing countries, are facing the devastating consequences of climate change. The main problems arises due to climate change are rising temperature, melting glaciers which in turn rises the sea level leading to inundation of coastal areas (IPCC 2007). Precipitation pattern also changes along with the increasing tendency of high intensity rainfall for short time leading to increased risk of either devastating floods or recurrent droughts. Expansion of pest attack due to unnatural weather condition may harm the biodiversity along with many public health related issues. Thus it is obvious that different ecosystems are going to be disturbed due to imbalance in their systems. People living in different ecosystems across the world have been recognized as being predominantly risk prone to the impacts of climate change (Pittock 2005). It is also true that ecosystems are under severe threat from poor and unsustainable resource management along with the impact of climate change (Sokona and Denton 2001).

Generally climate change negatively affects natural ecosystems leading to biodiversity loss which may be up to 40% before the end of the century in some areas. The Climate Change Vulnerability Index measures the current vulnerability condition of different countries to extreme climate-related events and how well it is prepared to combat the impacts of climate change. A report based on the study on 160 countries revealed that many countries mainly Africa, south and south-east Asia are at 'extreme risk' from the impacts of climate change (Maplecroft 2010). Poor nations with few natural resources, limited infrastructure and most importantly large scale population are the main victims of climate change. Thus African and south Asian countries like Somalia, Haiti, Afghanistan, Pakistan, Philippines, India and Indonesia are topping the list. In India, coastal areas and Indo-Gangetic plains are declared as the most vulnerable zones for extreme weather events.

Among all ecosystems, coastal ecosystem is the most productive but highly threatened ecosystems in the world. The ecosystem derived from marine habitats is greatly affected by human activities and large numbers of population pressure in the coastal areas (MEA 2005; McGranahan et al. 2007). Economic opportunities like highly fertile low lying delta areas with good transport facilities and easy access to sea food are the main reason for high population density near the coastal areas (Darwin and Tol 2001). Coastal ecosystems are very rich in species diversity (flora and fauna) like mangrove forests, coral reefs, sea fish and aquaculture, etc. Coastal areas have a relatively higher Gross Domestic Product share compared to other inland regions due to their productive nature (Dasgupta et al. 2007). During recent years the physical, biological and biogeochemical characteristics of the oceans and coasts are changing due to pollution and climate change impact which modify the ecological structure and functions.

Reviewing a large numbers of literatures and articles it can be stated that the two main concerning impacts of climate change for coastal ecosystem are:

(i) Sea level rise and
(ii) Extreme weather events such as cyclones, storms, heavy rainfall, etc.

4.4 Impact of Sea Level Rise on Coastal Ecosystem

The rise in sea level has a negative impact on the biophysical and socio economic characteristics of coastal ecosystem. According to a report, most of the destructive consequences of climate change are evolving around water-resource (Stern 2006). It is expected that sea levels may rise approximately 50 cm by the year 2100 compared to 1990 levels. The most alarming news according to them is 'some island nations like the Seychelles and Maldives are expected to be submerged into the sea within the next century if the present rate of sea level rises continue'. Approximately 1.28% of the total population live in developing countries will be affected if the sea level rises 1 m in height (Dasgupta et al. 2007). A study was conducted to estimate the potential impacts of sea level rise for different countries (developed, and developing) considering a 1–5 m rise in sea levels (TERI 1996). Few other researchers also tried to find out the impact of SLR at regional level (Nicholls and Mimura 1998; Darwin and Tol 2001; Dasgupta et al. 2007). Another study was conducted to estimate the climate change impact for 84 developing countries situated in the coastal region by grouping them into five world regions, namely, Latin America and the Caribbean countries; Middle East and North Africa; Sub-Saharan Africa; East Asia and South Asia (Dasgupta et al. 2007). Based on six indicators (land, population, GDP, urban extent, agricultural extent and wetlands), the researchers identified the most affected regions due to sea level rise are East Asia and South East Asia. If we consider the scenario of south Asian countries, researchers reported that sea level has risen at a rate of 2.5 mm per year along the Indian coastline since 1950s. India has also been identified in the top 27 most vulnerable countries list which are mostly affected by the impacts of climate change related accelerated sea level rise (UNEP 1989).

Report of IPCC claimed that through the twentieth century, global sea level rise contributed to increase in coastal inundation, erosion and ecosystem losses, but with considerable local and regional variations (IPCC 2007). Researchers across the world predicted that changes in water temperature, precipitation and oceanographic variables like, wind velocity, wave action and sea level rise can bring significant ecological and biological changes to coastal ecosystems. Loss of property and life due to increasing flood risk along with loss of habitats, infrastructure damage, loss of tourism, recreation and transportation functions and most importantly degradation of soil and water quality affecting agriculture and aquaculture, are the main issues on climate change (Nicholls and Lowe 2004). Coral reefs have significant importance in the livelihood of coastal people because these reefs are acting as a food source to marine fishes as well as it attracts the tourists too. Degradation of coastal ecosystems will definitely destroy those coral reefs which have a direct socioeconomic impact on the societies dependent on the coastal ecosystems for goods and services.

4.5 Extreme Weather Events and Their Impact

The coastal communities over the world are witnessing many extreme weather events like cyclone, storm, tornedo, etc., specially from beginning of twenty-first century and scientists are expecting a lot more to come. Reports claimed that nearly 120 million people across the globe are exposed to tropical cyclones annually, which killed 2,50,000 people from 1980 to 2000. Scientists predicted a 15% increase in the intensity of tropical cyclones which would significantly enhance the vulnerability of population living along the shoreline of India (Aggarwal and Lal 2008). Cyclonic storms affect the coast more adversely as it is associated with high tides along with heavy rainfall which is sufficient to destroy coastal habitats in few hours. For example, severe cyclonic storm 'Aila' formed at Bay of Bengal and hit the coast of West Bengal, India and the coasts of Bangladesh on 25th May, 2009 (Table 4.2). The cyclone hit the coast with 6.5 m high tidal surges and affected 11 coastal districts of India and Bangladesh. This surge of water damaged and washed away over 1743 km of embankments and forced many people to leave their villages. The actual calamities started after the cyclone when daily high tides, and particularly during periods of full moon, inundated the coastal area due to absence of embankments. These tidal waves cause intrusion of huge amount of saline water into the agricultural land and stays there for a long time. As a result, salinity of the soil increased highly which severely affected the agricultural production. Fresh water resources like ponds, wells are also contaminated with the saline water which caused severe drinking water scarcity in the area. Aila has altered the livelihood of the peoples by severely hampering the economy of this region for few years.

Table 4.2 Damage caused by the cyclone 'Aila'

Damages caused by the Aila	
Number of villages affected:	4249
Size of affected population:	25,62,442
Number of people missing:	8000
Number of deaths:	Official-70; Unofficial-300
Length of embankment breached:	400 km
Number of cattle lost:	2,12,8512,12,851
Total area of agricultural land affected:	1,25,872 ha
Estimated financial loss in agriculture:	Rs. 337 crore
Number of houses fully damaged:	1,94,390
Number of houses partially damaged:	1,94,701
Total loss:	Rs. 1495.63 crore

Adapted from Rudra (2010)

4.6 Impact of Climate Change on Agriculture

Climate is one of the most important factors for agricultural productivity. At a global scale, scientists offer different views at the impact of climate change on agriculture as they thought that there might be no significant losses on its production with changing climate (IPCC 2007; Mendelsohn 2008). They explained this point by the logic that production losses in drier African regions will be compensated by increased production in high latitude regions where global warming will raise temperatures and extend planting seasons by reducing the risk of frost (Hassan 2010). The overall impact of climate change on agriculture is given in Table 4.3. Though we are very much concerned about the impact of climate change on agriculture but it is also true that this sector itself contributes to global warming through carbon dioxide (CO_2), methane (CH_4) and nitrous oxide (N_2O) gas emissions. At a global scale, approximately 20% of the annual increase in greenhouse gas emissions was contributed by agricultural system (IPCC 2007).

Many researchers studied the impact of global warming on crop growth and production (Kurukulasuriya and Rosenthal 2003; Easterling et al. 2007; Lobell and Gourdji 2012). Most of the authors opined that the crop and livestock productivity may decline because of rising temperatures and drought-related stress, especially in

Table 4.3 Impact of climate change on agriculture

Climatic element	Expected changes by 2050's	Confidence in prediction	Effects on agriculture
CO_2	Increase from 360 ppm to 450–600 ppm (2005 levels now at 379 ppm)	Very high	Good for crops: increased photosynthesis; reduced water use
Sea level rise	Rise by 10–15 cm increased in south and offset in north by natural subsistence/rebound	Very high	Loss of land, coastal erosion, flooding, salinisation of groundwater
Temperature	Rise by 1–2 °C. Winters warming more than summers. Increased frequency of heat waves	High	Faster, shorter, earlier growing seasons, range moving north and to higher altitudes, heat stress risk, increased evapotranspiration
Precipitation	Seasonal changes by ±10%	Low	Impacts on drought risk' soil workability, water logging irrigation supply, transpiration
Storminess	Increased wind speeds, especially in north. More intense rainfall events.	Very low	Lodging, soil erosion, reduced infiltration of rainfall
Variability	Increases across most climatic variables. Predictions uncertain	Very low	Changing risk of damaging events (heat waves, frost, droughts floods) which effect crops and timing of farm operations

Adapted from Mahato (2014)

the tropical regions. Studies revealed that a 2.5–4.9 °C temperature rises in India may reduce the rice yields by 32–40% and wheat yields by 41–52% which will cause GDP to fall by 1.8–3.4% (GoI 2011; Guiteras 2007). Many authors predicted that depending on crop variety its productivity may also increase reasonably with the increment in mean ambient air temperature of up to 1–3 °C. Though this scenario can only be experienced at mid to high latitude regions. Same experiment at lower latitude regions produced negative result. Moderate increase in temperature will reduce crop yields as crops become intolerant of high temperatures. The intrusion of sea-water in the fertile land due to cyclone or water surge, the fertility of the soil can be destroyed (Figs. 4.2 and 4.3). For example, due to Aila, the production of rice reduced to 3000 Kg per ha compared to the 6000 Kg per ha in the coastal zone of West Bengal (Debnath 2013). Historical studies claimed that, at global level, a significant climate-associated yield reductions of 40 million tons per year was experienced from 1981 to 2002 for maize, wheat and other major crops (Lobell et al. 2011). Projecting the agricultural production under various climate scenarios of 2055, some scientists expected a 10% reduction in maize production for Africa and Latin America which may result a loss of US$2 billion per year (Jones and Thornton 2003).

Fig. 4.2 Sea water intrusion in the coastal region of West Bengal, India

Fig. 4.3 Visible impact of sea-water intrusion

4.7 Impact of Climate Change on Fisheries and Aquaculture

High tides during cyclonic storm or sea level rise not only contaminate the fresh water resources but also destroy the fresh water fish population through salinisation, coastal erosion and wetland flooding. Some freshwater fish species can't survive on saline water and may become extinct along the coastlines. Degradation of coral reefs due to the increasing oceanic temperature, coastal fish population has also declined considerably. Increase in oceanic temperature and destruction of coral reefs along the coasts of many Asian cities reduced the fish population near the coast and hampered the fishery sector severely. Many researchers found that marine fisheries are affected by the direct and indirect impacts of climate change (Allison et al. 2005; Vivekanandan 2006; Allison et al. 2009; Badjeck et al. 2010; Sumaila et al. 2011). Although it is also true that many non-climatic factors are also associated with this declination in fish catch. But there is no doubt that climate change will add to the problems in fishery sector that is already visualized in the coastal regions.

In a nutshell, the ecological impacts of climate change on fishery include:

(a) Change species distribution i.e. fish migration, increased variability of catches, changes in seasonality of production i.e. decrease in fishing season.
(b) Damage to infrastructure, damage to fishing gears, increased danger at sea and problem in navigation routes.
(c) Socio-economic impact, migration, rehabilitation, increase in fuel costs, reduced health due to diseases.

4.8 Climate Resilient Agriculture

It is the integral part of sustaining agriculture in the era of climate change involving proper management of resources used for agricultural operations. It aims at increase of farm production and productivity and to minimize the adverse effect of climate change. The natural resource management, soil health improvement, crop production enhancement and livestock management are the major components of any climate resilient agriculture system. Climate resilient agriculture offers both mitigation and adaptation measures to climate changes (FAO 2010a, b; Matthews et al. 2013). The adaptation measures vary a lot depending upon the region, agroclimatic situation and socioeconomic status of the farming community. They may be switching over to agro-forestry practices, establishment of orchards, nutrient recycling and use and rational use of water towards greater water productivity (Pretty et al. 2006). Diversification of crops in promotion of mixed farming and agroforestry will also help the farmer to fight weather extremes, disease epidemics and crop failures and sustain the farmers' livelihood support (Davis et al. 2012; Lin 2011).

One of the more talked about issues in climate change is carbon sequestration and orchards and agro-forestry shall pave the way for more carbon assimilation and delay the pace of climate change (Blanco-Canqui and Lal 2004). This is because diversity of ecological attributes arising from a large number of species also provides easier access to limited resources and may also reduce instability in the ecosystem processes through asynchronous responses of the different species to environmental fluctuations. As water is becoming scarce over time there will be a decline in rice areas and rice ecosystems are to be gradually shifted to others. At the same time promotion of fodder and animal husbandry along with poultry will indicate a better balance of sustainability in agriculture and welcome the way for greater organic interventions and internal resources in input use pattern (Ortmann and Machethe 2003; Zheng et al. 2014). This in conjunction with mixed farming becomes important very much because in the light of greater use of farm supported energetic in crop production and cutting down of fossil fuel use in manufacture of inputs (Figs. 4.4 and 4.5).

Domestic waste water if at all used to support and get a harvest out of a homestead garden can contribute to the awareness of the value of water and its rational use. The models of agriculture supported by mixed cropping and animal husbandry is not only climate friendly but also it creates man-days and takes care of farm employment

Fig. 4.4 Intensified farming through utilization of banks of pond for vegetable cultivation

Fig. 4.5 Vegetable cultivation in the bunds of rice field

which also contributes to sustainability in a greater way (Magdoff 2007). Conservation tillage will be an important strategy to combat the climate change also. It not only means less fossil fuel spent on the mechanical operations on part of the farmer or enterprise, it also means less compaction of the farm lands due to lessened burden of traffic. Further opening the entire land by cultivators more carbon from below the surface gets oxidised and escapes in the atmosphere. Less compaction also results in more biological tillage by worms and less disturbed micro-flora and drives the system towards sustainability. Organic residue of rice is normally burned in many areas which promotes carbon emissions. Organic residues of the crop when incorporated it helps to attain the soil quality and water balance improvements that are needed to realise the benefits which offer resilience to climate change.

Different types of livelihood assets (physical, financial, social, human and natural capital) form the basis for households' choices of livelihood strategies, including agricultural practices, which in turn influence their food security status and level of well-being. We find empirical evidence that land (natural capital) and household asset and ownership (physical capital) are positively linked to household food security levels. Relevant information available, their integration, capacity building allows further receptivity and makes education (human capital) is closely associated with adaptation by the vulnerable households. The results also show that households that diversify what they choose to produce, as well as to sell, are pursuing key livelihood strategies that make them more food secure, and more able to take up new agricultural practices to deal with changing circumstances.

To bridge the gap between science and field application, there is a need for 'translators' of climate information to assist communities and planners to understand the implications of results for their immediate planning decisions. Enhanced communication between producers and users of climate science is clearly a requirement. New climatological information that is too coarse in resolution for application to crop decision-making must be downscaled for use at the field decision-making level. Developing tools and documents for mainstreaming climate information to agricultural decision making, with the building of a large-scale system of support for the operational use of seasonal climate information for the countries are needed. It will be important to devise best practices for integrating local knowledge with scientific knowledge in the formulation of adaptation strategies. While local knowledge has much to offer in terms of informing adaptation strategies, combining the two has proven challenging to date. Participatory mechanisms for bringing farmers together to disseminate expert knowledge and weather information are necessary. Active collaboration between climate forecasters, agrometeorologists, agricultural research and extension agencies in developing appropriate products for farming communities is essential.

In India, considering the importance of climate resilient agriculture, National Initiative on Climate Resilient Agriculture (NICRA) was launched in the year 2011 by Indian Council of Agricultural Research (ICAR) with the funding from Ministry of Agriculture, Government of India. The mega project has three major objectives of strategic research, technology demonstrations and capacity building (ICAR 2016). Under strategic research, assessment of the impact of climate change simultaneous with formulation of adaptive strategies is the main approach. The climate resilient agricultural technologies aim at increasing farm production and productivity and managing natural resource in a better way. The objectives of NICRA are to enhance the resilience of Indian agriculture covering crops, livestock and fisheries to climatic variability and climate change through development and application of improved production and risk management technologies; to demonstrate site specific technology packages on farmers' fields for adapting to current climate risks and to enhance the capacity of scientists and other stakeholders in climate resilient agricultural research and its application. Presently the technology demonstration component is going on in 100 vulnerable villages through Krishi Vigyan Kendra i.e., Farmers' Scientific Centre (Figs. 4.6 and 4.7).

Under NICRA, the village level interventions towards climate resilient agriculture are as follows:

(a) Maintaining soil health: Resilience of crop production under changing climate is mainly dependent on soil health. A number of interventions are made to build soil carbon, control soil loss due to erosion and enhance water holding capacity of soils, all of which build resilience in soil. Improved methods of fertilizer application should be practiced matching with crop requirement to reduce nitrous oxide emission.
(b) Adapted cultivars and cropping systems: Improved, early duration drought, heat and flood tolerant varieties should be adopted as per climatic situation of a village. Crops and varieties should be chosen in such a way that optimum yields can be achieved despite climatic stresses.
(c) Agromet Advisory service system: Village level weather based agro advisories and contingency crop planning should be prepared on regular basis.
(d) Rainwater harvesting and recycling rainwater: Rainwater harvesting and recycling through farm ponds, restoration of old ponds in dryland/rainfed areas, percolation ponds for recharging of open wells, bore wells and injection wells for recharging ground water are taken up for enhancing farm level water storage.
(e) Water saving technologies: Deficit irrigation, water saving technologies like direct seeded rice, zero tillage and other resource conservation practices should be introduced at drought prone villages.
(f) Livestock and fishery interventions: Use of community lands for fodder production during droughts/floods, improved fodder/feed storage methods, feed supplements, micronutrient use to enhance adaptation to heat stress, vaccination, improved shelters for reducing heat/cold stress in livestock, management of fish ponds/tanks during water scarcity and excess water are some key interventions in livestock/fishery sector.
(g) Formation of cooperative and Institutional interventions: Seed bank, fodder bank, commodity groups, custom hiring centre, collective marketing, introduction of weather index based insurance and climate literacy through a village level weather station should be introduced to ensure effective adaptation of all other interventions.

Fig. 4.6 Farmers' Awareness Program for technology demonstration

Fig. 4.7 Farmers-scientists interaction program in one vulnerable village

4.9 Conclusion

The coastal zone is the most vulnerable region due to climate change. The climate resilient agriculture operation must be implemented for the sustainable development of agricultural system in the coastal area. In the early phase, the major focus should be on assessing vulnerabilities and identifying adaptation options for coastal ecosystems. Designing optimum adaptation strategies are extremely difficult due to its multidisciplinary nature and multi-sector, multi-stakeholder interests. Successful implementation of farm level climate resilient agriculture depends on how different stakeholders play their roles. There are many stakeholders at various levels like national, regional and international organizations, dealing with various elements of climate change adaptation. Additionally, various NGOs operating at regional levels should be involved in community-based interventions to improve the livelihoods of farmers, water and food security. In the coastal zone of developing countries, major emphasis should be given on poverty reduction and disaster mitigation.

Acknowledgement The help and encouragement of Director of Research and Honorable Vice Chancellor, BCKV are duly acknowledged.

References

Aggarwal D, Lal M (2008) Vulnerability of the Indian coastline to sea level rise, SURVAS (Flood Hazard Research Centre). http://www.survas.mdx.ac.uk/pdfs/3dikshas.pdf

Allison EH, Adger WN, Badjeck MC, Brown K, Conway D, Dulvy NK, Halls Perry A, Reynolds JD (2005) Effects of climate change on the sustainability of capture and enhancement fishery important to the poor: analysis of the vulnerability and adaptability of fisherfolk living in poverty, Project No. R4778J, Fishery management science programme, Department for International Development (DFID). http://r4d.dfid.gov.uk/Output/122513/

Allison E, Allison LP, Badjeck M, Adger WN, Brown K, Conway D, Ashley SH, Pillin GM, Reynolds JD, Andrew NL, Dulvy NK (2009) Vulnerability of national economies to the impacts of climate change on fishery. Fish Fish 10:1–24. https://doi.org/10.1111/j.1467-2979.2008.00310.x

Badjeck MC, Allison EH, Halls AS, Dulvy NK (2010) Impacts of climate variability and change on fishery-based livelihoods. Mar Policy 34:375–383. https://doi.org/10.1016/j.marpol.2009.08.007

Bandyopadhyay BK, Maji B, Sen HS, Tyagi NK (2003) Coastal soils of West Bengal – their nature, distribution and characteristics, Bulletin No. 1/2003. Canning Town, West Bengal, India: Central Soil Salinity Research Institute, Regional Research Station, p 62

Blanco-Canqui H, Lal R (2004) Mechanisms of carbon sequestration in soil aggregates. Crit Rev Plant Sci 23:481–504. https://doi.org/10.1080/07352680490886842

Burke LA, Kura Y, Kaseem K, Ravenga C, Spalding M, McAllister D (2000) Pilot analysis of global ecosystems: coastal ecosystems. World Resource Institute, Washington, DC

Costanza R, Arge R, Groot R, Farber S, Grasso M, Hannon B, Limburg K, Naeem S, O'Neill RV, Paruelo J, Raskin RG, Sutton P, Belt M (1997) The value of the world's ecosystem services and natural capital. Nature 387:253–260

Cronk JK, Fennessy MS (2001) Wetland plants: biology and ecology. Lewis Publishers, Washington, DC

Darwin RF, Tol RSJ (2001) Estimates of the economic cost of sea level rise. Environ Resour Econ 19(2):113–129. https://doi.org/10.1023/A:1011136417375

Dasgupta S, Laplante B, Meisner C, Wheeler D, Yan J (2007) The impact of sea level rise on developing countries: a comparative analysis. World Bank policy research working paper 4136

Davis AS, Hill JD, Chase CA, Johanns AM, Liebman M (2012) Increasing cropping system diversity balances productivity, profitability and environmental health. PLoS One 7(10): e47149. https://doi.org/10.1371/journal.pone.0047149

Debnath A (2013) Condition of agricultural productivity of Gosaba C.D. Block, South24 Parganas, West Bengal, India after severe cyclone Aila. Int J Sci Res Publ 3(7):1–4

Dickson ME, Walkden MJ, Hall JW (2007) Systemic impacts of climate change on an eroding coastal region over the twenty-first century. Clim Chang 84(2):141–166. https://doi.org/10.1007/s10584-006-9200-9

Easterling WE, Aggarwal PK, Batima P, Brander KM, Erda L, Howden SM, Kirilenko A, Morton J, Soussana JF, Schmidhuber J, Tubiello JN (2007) Food, fiber and forest products. In: Parry ML (ed) Climate change 2007: impacts, adaptation and vulnerability: contribution of working group II to the fourth assessment report of the intergovernmental panel on climate change. Cambridge University Press, Cambridge, pp 273–313

FAO (2010a) Climate-smart agriculture policies, practices and financing for food security. Adaptation and mitigation. http://www.fao.org/docrep/013/i1881e/i1881e00.pdf

FAO (2010b) The state of world fishery and aquaculture. The United Nations/Food and Agriculture Organization, Rome

GOI (2011) Census of India 2011, Government of India (GOI), New Delhi

Guiteras R (2007) The impact of climate change on Indian agriculture. Working paper, Department of Economics, MIT

Hassan RM (2010) Implications of climate change for agricultural sector performance in Africa: policy challenges and research agenda (dagger). J Afr Econ 19(Supplement 2):77–105. https://doi.org/10.1093/jae/ejp026

Hoorweg J, Muthiga N (2009) Advances in coastal ecology people, processes and ecosystems in Kenya. African studies collection (20). African Studies Centre, Leiden

ICAR (2016) Climate resilient agriculture-NICRA. Accessed through www.nicra-icar.in

IPCC (2007) Climate change 2007: impacts, adaptation and vulnerability, contribution of working group II to the fourth assessment report of the intergovernmental panel on climate change. Cambridge University Press, Cambridge

Islam S (2003) Sustainable eco-tourism as a practical site management policy. AH Development Publishing House, Dhaka

Jena KC, Mishra MR (2011) Effect of global warming and climate change on coastal zones and sea level. Orissa Rev 10(8):82–94

Jones PG, Thornton PK (2003) The potential impacts of climate change on maize production in Africa and Latin America in 2055. Glob Environ Chang 13:51–59. https://doi.org/10.1016/S0959-3780(02)00090-0

Kurukulasuriya P, Rosenthal S (2003) Climate change and agriculture a review of impacts and adaptations. Climate change series paper, (91). World Bank, Washington

Lal R (2005) Climate change, soil carbon dynamics, and global food security. In: Lal R, Stewart B, Uphoff N (eds) Climate change and global food security. CRC Press, Boca Raton, pp 113–143

Lem A (2001) International trade in ornamental fish. In: Proceedings of 2nd international conference on marine ornamentals: collection, culture and conservation. Florida, USA. November 26 – December 01, 2001

Lin BB (2011) Resilience in agriculture through crop diversification: adaptive management for environmental change. BioScience 61:183–193. https://doi.org/10.1525/bio.2011.61.3.4

Lobell DB, Gourdji SM (2012) The influence of climate change on global crop productivity. Plant Physiol 160:1686–1697. https://doi.org/10.1104/pp.112.208298

Lobell DB, Schlenker W, Costa-Roberts J (2011) Climate trends and global crop production since 1980. Science 333:616–620. https://doi.org/10.1126/science.1204531

Magadza CHD (2000) Climate change impacts and human settlements in Africa: prospects for adaptation. Environ Monit Assess 61(1):193–205. https://doi.org/10.1023/A:1006355210516

Magdoff F (2007) Ecological agriculture: principles, practices, and constraints. Renewable Agric Food Syst 22(2):109–117. https://doi.org/10.1017/S1742170507001846

Mahato A (2014) Climate change and its impact on agriculture. Int J Sci Res Publ 4(4):1-–11

Maplecroft (2010) Climate change risk atlas 2010. https://maplecroft.com/about/news/climate_change_risk_list_highlights_vulnerable_nations_and_safe_havens_05.html

Matthews RB, Rivington M, Muhammed S, Newton AC, Hallet PD (2013) Adapting crops and cropping systems to future climates to ensure food security: the role of crop modelling. Glob Food Sec 2:24–28. https://doi.org/10.1016/j.gfs.2012.11.009

McGranahan G, Balk D, Anderson B (2007) The rising tide: assessing the risks of climate change and human settlements in low-elevation coastal zones. Environ Urban 19(1):17–37. https://doi.org/10.1177/0956247807076960

MEA (2005) Ecosystems and human well-being: current state and trends. In: Hassan R, Scholes R, Ash N (eds) Findings of the condition and trends working group millennium ecosystem assessment. Island Press, Washington, DC. http://www.maweb.org/en/Condition.aspx

Mendelsohn R (2008) The impact of climate change on agriculture in developing countries. J Nat Resour Policy Res 1(1):5–19. https://doi.org/10.1080/19390450802495882

Mertz O, Halsnæs K, Olesen JE, Rasmussen K (2009) Adaptation to climate change in developing countries. Environ Manag 43:743–752. https://doi.org/10.1007/s00267-008-9259-3

Nicholls RJ, Lowe JA (2004) Benefits of mitigation of climate change for coastal areas. Glob Environ Chang 14:229–244. https://doi.org/10.1016/j.gloenvcha.2004.04.005

Nicholls RJ, Mimura N (1998) Regional issues raised by sea level rise and their policy implications. Clim Res 11:5–18

Ortmann G, Machethe C (2003) Problems and opportunities in south African agriculture. In: Nieuwoudt L, Groenewald J (eds) The challenge of change: agriculture, land and the South African economy. University Natal Press, Pietermaritzburg, pp 47–62

Pittock B (2005) Climate change – turning up the heat. Earthscan/CSIRO Publishing, Melbourne

Pretty J, Noble AD, Bossio D, Dixon J, Hine RE, Penning de Vries FW, Morison JI (2006) Resource- conserving agriculture increases yields in developing countries. Environ Sci Technol 40:1114–1119. https://doi.org/10.1021/es051670d

Raha A, Das S, Banerjee K, Mitra A (2012) Climate change impacts on Indian Sunderbans: a time series analysis (1924–2008). Biodivers Conserv 21:1289–1307. https://doi.org/10.1007/s10531-012-0260-z

Rudra K (2010) Aila: unpublished records of the Government of West Bengal. A South Asian journal on forced mitigation. MCRG, Kolkata, pp 86–93

Senapati S, Gupta V (2014) Climate change and coastal ecosystem in India: issues in perspectives. Int J Environ Sci 5(3):530–543. https://doi.org/10.6088/ijes.2014050100047

Sokona Y, Denton F (2001) Climate change impacts: can Africa cope with the challenges? Clim Pol 1(1):117–123. https://doi.org/10.1016/S1469-3062(00)00009-7

Sovacool BK (2014) Environmental issues, climate changes, and energy security in developing Asia. ADB economics working paper series, No. 399. Asian Development Bank, Philippines

Stern N (2006) What is the economics of climate change? World Econ 7:1–10. https://doi.org/10.1257/aer.98.2.1

Sumaila UR, Cheung WWL, Lam VWY, Pauly D, Herrick S (2011) Climate change impacts on the biophysics and economics of world fishery. Nat Clim Chang 1301:1–8. https://doi.org/10.1038/NCLIMATE1301

TERI (1996) The economic impact of a one metre sea level rise on the Indian coastline: method and case studies, report submitted to the ford foundation. Tata Energy Research Institute, New Delhi

Tompkins EL, Adger WN (2005) Defining a response capacity for climate change. Environ Sci Policy 8:562–571. https://doi.org/10.1016/j.envsci.2005.06.012

UNEP (1989) Criteria for assessing vulnerability to sea level rise: a global inventory to high risk area (Delft: Delft Hydraulics), 51, United Nations environment Programme (UNEP)

Vidthayanon C, Premcharoen S (2002) The status of estuarine fish diversity in Thailand. Mar Freshw Res 53:471–478

Vivekanandan E (2006) Impact of climate change on marine fishery, Central Marine Fishery Research Institute, CMFRI Newsletter, No 112

Weiner J (2003) Ecology – the science of agriculture in the 21st century. J Agric Sci 141:371–377. https://doi.org/10.1017/S0021859603003605

Wezel A, Bellon S, Dore T, Francis C, Vallod D, David C (2009) Agroecology as a science, a movement or a practice. A review. Agron Sustain Dev 29(4):503–515

Zheng C, Jiang Y, Chen C, Sun Y, Feng J, Deng A, Song Z et al (2014) The impacts of conservation agriculture on crop yield in China depend on specific practices, crops and cropping regions. Crop J 2:289–296. https://doi.org/10.1016/j.cj.2014.06.006

Chapter 5
Methanogenesis and Methane Emission in Rice / Paddy Fields

N. K. Singh, D. B. Patel, and G. D. Khalekar

Abstract Rice fields are a major source of atmospheric methane (CH_4), a greenhouse gas. CH_4 emissions from wetland rice fields represents globally 15–20% of the annual anthropogenic CH_4 emissions, and about 4% of the global CH_4 emissions. Methane emission from rice cultivation may increase from the 1990 level of 97 Tg/year to 145 Tg/year by 2025 due to the increase in acreage and intensification of paddy cultivation. Here we review the role of anaerobic methanogenic bacteria in methane emission. We discuss the factors that influence methane emissions from rice fields, such as water regime, cropping season, soil temperature, fertilizer application, soil physico-chemical properties, crop cultivation, agricultural practices, soil type, soil profile and crop management practices. These practices control soil bacterial communities. Other influencing factors include intercultural operations such as ploughing, puddling and frequent mixing of soil during the paddy field preparation. Methane emission from paddy field follows a seasonal pattern of variation due to influence of climatic factors like temperature, sunlight, and precipitation. Algae, microphytes, macrophytes and anoxygenic photosynthetic bacteria significantly reduce CH_4 emissions when they grow actively under illuminated condition. Methane emission is limited by alternate flooding-drying; cultivars with few unproductive tillers, small root system, high oxidative ability, and high harvest index; excessive application of organic amendments; application of potassium, biochar, nitrate, sulfate and ferric iron; and urease and nitrification inhibitors.

Keywords Archaea · Biochar · Fertilizers · Methanogens · Soil properties

N. K. Singh (✉) · D. B. Patel
Department of Microbiology, C.P. College of Agriculture, S.D. Agricultural University, Sardarkrushinagar, Gujarat, India

G. D. Khalekar
Department of Plant Molecular Biology & Biotechnology, C.P. College of Agriculture, S.D. Agricultural University, Sardarkrushinagar, Gujarat, India

E. Lichtfouse (ed.), *Sustainable Agriculture Reviews 33*, Sustainable Agriculture Reviews 33, https://doi.org/10.1007/978-3-319-99076-7_5

5.1 Introduction

Methane, the second most important greenhouse gas after CO_2 in the atmosphere, is increasing at an average rate of 0.8% per annum (Prinn 1995). Since pre-industrial era, the atmospheric methane concentration has increased almost three fold (Keppler et al. 2006; Zou et al. 2005). The tropospheric CH_4 concentration (1.8 ppmV), although very low as compared to CO_2 (357 ppmV), accounts for about 15–20% to global warming (IPCC 1996; Schimel 2000). About 1 kg of atmospheric methane is considered 21 times more destructive to the radiation balance (Denier van der Gon et al. 2002) and has 23 times greater global warming potential on mass balance basis (Towprayoon et al. 2005) than does an equal amount of CO_2 when released into the atmosphere. The greater global warming potential of methane may be due to its high absorption potential for the infrared radiations.

Nearly 70% (550 Tg y^{-1}) of the total methane emission has anthropogenic origin (Hattori et al. 2001; Sass and Fisher 1994; Zou et al. 2005). Global estimates of emission rates from wetland rice fields range from 60 to 150 Tg year−1 (Aulakh et al. 2000). More recent findings estimate the global CH_4 emission from rice fields to be 25.6 Tg y^{-1}, accounting for about 4% of the total global CH_4 emission (IPCC 2007; Yan et al. 2009). However, methane emission from rice cultivation may increase from the 1990 level of 97–145 Tg y^{-1} by 2025 (Anastasi et al. 1992). The flooded paddy fields represent a unique anoxic environment where methanogenic archaea produce methane in the terminal step of anaerobic microbial decomposition of organic material.

Paddy is grown over more than 150 million hectares of land and is consumed more than any other cereal globally (Bloom and Swisher 2010). The rice producing area across the globe has expanded in the past from 116 million hectares in 1961 to 153 million hectares in 2004 in order to fulfil the food demand of growing population (IRRI 2006). Table 5.1 represents the rice cultivation area under main geographical regions and countries (FAOSTAT Database 2008). Nevertheless, rice cultivation has to be intensified in order to support the booming human population and to meet the increasing food demand; especially in the tropical and subtropical regions. Such intensification will require increased fertilizer dose which may exacerbate the methane problem from this anoxic agroecosystem. This may lead to further intensification of CH_4 emission and increased concern among policy makers (Dubey 2001; Inubushi et al. 2001; Neue and Roger 2000). Although, microbiology of flooded rice field has been reviewed by few workers (Kimura 2000; Liesack et al. 2000), the present article tries to focus on the factors that influence CH_4 production and emission from paddy fields. A broader knowledge of the factors affecting CH_4 emissions from rice fields is essential so that feasible and cost-effective technologies and methods can be developed in order to mitigate CH_4 emissions from paddy fields without affecting the productivity.

Table 5.1 Major rice growing geographical regions and countries

Country and geographical region	Area (000 ha)	Country and geographical region	Area (000 ha)
World	156,688	Egypt	668
Asia	140,036	Tanzania	665
Africa	9386	Malaysia	660
South America	4806	Iran	630
N&C America	1826	Sierra Leone	630
Europe	606	Korea, DPR	575
Oceania (including Australia)	27	Congo, Dem Rep	418
India	44,000	Mali	377
China	29,230	Colombia	360
Indonesia	12,166	Côte d'Ivoire	345
Bangladesh	11,200	Peru	339
Thailand	10,360	Ecuador	325
Myanmar	8200	Italy	233
Vietnam	7305	Mozambique	204
Philippines	4250	Uruguay	180
Nigeria	3000	Argentina	164
Brazil	2901	Russia	163
Pakistan	2600	Cuba	160
Cambodia	2542	Venezuela	160
Japan	1678	Bolivia	145
Nepal	1440	Dominican Rep	145
Madagascar	1300	Panama	130
USA	1112	Liberia	120
Korea, Rep	950	Ghana	120
Laos	820	Uganda	119
Sri Lanka	796	Guyana	105
Guinea	789	Spain	102

Source: FAOSTAT Database 2008. FAO, Rome. 22 Sep 2008. http://beta.irri.org/solutions/index.php?option=com_content&task=view&id=250

5.2 Mechanism of Methane Production in Paddy Fields

The methanogenic archaea responsible for methane production may be obligate chemolithotrophic or quasi-chemolithotrophic or methylotrophic, transform organic carbon into methane, and perform better under anaerobic conditions (Bloom and Swisher 2010). Crop residues and exudates from previous crops and rice plants accumulated in the soil and during prolonged inundation get decomposed and act as main substrates for methanogenesis (Naser et al. 2007). Firstly, the hydrolytic microorganisms convert the complex organic molecules into monomers and oligomers. The cellulolytic and saccharolytic organisms then ferment the above compounds and lead to production of various organic acids, alcohols, H_2, and CO_2;

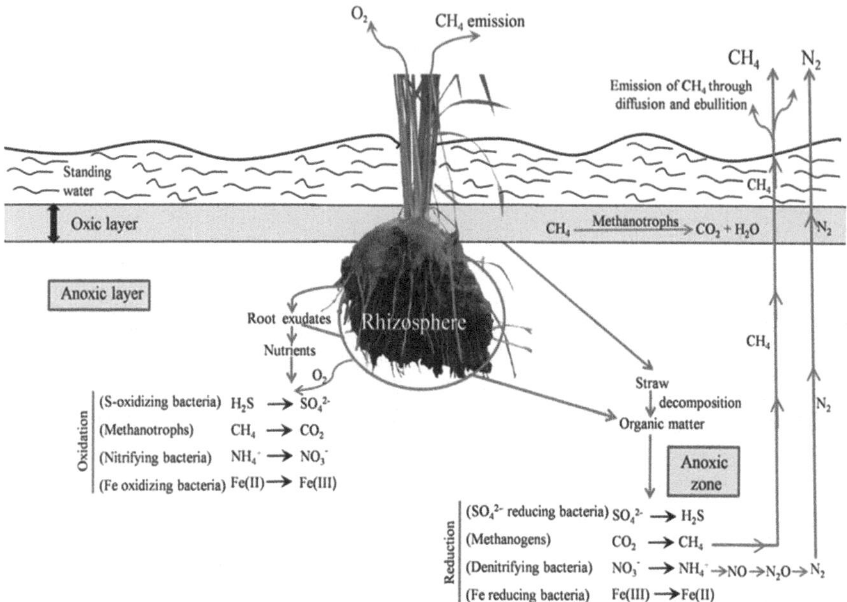

Fig. 5.1 Schematic diagram of methane production, oxidation and emission from paddy field [Note the oxidation of root exudates in the vicinity of rhizosphere and in the oxic layer prevailing at the soil water interface. Reduction of organic matter originating from the leached down root exudates, rice straw, and other sources takes place due to the presence of various microorganisms in the anoxic layer of the lowland paddy field ecosystem]

whereas, homoacetogenenic bacteria exclusively convert simple sugars to acetate and syntrophic bacteria convert organic acids/alcohols to acetate, H_2, and CO_2. Lastly, the anaerobic acetoclastic methanogenic archaea convert acetate to CH_4 and CO_2. However, the sulfate and iron reducing bacterial populations in presence of sulfate or ferric iron compete with the methanogenic archaea for reducing equivalents and convert CH_4 to CO_2 (Chidthaisong and Conrad 2000). The methane thus produced in the paddy fields gets released into the atmosphere (Fig. 5.1).

5.3 Factors Affecting CH_4 Production and Emission from Paddy Fields

The important variables that affect CH_4 production and emission from paddy fields include climatic factors; physical, chemical and biochemical properties of soil; temperature; water management; rice cultivar; application of manure, fertilizers, urease and nitrification inhibitors and other organic carbon or nitrogen sources (Aulakh et al. 2001; Neue and Roger 2000). These variables affect one or more of the principal processes of CH_4 production, its transport and oxidation; which control

CH_4 emission from paddy fields. However, production, transport and oxidation of CH_4 are the result of a large number of complex processes that operate on microscopic scale, which ultimately control the CH_4 flux form rice fields to the atmosphere. Since the first study on methane production from paddy field in 1981 (Cicerone and Shetter 1981), a large number of experiments have been conducted in the last three decades, the results of which are of tremendous importance in understanding of the complexity of processes which control and affect methane emission from paddy fields (Table 5.2).

5.3.1 Influence of Soil Properties

Soils from different location may differ in their physical, chemical and biochemical properties like, soil texture and structure, aeration, porosity, density, dynamic water content, water holding capacity, clay percent, mineral content and their availability, nutrient and organic matter status; all of which may affect the soil bacterial community and CH_4 production, oxidation and transport. Apart from carbon supply and water regime, soil conditions responsible for higher methane production in wetland paddy soils are temperature, texture and mineralogy, Eh/PH buffer, and salinity (Neue and Roger 2000). The CH_4 emission rates also differ markedly with the soil type (Yagi and Minami 1990).

5.3.1.1 Soil Type

Soil type is a complex factor, influenced by soil organic matter, decomposition of organic fertilizer, soil pH, Eh and sand percentage. It has been reported that soil type and management practice largely determine the structure of bacterial communities (Clegg et al. 2003). Wang et al. (2010) investigated the community structures of methanogenic archaea by polymerase chain reaction-denaturing gradient gel electrophoresis (PCR-DGGE), sequencing and real-time PCR and concluded that the methanogenic community structure was influenced by the soil types and sampling locations. Ramakrishnan et al. (2001) also showed differences in the methanogenic archaeal communities in rice soils among distinct geographical regions. The methanogenic archaeal communities in paddy field soils were reported stable throughout the year and were slightly influenced by the sampling period and fertilization; but distinctly influenced by the soil type and sampling region (Watanabe et al. 2006). Above findings suggest that soil type is known to affect the methanogenic communities and it is the methanogens which affect the methane production. However, Sass and Fisher (1992) reported a clear inverse relation of the seasonal CH_4 emissions in Texas with that of the soil clay content.

Table 5.2 Factors affecting bacterial methane emission from paddy fields

Factors		Impact	References
Season	Seasonal pattern	Emission records from paddy field show a distinct seasonal pattern: an early flush of CH_4 before transplanting, an increasing trend in emission rates reaching maximum toward grain ripening, and a second flush after water is withdrawn prior to harvesting	Corton et al. (2000)
Cultivars	Biomass	The cultivar producing higher biomass (Pusa Basmati) show higher CH_4 emissions than the cultivar producing comparatively lower biomass (IR72)	Jain et al. (2000)
Sowing method	direct seeding	The practice of direct seeding instead of transplanting result in 16–54% reduction in CH_4 emission	Corton et al. (2000)
Presence of other organisms	Indigenous phototrophs	Methanogenesis in paddy soil occurs in the soil-floodwater interface if plant residues like rice straw exist. However, such methanogenesis is likely to be suppressed by the growth of indigenous phototrophs under illumination	Harada et al. (2005)
	Azolla	Growing *Azolla* at the soil-floodwater interface in low-land rice filed exert a moderating effect on the CH_4 flux by increasing the dissolved oxygen concentration	Bharati et al. (2000)
Fertilizers	NH_4^+ containing N fertilizers	Such N fertilizers often result in increased CH_4 emission either by stimulation of plant growth and methanogenic activity or inhibition of methane-oxidizing activity	Kruger and Frenzel (2003)
	Urea	Seasonal flux of CH_4 increased by 94% following application of urea @120KgNha^{-1}	Adhya et al. (2000)
	Phosphorus	Low phosphate supply to rice plants result in the enhancement of CH_4 emission whereas application of P	Adhya et al. (1998), Conrad et al. (2000), Lu et al. (1999)

(continued)

Table 5.2 (continued)

Factors		Impact	References
		inhibit acetotrophic methanogenesis. Application of phosphorus as single superphosphate (SSP) inhibits CH_4 emission from flooded rice planted plot	
	Phosphogypsum	Phosphogypsum reduce CH_4 emission by 72% when applied in combination with urea fertilizer	Corton et al. (2000)
	Muriate of potash (K_2O)	Application of potassium @30 kg K ha – 1 (through muriate of potash) in paddy field show 49% reduction in CH_4 emission as compared to control plots (without K application). Potassium fertilizers act by preventing reduction of redox potential, decrease the content of reducing substances and Fe_2^+ content in the rhizosphere, inhibit methanogenic bacteria, stimulate methanotrophic bacterial population, and produce higher plant biomass and grain yield.	Jagadeesh Babu et al. (2006)
	Sulfate	Presence of sulfate suppresses CH_4 formation	Saenjan and Wada (1990)
		Incorporation of ammonium sulfate as N fertilizer in place of urea results in 25–36% reduction in CH_4 emissions	Corton et al. (2000)
		One mol of SO_2^{-4} is needed by sulfate reducing bacteria to inhibit the production of one mol of CH4. The cost of SO_2^{-4}-containing fertilizer as a mitigation option to reduce CH_4 emissions in rice fields is estimated at 5–10 US dollar per Mg CO_2-equivalent	Denier van der Gon et al. (2001)
	Slag silicate	Application of slag silicate fertilizer (@ 4 $Mgha^{-1}$) in tillage and no-tillage plots, reduced the total seasonal CH_4 flux by about 20% and 36%, respectively, while	Ali et al. (2009)

(continued)

Table 5.2 (continued)

Factors		Impact	References
		grain yields increased by about 18% and 13%, respectively	
	Nitrate, ferric iron, and sulfate	Reductions of nitrate, ferric iron, and sulfate play important roles in the mineralization process, especially during the early phase of flooding of rice fields when nitrate, ferric iron, and sulfate have not been completely depleted	Yao et al. (1999)
Organic amendments	*Sesbania*, *Azolla* and compost	Application of *Sesbania*, *Azolla* and compost result in emission of 132, 65 and 68 kg CH_4 ha^{-1} in the wet season while pure urea application result in 42 Kg CH_4 ha^{-1}	Adhya et al. (2000)
	Rice straw	Addition of rice straw compost increase CH_4 emission by only 23–30% as compared to 162–250% increase in emissions with the use of fresh rice straw	Corton et al. (2000)
		Rice straw stimulated the immediate accumulation of fermentation products during the incubation period. In the detected organic acids, acetate was in most abundance during the decomposition of rice straw, whereas propionate showed the lowest concentration	He et al. (2008)
		Propionate is an important intermediate among fermentation products following acetate in paddy field receiving high amount of paddy straw	Glissmann and Conrad (2000)
		Application of comparatively higher carbon substrate stimulate the growth of *dsr*AB [dissimilatory (bi)sulfite reductase genes]-containing sulfate-reducing prokaryotes. Sulfate is not the limiting factor to growth of sulfate-	Hadas and Pinkas (1995), He et al. (2008), Oude Elferink et al. (1994)

(continued)

Table 5.2 (continued)

Factors		Impact	References
		reducing prokaryotes and these bacteria can survive by fermentation in syntrophic interaction with methanogens under low-sulfate condition	
	Organic carbon	Application of organic carbon into freshwater wetlands stimulates sulfate reduction, production of organic acids, and CH_4 emission	Feng and Hsieh (1998), Glissmann and Conrad (2000)
	Biochar	Almost complete suppression of CH_4 emissions in biochar (ground to <1 mm) amended soils (@20 g Kg^{-1}) has been reported in the Eastern Colombian Plains	Rondon et al. (2005)
		Soil amendment with high dose of bamboo char and straw char reduce CH_4 emissions from the paddy field respectively by 51.1% and 91.2% compared to control (without biochar)	Liu et al. (2011)
	C/N ratio of organic amendments	Soil amendment with organic products like animal manure, rice straw, compost, and green manure having C/N ratio of 100, 51, 12, and 10 kg C (kg N)$^{-1}$ respectively show that amendment having least C/N ration (green manure) is most desirable due to their role in increased grain yield and comparatively less contribution to increased CH_4 emissions	Matthews et al. (2000)
Water management	Drainage	Midseason drainage reduces CH_4 emission by 43% which can be explained by the influx of oxygen into the soil	Corton et al. (2000)
		The mean CH_4 flux from permanently flooded rice fields in fallow season is about 5–6 times higher than that from the rice fields drained intermittently during rice season	Xu et al. (2000)

(continued)

Table 5.2 (continued)

Factors		Impact	References
		Drainage of rice fields in the fallow season reduces annual total CH_4 emission by 56–59%	Jiang et al. (2006)
		Drainage in the fallow season reduce CH_4 emission during rice growing season by 13–42% and annual total CH_4 emission by 48–68%	Cai et al. (2003)
		The rate of CH_4 production in the paddy soil and in rice roots during the rice growing season are respectively 42–61 and 56% lower in the flooded condition than in drained rice field	Zhang et al. (2011)
		Keeping soils drained as much as possible during winter seems to be a feasible option to reduce CH_4 emissions during the following rice growing seasons	Kang et al. (2002)
	Intermittent irrigation	Mean CH_4 emission rate during the 4 year period amount to 25.57 and 18.33 kg ha^{-1} respectively under intermittent and continuous flooding showing a reduction of 28% by adopting the practice of intermittent irrigation over continuous flooding. However, this reduction is accompanied with a slight reduction (3.2%) in grain yields	Jain et al. (2000)
	Water regimes and aeration	In comparison to local practice of normal aeration, early aeration reduce CH_4 emission by 13.3–16.2% and increase N_2O emission by 19.1–68.8% while delayed aeration reduce N_2O emission by 6.8–26.0% and increase CH_4 emission by 22.1–47.3%	Li et al. (2011)
Nitrification inhibitor	Dicyandiamide	Dicyandiamide (@ 30 Kg ha-1) reduce CH_4 emission by 13%	Adhya et al. (2000)

(continued)

Table 5.2 (continued)

Factors		Impact	References
	Nimin	Nimin (@ 1% of urea-N) increase CH_4 flux by 9.6% over that of urea	Adhya et al. (2000)
	Calcium carbide and nitrapyrin	Calcium carbide and nitrapyrin inhibit CH_4 emission from flooded soil planted to rice besides controlling N losses	Bronson and Mosier (1991), Keerthisinghe et al. (1993)
	Sodium azide, dicyandiamide, pyridine, aminopurine, ammonium thiosulfate, thiourea	These compounds act as nitrification inhibitor, prevent nitrogen loss, and suppress rapid release of methane from paddy fields. The nitrification inhibitor-based inhibition of CH_4 production follows the order: sodium azide > dicyandiamide > pyridine > aminopurine > ammonium thiosulfate > Thiourea	Bharati et al. (2000)
	Organic ammendments	Application of *Sesbania*, *Azolla* and compost respectively result in emission of about 132, 65 and 68 kg CH_4 ha^{-1} in the wet season while pure urea application result in 42 KgCH_4 ha^{-1}	Adhya et al. (2000)

Sandy soils having high organic carbon are reported to produce more methane than clay soils with similar organic carbon content (Yao et al. 1999). Among the three soil types tested, peaty soil showed highest emission rate followed by the alluvial and the Andosol. Higher percolation rates of soil indirectly result in low emissions of CH_4 from paddy fields (Inubushi et al. 1992; Jain et al. 2000). Moreover, leached sandy soil in China also resulted in lower emissions from the paddy fields (Mingxing and Jing 2002). Aggregate size of soil is supposed to affect CH_4 uptake and its oxidation in soil. At high CH_4 concentrations (16.5% v/v), the uptake rate of CH_4 and O_2 linearly decreased with aggregate size of soil between 2 and 10 mm. However, at low CH_4 concentration, CH_4 (1.8 ppmv) was consumed in soil aggregates <6 mm but soil aggregates >6 mm released CH_4 into the atmosphere. The atmospheric CH_4 uptake rate also increased threefold when the loamy soil was amended with sterile quartz sand (Jackel et al. 2001). High percolation rate and frequent water replenishment may result in constant inflow of oxygen and downward discharge of methanogenic substrate into the soil, and therefore, low methane production and emission.

5.3.1.2 Water Holding Capacity

Water holding capacity of soil depends upon the soil type, especially texture and is known to influence the soil microflora and hence the process of methanogenesis. Soil water affects the seasonal variation pattern of CH_4 flux and soil redox potential. The higher the soil water content, the quicker soil Eh declined and the earlier CH_4 emission initiated after rice transplantation (Xu et al. 2003). The CH_4 emission rates from paddy fields increased until the water content reached to about 75–82% of water holding capacity (Jackel et al. 2001). However, Yang and Chang (1998) observed a increase in methane production with water content of soil with a maximum at 66.7% water. There was a positive linear correlation between total methane production and water content of soil from 16.7% to 66.7%. The water-history-induced change of soil organic C content may affect the soil reduction rate, and then CH_4 production and emission within the rice-growing season (Xu et al. 2003). At 68% of water holding capacity, maximum CH_4 oxidation was observed with biphasic kinetics. It is worthwhile to mention that activity with low K_m allows oxidation of atmospheric CH_4. However, when the soil moisture reduced below 20%, the soil turned from net uptake to net release of atmospheric CH_4 (Jackel et al. 2001). They showed that the methanotrophic microorganisms get inactivated at an earlier stage of drainage than the methanogenic microorganisms. Hence, it is unlikely that the rice fields can act as a net sink for atmospheric CH_4 even when drained.

5.3.1.3 pH of Soil

Soil pH has been regarded as one of the most important factor in determining soil bacterial community structure (Fierer and Jackson 2006; He et al. 2008). The optimal pH for several species of methanogens is near neutral *i.e.*, between 6.4 and 7.1 (Liu and Wu 2004) and a soil pH less than 6.0 may inhibit the growth of methanogens (Wang et al. 1993). The relationship between the pH and the methane production rate can be described by a parabolic curve (Liu and Wu 2004). The correlation coefficient between methane emission rate and water pH was recorded less than 0.25, and between methane emission rate and soil pH it was less than 0.30 (Yang and Chang 1999). The DGGE banding patterns of 16S rDNA PCR products of methanogens from two black calcareous soil samples (BC-I, BC-II) of NE China were different from those in the other soil samples (Wang et al. 2010). These two soil samples were collected from soils having pH > 8.0, which was comparatively higher than the soil pH of the other samples. Therefore, it is likely that soil pH might be a factor influencing the methanogenic archaeal community.

5.3.1.4 Redox Potential

The methanogenic archaea are obligate anaerobe, need low redox potential (less than – 0.33 V) for their growth, and are known to survive in paddy field soil even under drained conditions. These bacteria can metabolize only in strictly anaerobic environments and the Eh must be below −200 mV to cause significant CH_4 production (Kludze et al. 1993; Yamane and Sato 1964). The correlation coefficient between methane emission rate and soil redox potential was less than 0.35 (Yang and Chang 1999). The flooding leads to a low redox potential, establishes an anaerobic soil environment for the mathanogens, and increases the rate of CH_4 production (Sebacher et al. 1986). Most of the Asian paddy fields are flooded during rice cultivation period in summer and remain under fallow or cropped with wheat or barley under drained condition in winter. This may be a reason for prevalence of a wide seasonal variation in redox potentials (+0.6 to −0.3 V) of paddy field soils (Takai and Kamura 1966). Therefore, the archaeal community structure from drained soil did not reflect the methane flux from paddy fields under flooded conditions (Yagi and Minami 1990); and the communities of methanogenic archaea determined in such case may not correspond to the active populations of methanogenic archaea in paddy field soil (Watanabe et al. 2007). Soils containing higher amounts of readily decomposable organic substrates (acetate, formate, methanol, methylated amines, etc.) and low amounts of electron acceptors (NO_3^-, Fe^{3+}, Mn^{4+}, and SO_4^{2-}) are likely to show high amount of CH_4 production. During oxidation–reduction reaction O_2 is the first to get reduced at an Eh of about +30 mV; followed sequentially by NO_3^- and Mn^{4+} at 250 mV, Fe^{3+} at +125 mV and SO_4^{2-} at −150 mV (Patrick 1981). Re-flooding of drained soils require longer duration to sufficiently lower the soils redox potential (Eh), and hence, CH_4 production and fluxes is delayed in more oxidized soil, which increase only slowly after transplanting of seedlings (Xu 2001). However, if flooding is continued between two rice cropping seasons, Eh of the soil remains low and CH_4 production and its fluxes starts immediately after rice transplanting.

5.3.2 *Spatial Variation with Soil Depth*

Flooded rice fields are characterized by O_2 and nutrition gradients and hence, different physical–chemical characteristics of the soil profile. Ratering and Schnell (2000) reported highest concentrations of oxygen at the surface and a depletion of its level below 3 mm depth. They observed zero CH_4 production at the soil surface but an increased with the depth. However, in soil below 4 mm depth where iron(III) concentrations decreased, higher methane production rates were recorded. Wang et al. (1997) observed a significant decrease in kinetic parameters: Vmax and Km from top to bottom in the paddy rice soil profile, ranging from 12.5 to 1.2 μg h^{-1} g^{-1} and 165 to 4.1 μg g^{-1}, respectively. Oxygen gradient affect the diversity and succession of bacterial community structure in paddy soil (Noll et al. 2005). Only a slight change in soil depth affects the relative abundance of the hydrogenotrophic Methanomicrobiales or RC-I relative to the acetoclastic Methanosaetaceae (Wu et al. 2009). The diversity and composition of bacterial communities in different habitats of paddy field such as floodwater, plow-layer soil, rice straw and its compost incorporated in soil differed considerably (Asakawa and Kimura 2008; Nakayama et al. 2006; Okabe et al. 2000). The relative abundance of methanogenic bacterial population composition was also different in the rhizosphere than the bulk soils (Kruger et al. 2005). Therefore, it is quite clear that spatial variation in the soil profile influence the structure of methanogenic community and hence the process of methanogenesis in rice soil. However, due to ploughing, puddling and the frequent mixing of soil during the field preparation for planting, the effect of soil profile may not be always observed.

5.3.3 Seasonal Variations in Methane Emissions

Methane emissions from paddy fields follow a seasonal pattern of variations depending upon the availability of moisture and an inverse correlation exist between CH_4 emissions and grain yields (Corton et al. 2000). A seasonal variation of 36.3, 18.4, and 20.1 gm^{-2} in daily rate of methane emissions has been observed respectively from flooded paddy field, field planted with weeds, and unplanted field (Holzapfel-Pschorn et al. 1986). Low solar radiation during pre-anthesis period result in reduced supply of assimilate to the spikelets which may reduce the yields and harvest index during wet season whereas, during dry season the yields and harvest index are comparatively higher (Denier van der Gon et al. 2002). During wet season sufficient photosynthates are produced but comparatively lower flowers and spikelets frequency are unable to make use of this carbon for grain production and therefore result in lower yield. The excess carbon gets into the soil as rhizodeposition and as leaf litter which serve as raw materials for methanogenesis and thus higher methane emissions in wet season. Whereas, much of the carbon during dry season is more effectively used for seeding and active rice production and hence a lower methane emission and higher rice yield (Denier van der Gon et al. 2002; Sass et al. 1990). Thus it clear that submerged rice fields in wet season show increased methanogenesis and lower rice yield harvest index as compared to the dry season crops or crops obtained with intermittent irrigation and drainage.

5.3.4 *Climatic Factors*

The climatic factors exert a natural selection pressure on the paddy fields microflora and greatly influence their structural and functional diversity. Temperature, sunlight, and precipitation are the few climatic factors that mainly affect CH_4 emission from paddy fields either directly or indirectly. These factors affect the growth and activity of both the microorganisms and the rice plants besides affecting many physical and chemical processes that occur in the paddy field ecosystems. It affects mineralization of nutrients, process of methanogenesis, methane oxidation, transport and emission in the atmosphere. The effect of sunlight on CH_4 production and emission is due to its influence on the temperature and growth of the rice plants. The relation between solar radiation and grain yield is significant. Daily variations of CH_4 flux have showed good correlation with soil temperature at specific depth in some paddy fields (Schuetz et al. 1989). Dual cropped paddy fields generally show greater CH_4 flux for late rice than for early rice. Moreover, seasonal variation of CH_4 emission is more coincident with the temperature pattern than early rice. With increase in soil temperature (at 5-cm depth) from 20 to 35 °C in Italy and 18–31 °C in China, methane emissions measured from rice paddies increased rapidly (Holzapfel-Pschorn et al. 1986; Khalil and Rasmussen 1991).

Elevated methane production and emission in dry season has been related to a better rice growth and higher solar radiation (Neue and Roger 2000). Moreover, daily pattern of CH4 emission from paddy field is not controlled by plant metabolism, as proved by cutting rice shoots above the water level (Neue et al. 1992). However, precipitation seems to play a more important role on CH_4 emission because the hydrological condition of the field changes significantly in different years at different places.

5.3.4.1 Temperature

Sensitivity analysis reveals temperature as the most important parameter governing the activity of methanogens and methane emission rate (Chin and Conrad 1995; Liu and Wu 2004). Temperature is positively correlated with both diurnal and seasonal variations in CH4 flux (Schutz et al. 1989). 30–40 °C is regarded as the optimal temperature for the activity of majority of the methanogens (Neue and Scharpenseel 1984). CH4 production gets doubled as the temperature increased from 20 to 25 °C and raising of temperature by 10 °C in the temperature range of 15–30 °C increases the methane production rate by a factor of 2.5–3.5 (Schutz et al. 1989). The methanogenic bacterial population level increase as soil temperature rise gradually. Some studies have correlated variations in soil temperature to CH_4 emission during growing season (Schutz et al. 1989; Yagi and Minami 1990). Three seasonal maxima for methanogenesis have been reported- the first shortly after flooding, second during vegetative stage of rice plants, and third during the grain filling and maturation stage. Continuously flooded paddy field with N-application through urea showed significant positive correlation of CH_4 emission with soil temperature ($R^2 = 0.281$, $p < 0.05$) and variation of CH_4 flux with population of methanogens ($R^2 = 0.82$, $p < 0.05$) (Yue et al. 2005).

Reduction of incubation temperature of methanogenic rice soil from 30 to 15 °C reduced the CH_4 production rate and changes the degradation pathway of organic matter. Lower temperature decreases the steady-state partial pressure of H_2 and lead to transient accumulation of acetate, propionate, caproate, lactate, and isopropanol (Chin and Conrad 1995). At reduced temperature, acetate become increasingly more important methanogenic precursor and allow proliferation of fast growing acetoclastic *Methanosarcinaceae,* while low acetate concentrations at higher temperatures favours slow-growing but the better-adapted *Methanosaetaceae* (Chin et al. 1999; Fey and Conrad 2000). Incubation of anoxic rice field soils at 30 °C for 1 week showed abundance of *Crenarchaeota* whereas; relative abundance of *Methanosarcinaceae* further increased when the soil was incubated for another 2 weeks (Chin et al. 1999). Brief incubation rice field soil at 50 °C resulted in a pronounced inhibition of acetoclastic methanogenesis, but only in a relatively small group of acetoclastic methanogenic populations (Wu et al. 2001). However, Fey et al. (2001) showed that prolonged incubation at 50 °C caused a drastic change in the methanogenic microbial community and resulted in the dominance of Rice cluster I methanogens and production of CH_4 exclusively from H_2/CO_2.

5.3.4.2 Illumination

The closed chamber method showed practically no or only minor effects of illumination on CH_4 emissions when rice straw was either not applied or was incorporated in the soil. However, when the rice straw was surface-applied, illumination significantly reduced CH_4 emissions (Harada et al. 2005). Further, the amount of methanogenesis in the rice straw incubated under light was significantly lower than that incubated in the dark. Earlier, Yang and Chang (1998) reported suppressive effects of light exposure on CH_4 emission from soil slurries was due to the growth of algae in water, which reduced the CH_4 emission from paddy field. Actively growing Azolla was also reported to exert moderating effect on CH_4 flux from flooded soil by increasing the dissolved oxygen concentration at the soil-floodwater interface (Bharati et al. 2000). Earlier, Yang and Chang (1998) reported that suppressive effects of light exposure on CH_4 emission from soil slurries was due to the growth of algae in water, which reduced the CH_4 emission from paddy field. Actively growing Azolla was also reported to exert moderating effect on CH_4 flux from flooded soil by increasing the dissolved oxygen concentration at the soil-floodwater interface (Bharati et al. 2000). Methanogenesis in fact occurs even in the soil-floodwater interface in the presence of plant residues like rice straw. However, under such condition, methanogenesis is suppressed by the growth of indigenous phototrophs (algae, microphytes, macrophytes and anoxygenic photosynthetic bacteria) growing actively under illuminated condition. They increase the dissolved oxygen status of soil and thus play an important role in controlling CH_4 emission from rice paddies by suppressing the activity of methanogenic archaea and stimulating the methanotrophic microorganisms.

5.3.4.3 Water Regime

Wang et al. (2000) compared three different water regimes: local practice (drying of fields at 50–68 DAT and at 112–138 DAT); alternate flooding/drying (7 times drying: 12–16, 25–32, 44–50, 59–64, 73–78, 86–91, and 100–135 DAT); and continuous flooding (dry only at 32 day before harvest). The practice of alternate flooding/drying showed least CH_4 emission whereas; the local practice of mid-season drainage reported highest biomass and grain yield. However, the difference in grain yield was not significant although, it was at par with that of alternate flooding/drying. Aeration status of soil aggregates is a function of soil moisture. A comparative view of the rhizosphere of paddy plants grown under different water regimes like waterlogged, drained soil, and upland rice has been represented in the Fig. 5.2. Li et al. (2011) compared four different water regimes (early aeration, normal aeration, delayed aeration, and prolonged aeration) with respect to global warming potential of CH_4 and N_2O emissions and rice yields. In comparison to normal aeration of local practice, early aeration reduced CH_4 emission by 13.3–16.2% and increased N_2O emission by 19.1–68.8%; while delayed aeration reduced N_2O emission by 6.8–26.0% and increased CH_4 emission by 22.1–47.3%. They reported least emission of CH_4 and N_2O with prolonged aeration treatment; however, grain yield got reduced by 15.3% as compared to the normal practice. Midseason aeration around 1 month after seedling transplantation would optimize rice yields and may simultaneously limit global warming potential of CH_4 and N_2O. Kang et al. (2002) concluded that CH_4 emission from rice fields could also be mitigated by improving soil water regime between rice crops. Therefore,

Fig. 5.2 Rhizospheric regions of paddy plants from (**a**) waterlogged soil, (**b**) drained soil, and (**c**) upland rice [Note the prolific growth of roots of paddy from lowland rice, more numerous and thinner roots from drained soil and poor roots growth in case of upland rice. The green coloured cyanobacterial mats are present on the surface of root zone of paddy plants from lowland; whereas, it is absent in the case of drained and upland rice plants.]

preventing water logging, keeping rice fields well-drained and drying soils as much as possible in the winter would be options to mitigate CH_4 fluxes from rice fields.

5.3.5 Intermittent Irrigation

As mentioned earlier, anaerobic condition is a prerequisite for CH_4 production in paddy fields. However, drainage may eliminate this prerequisite and eventually may reduce CH_4 emission (Watanabe et al. 2010; Zhang et al. 2011). In case of permanently flooded field, drainage is effective in preventing CH_4 emission not only from rice fields in the fallow season, but also during the following rice season. The total CH4 emission from permanently flooded rice fields during the rice season were reported 1.2–4.8 times that of the rice fields drained in the previous fallow season (Cai et al. 2000; Kang et al. 2002). Drainage of the paddy field in fallow season reduced CH_4 emission during the rice season by 13–42% and the annual total CH_4 emission by 48–68% (Cai et al. 2003; Jiang et al. 2006). However, Zhang et al. (2011) reported respectively 42–61% and 56% reduction of mean CH4 production from the rice soil and rice roots during the rice growing season in drained field as comparison to the flooded field.

The plant photosynthates and decaying rice roots are important carbon sources for CH4 production (Conrad and Klose 2005; Dannenberg and Conrad 1999). Under reduced soil water content in the fallow season and the resultant lack of anaerobic condition, the methane production from paddy fields also gets reduced (Xu et al. 2003). CH_4 production rate in rice roots was also significantly lower in drained fields than in permanently flooded rice fields at four rice growth stages namely; rice tillering, booting, grain filling and ripening stages. The root exudate and plant debris (organic carbon) supplied by the rice plants to the soil, especially at the reproductive and ripening stages play a significant role in second or third peaks of the seasonal variation of CH_4 emission (Wang 2001). During rice growing season, soils in the paddy plots remain constantly under strong reductive condition, which may increase the anoxic condition of soils around the rice roots. Under such condition growth and activity of rice roots would be limited and roots would begin to age and decay comparatively earlier (Chen et al. 2007; He et al. 2008).

Water-saving practice like alternate wetting and drying brings about changes in both the community structures and transcriptional activities of methanogenic archaea (Watanabe et al. 2010). The principal component analysis and sequencing analysis of 16S rDNA indicated the dominance of members of *Methanosarcinales* in continuously flooded field and *Methanocellales* in the field under alternate wetting and drying. Intermittent irrigation shows a reduction in the number of methanogenic archaeal community (Yue et al. 2005) due to abundant entry of atmospheric O_2 into the surface layer of submerged paddy soils, while methanotrophic bacteria population increase and result in reduced CH_4 emission. Hence, more reduced nature of soils in permanently flooded rice fields than in drained fields over the rice season leads to greater decomposition rates in permanently flooded rice fields than in drained fields, causing higher CH_4 production. Therefore, reduction of CH_4 production in the rice roots in drained fields in the fallow season decreases the CH_4 emission to some extent in comparison to that of the permanently flooded rice fields.

5.3.6 Effects of Cultivars on Methanogenesis

Over 80,000 rice cultivars are known throughout the world which varies in genotype and phenotype (Jia et al. 2002). Lower CH_4 emission was observed in the fields of Sichuan province planted with hybrid rice than that with normal variety (Schuetz et al. 1989; Wang 2001). A comparison of three temperate rice cultivars reported higher CH_4 fluxes with the cultivars Jingyou (japonica hybrid) and Zhonghua (tall japonica), whereas CH_4 emission from Zhongzhuo (modern japonica) was comparatively lower. Among these temperate varieties, Zhongzhuo had the lowest emission rates and the highest yield (Wang et al. 2000). The differentiation in the CH_4 emission potential of different rice cultivars gets reflected only in the second week after transplanting, when CH_4 emission start to increase. Cultivar level variation is also influenced by the CH_4-oxidizing activity of paddy fields. Rice cultivar IR65598 showed comparatively higher CH_4 oxidizing activity as compared to the cultivars IR72 and Chiyonishiki (Wang and Adachi 2000). They observed a significant difference in the population level of methanogenic bacteria in soil grown (at booting and ripening stages) and methanotrophic bacterial population in rice roots (at ripening stage) to different rice cultivars.

Rice plant is supposed to affect CH_4 emission due to their influence on the processes of CH_4 production, consumption and transport. With no organic amendment, root exudation and root death (rhizodeposition) contribute about 380 kg C ha $^{-1}$ of methanogenic substrate over the season; representing 37% of the total methanogenic substrate from all sources (Matthews et al. 2000). Additionally, about 225 kg C ha^{-1} (22%) is predicted to come from residues from previous crop. Thus rice crop contribute a sum total of about 59%, while the remaining 41% comes from the humic fraction of the soil organic matter. However, the cultivar based differences in CH_4 emission rates become significant only in the middle and late growth stages of rice plants, and accordingly, the production of root exudates varied significantly among the cultivars tested (Wang and Adachi 2000). Significant positive correlation of root exudates to root dry matter production was also observed among rice cultivars. The root exudate and plant debris (organic carbon) supplied by the rice plants to the soil is supposed to be the predominant factor responsible for peaks in CH_4 production during the reproductive and ripening stages of the rice (Lu et al. 2000; Wang 2001). Various cultivars differ in their ability to produce biomass due to variation in their morphological, physiological, and biochemical attributes; which might be one of the reasons for higher CH_4 emission potential with some cultivars. Root morphology and physiology vary significantly among rice cultivars; hence, different niche conditions may develop around the roots of different cultivars. Moreover, rice cultivars with few or practically negligible unproductive tillers, small root system, high root oxidative activity, and high harvest index are supposed to be ideal for mitigating CH_4 emission in rice fields.

5.3.7 Influence of Readily Mineralizable Organic Amendments

Excessive application of nitrogen-rich organic amendments may however, supply available carbon and hence enhances the risk of global warming (Qin et al. 2010; Win et al. 2010). It is noteworthy to mention that Azolla and a large number of microalgae grow naturally and are also used for better crop growth and yield (Singh and Dhar 2011; Singh and Patel 2012). These microphytes besides having high ptotosynthetic efficiency fix atmospheric nitrogen and serve as a good source of readily mineralizable nitrogen, phosphates, potash and a large number of micronutrients including growth factors (Singh and Dhar 2006, 2007). These cyanobacteria provide multiple benefits, depict a promising multifaceted bioinoculants in organic farming practices, and are useful pointers for improving cultivation practices and establishment of plants in inhospitable habitats (Singh et al. 2016a). Although, organic amendments in general enhance CH_4 emission, application of compost increases CH_4 emission comparatively lower than does rice straw (Yagi and Minami 1990). Applications of fermented organic fertilizers from biogas generators also produce less CH_4 than the fresh manure (Wang et al. 1993). Amendment with readily mineralizable soil organic matter can affect an immediate increase in CH_4 emission up to 400 $mgCH_4m^{-2}$ day^{-1} which gets reduced after 10 days of incubation to below 100 $mgCH_4m^{-2}day^{-1}$; but remained consistently at a higher level as compared to the treatments with inorganic fertilizers (Adhya et al. 2000). Fertilizer from biogas pits often contains comparatively less amount of easily decomposable carbon than the fresh manure (Wang et al. 2000). The organic matter having high amount of readily mineralizable organic carbon serves as the main source of fermentation products, which may be driven to CH_4 by strict methanogens in flooded soils and sediments.

Organic amendments with low C/N are supposed more beneficial in terms of enhancing crop yields and reducing CH_4 emissions (Matthews et al. 2000). This may be due to higher rates of C immobilization into microbial biomass, removing it temporarily as a methanogenic substrate. Use of biofertilizers and readily utilizable organic amendments are advocated to sustain crop productivity and soil health (Singh et al. 2016b). Of animal manure, rice straw, compost, and green manure having C/N ratio of 100, 51, 12, and 10, respectively; green manure proved most desirable for soil amendment in paddy fields. This may be due to increase in grain yield and comparatively less increase in CH_4 emissions than rest of the treatments. However, animal manure appeared to be the worst option in terms of reducing CH_4 emissions (Matthews et al. 2000). Seasonal CH_4 fluxes (cumulative) from pig manured plots exceeded over the plots supplemented with ammonium sulfate fertilizer by a factor of 15–35 (Wang et al. 2000). The average fluxes observed from the paddy fields were 139, 31, 102, and 04 mg CH_4 m^{-2} day^{-1} respectively, in plots treated with pig manure, cattle manure, rice straw, and pure mineral fertilizer.

Organic manure greatly promotes CH4 emissions from paddy field as compared with mineral fertilizers. Matthews et al. (2000) compared methane emission from paddy fields when fertilized with urea alone and urea along with each of *Sesbania, Azolla*, and compost. They observed maximum methane emission with urea +*Sesbania* (212% increase); followed by urea+*Azolla* (61% increase), urea+compost (54% increase), and urea alone. Incorporation of vetch as a green manure has been regarded effective in improving soil fertility and rice productivity with a yield comparable to that of recommended chemical fertilization (Lee et al. 2010, 2011). Organic amendment increases CH_4 oxidation potential of the paddy field soil while nitrogenous fertilizer inhibits the process. Thus accelerating the process of CH_4 oxidation can be a feasible approach to mitigate CH_4 emission from wetland.

5.3.8 Application of Rice Straw and Methane Production

Application of rice straw improves the soil fertility and increases N_2 fixation, N uptake and crop yield (Takahashi et al. 2003; Tanaka et al. 2010). However, it also increase methane emission and influence successions of bacterial population in the paddy fields (Asari et al. 2007; Rui et al. 2009). Photosynthates released from rice roots (rhizodeposition) that are immediately utilized by the rhizospheric microorganisms, organic materials from rice roots, and rice straws incorporated in soil are the major sources of methane emitted from paddy fields (Kimura et al. 2004). Almost half amount of plant materials in soil gets decomposed before they reach a size less than 1 mm. Harada et al. (2005) observed a more active methanogenesis at the soil-floodwater interface in the surface-applied rice straw and the surrounding soil compared to the soil incubated without rice straw. In straw amended paddy wetland about 84–89% of the released methane comes from acetoclastic methanogenesis whereas, hydrogenotrophic methanogenesis accounts for about 11–27% (Glissmann and Conrad 2000). Methane production in straw amended fields starts after 8 days of incubation and reaches a stable state only after 20 days (Weber et al. 2001). Soil amendment with rice straw (@ 2 t ha^{-1}) significantly increased CH_4 production under both continuously flooded and intermittently flooded field plots (Adhya et al. 2000). However, higher grain yield and less amount of CH_4 released per tonne of grain yield were noticed under rice straw-amended intermittently flooded paddy field plots.

Application of paddy straw influence successions of bacterial population with fast-growing bacteria dominating at the beginning while the slow-growing one at the later stages (Asari et al. 2007; Rui et al. 2009; Sugano et al. 2005). The most representative methanogenic bacteria that colonize the paddy straw belong to the families Methanosaetaceae and Methanomicrobiaceae. The methanogenic bacterial diversity in the fertilized paddy soil was less than the unfertilized one (Weber et al. 2001). This may be due to the fact that paddy straw is predominately a habitat for fermentative bacteria whereas; methanogenic bacteria are more abundant in the soils around fertilizers, where the fermentation products get concentrated and serve as methanogenic substrate. The change in microbial communities of these habitats may depend upon the plant growth, decomposition of rice straws, plant residues and change in symbiotic and competitive relationships between methanogenic archaea and eubacteria during the cultivation period and thereafter.

5.3.9 Application of Biochar to Soil

Biochar is suggested to play a significant role in reducing greenhouse gases emissions from agricultural soils (Renner 2007; Yanai et al. 2007), improving sorption and desorption of pesticides (Wang et al. 2010; Zhang et al. 2010b), reducing leaching loss of nutrients (Hua et al. 2009), improving fertility status of soil (Major et al. 2005; Steiner et al. 2007) and boosting plant growth and crop yield (Steiner et al. 2007). Almost complete suppression of CH_4 emissions in ground biochar amended soils (@20 g Kg^{-1}) has been reported from the Eastern Colombian Plains (Rondon et al. 2005). CH_4 emissions from the paddy field soil amended with bamboo char and straw char at high rate reduced respectively by 51.1% and 91.2%, compared to control (without biochar) (Liu et al. 2011). The reduction of CH4 emissions from paddy soil with biochar may be due to the inhibition of methanogenic activity or stimulation of methylotrophic activity during incubation. Biochar amendment of paddy soil also resulted in significant reduction of CO_2 emission (Liu et al. 2011; Zhang et al. 2010a). Rondon et al. (2006) also demonstrated a net reduction in annual emissions of CH_4 and increase in soil carbon from a non-fertile tropical soil with the use of biochar derived from mango trees.

The effect of biochar on CH_4 emissions from paddy fields depend on the physico-chemical properties of the biochar, types of soils, microbiological considerations, and water and nutrients management (Cai et al. 1997; Xiong et al. 2007; Zou et al. 2007; Zwieten et al. 2009). Soil pH is one of the most important parameter that affects CH_4 emission rates from paddy soil. The methanogenic archaea prefer a near neutral pH (6.5–7.5). Hence, addition of bamboo char (pH 9.81) or straw char (pH 10.2) may result in reduced methanogenesis and hence less emission of CO_2 from the paddy soil (Liu et al. 2011). Moreover, higher C/N ratio of biochar may stabilize microbial biomass which may be the reason for reduced mineralization rates in soil supplemented with biochar. Therefore, CH_4 emission can be avoided by converting straw to biochar instead of directly using it. Moreover, the nature of biochar, the local soil condition, and the environmental factors must be considered before using this input as a mitigation strategy for greenhouse gases.

5.3.10 *Application of Mineral Fertilizers*

Soil organic carbon, available NPK, and micronutrients are regarded predominant factors affecting both crop productivity and bacterial community structure (Wu et al. 2011). Some researchers have focused on the responses of some functional microbes like denitrifiers, methanotrophs, and diazotrophic bacteria to fertilization in paddy soil (Chen et al. 2010; Qiu et al. 2008). Long-term fertilization is reported to significantly affect the population of free-living diazotrophs and methanotrophs in paddy fields (Islam et al. 2010; Zheng et al. 2008). Although application of some mineral fertilizers have been helpful in reducing CH_4 emission from paddy fields, they do not show a clear cut pattern and varying results have been reported (Lu et al. 2000; Wang 2001; Wang et al. 2000). Nitrate, sulfate, and ferric iron favours respectively the nitrate reducers, sulfate reducers, and iron reducers which compete successfully for the methanogenic substrates and suppress methanogenesis.

5.3.10.1 Nitrogenous Fertilizers

Nitrogen is the most important nutrient required by plants for proper growth and development. It serves as building blocks of several carbohydrates, lipids, proteins and many other products. Nitrogen fertilizer influence the root physiology of paddy plants by changing the exudation pattern of organic substance and H^+/OH^- ions and therefore, the substrates and pH. With increase of N fertilizer in paddy field, an increase in the relative abundance of hydrogenotrophic methanogens on the rice roots has been observed (Wu et al. 2009). Application of N fertilizers, especially ammonical fertilizer can inhibit the process of CH_4 oxidation. Moreover, $(NH_4)_2SO_4$ is reported to reduce CH_4 emission more efficiently than NH_4HCO_3 and $(NH_4)_2HPO_4$ (Wang et al. 2000). Addition of N-compounds (nitrate and its denitrification products- nitrite, NO, N_2O) result in a largely reversible inhibition of methanogenesis. These reduce the H_2 partial pressure well below the threshold limit required by methanogens, and hence, do not allow exergonic production of methane (Kluber and Conrad 1998). Addition of nitrate and N_2O to paddy fields leads to oxidation of reduced iron and sulfur species and production of electron acceptors, which may be helpful in CH_4 oxidation. As a result, the methanogenic activity do not resume until all electron acceptors gets reduced, and hence, H_2 reach the threshold level for methanogenic archaea. Thus, competitions of methanogens with denitrifying bacteria, iron- and sulfate-reducing bacteria not only out-compete methanogens for H_2 and acetate but also generate N_2O and NO toxic to methanogens (Kluber and Conrad 1998).

Methanogens associated with rice roots may encounter nitrogen limitation due to severe competition between paddy plants and microbes for nutrients (Bodelier et al. 2000). Application of ammonium-based fertilizers at the commonly-adopted levels in China (150 or 250 kg N ha^{-1}) generally inhibited accumulative CH4 emission during rice season by about 28–30% as compared to no N addition across various climate zones. However, an increase in the application of ammonium-based fertilizers from the moderate level of 150 kg N ha^{-1} to the high rate of 250 kg N ha^{-1} did not significantly modify CH4 emission (Xie et al. 2010). The use of anaerobically digested cattle slurry for mitigation of ammonia volatilization also reduces the risk of nutrients being leached down. It did not increase the CH4 emission, and thereby, did not affect the C balance in the paddy field and the nutrients can effectively be taken up by plants (Win et al. 2010). Therefore, the application of N fertilizer may stimulate the activity of hydrogenotrophic methanogens associated with rice roots.

5.3.10.2 Phosphatic Fertilizers

Nirmal Kumar and Viyol (2009) suggested that greater phosphate content of the soil suppressed the methane flux from rice field. They observed a higher negative correlation ($r = -0.416$) of methane flux with phosphate content of the paddy field. High phosphate concentration is known to delay the development of methanosarcinal populations and the acetoclastic methanogenesis (Chin et al. 2004; Conrad et al. 2000). However, the exact reason behind such reduced methanogenesis is unknown and a confusion exist; whether this effect is only due to the inhibition of enzyme activity or also due to inhibition of growth of methanogens. It is further assumed that phosphate addition to rice paddies will decrease but do not stop the CH_4 production and the extent of such suppression depends on the concentrations of both the acetate and phosphate.

5.3.10.3 Application of K-fertilizers

K is regarded as one of the three primary nutrients and a deficiency of it in soil can adversely affect the plant performance. Kirk and Bajita (1995) have reported the utility of adequate K nutrition in maintaining the oxidizing power of rice roots. Potassium application is known to alleviate the extreme reducing conditions and associated imbalances in rice plants including iron toxicity (Chen et al. 1997). It reduces CH_4 emission in paddy fields by preventing a drop in soil redox potential, inhibiting methanogenic bacteria and simultaneously stimulating methanotrophic bacterial population (Jagadeesh Babu et al. 2006). Supplementation of K-deficient soils with K promoted root growth and enhanced α-naphthylamine oxidase activity (Chen et al. 1997). The α-naphthylamine oxidase activity of rice roots is used as an index of the oxidation status of the rhizosphere region and is significantly correlated with CH_4 efflux (Satpathy et al. 1998). Jagadeesh Babu et al. (2006) observed comparatively higher overall root oxidase activity with K application. The low CH_4 flux from K-amended plots may also be due to an impact of higher α-naphthylamine oxidase activity, indicating an overall higher oxidation status. Therefore, K amendment can correct nutritional imbalance (especially in K-deficient soils) and increase grain yield besides being an effective mitigation option for CH_4 emission due to higher α-naphthylamine oxidase activity.

5.3.10.4 Use of Sulfate Containing Fertilizers

Use of sulfate containing fertilizers like ammonium sulfate [$(NH_4)_2SO_4$] or soil amendments like gypsum ($CaSO_4$) are considered viable mitigation options to reduce CH_4 emission from paddy fields (Masscheleyn et al. 1993). The strictly anaerobic methanogenic bacteria mainly use acetate (CH_3COO^-) and H_2/CO_2 as substrates (Kristjansson et al. 1982) but have to compete with other microorganisms that use the same substrates as electron donor but have the capability to use sulphate as alternative electron acceptor. Sulfate reducing bacteria have slightly higher affinity for acetate than methanogens. They reduce the steady-state concentrations of H_2 to a level that is too low for methanogens to remain active, and hence, suppress the CH_4 production in paddy fields, but do not completely out-compete the methanogens for substrate (Denier van der Gon et al. 2001; Kristjansson et al. 1982). However, after reduction of SO_4^{2-} methanogens will start producing methane, which might be the reason for higher methane emission from rice fields at lower content of sulphate. Sulfate reducing bacteria need 1 mol of SO_4^{2-} to inhibit the production of 1 mol of CH_4 (Denier van der Gon et al. 2001). Methane emission from paddy field showed higher negative correlation ($r = -0.476$) with sulphate content of soil (Nirmal Kumar and Viyol 2009). Therefore, SO_4^{2-} addition to rice paddies reduces but do not stop the CH_4 production and the extent of such suppression depends on the concentrations of both the acetate and the SO_4^{2-}.

5.3.10.5 Application of Slag-Type Silicate Fertilizer

Silicate fertilizers are the by-products of steel industry, release ferric ions in soil solution, act both as oxidizing agents and electron acceptors, control the production of organic acids, enhanced methane oxidation and suppress the CH_4 emission (Furukawa and Inubushi 2002). Application of silicate fertilizer (@ 4 $Mgha^{-1}$) in tillage and no-tillage plots, reduced the total seasonal CH4 flux by about 20% and 36%, respectively; while grain yields increased by about 18% and 13%, respectively (Ali et al. 2009). However, Lee et al. (2011) observed that silicate fertilizers application @ 2.3 $Mgha^{-1}$ in rice field reduced the seasonal CH_4 flux by about 14.5% and increased rice yield by about 15.7% over control. However, the ratio of easily oxidizable organic matter to easily reducible Fe must be managed in order to get the best result (Conrad 2002). Silicate fertilizers improve soil nutrient balance by significantly increasing the soil pH, available phosphate and silicate, exchangeable calcium concentration in soil, and resistance to biotic and abiotic stresses (Ali et al. 2008; Takahashi et al. 1990) and thus could be good soil amendment to reduce CH_4 emission as well as sustaining rice productivity. However, increase of yield may partly be due to release of inorganic nutrients from the furnace slag and partly due to nutrients from the soil. Silicate fertilizers increase the soil pH, and hence, the availability of phosphate in acidic soil, which in turn further suppress the methane flux.

5.3.10.6 Availability of Fe(III) as Alternative Electron Acceptor

The inhibition of methane production in sediments in presence of poorly crystalline Fe(III) oxides is reported to inhibit methane production (Chidthaisong and Conrad 2000; Frenzel et al. 1999). Such inhibition is associated with reduction in the concentrations of acetate and hydrogen, the two primary electron donors for methanogenesis in sediments (Conrad 1999). However, addition of acetate or hydrogen to the sediments may result in resumption of the process of methane production. Fe(III), as such, is not considered toxic to the methanogens, but the inhibition of methanogenesis was due to the resurgence of Fe(III)-reducing microorganisms that maintained hydrogen and acetate at levels too low for methanogenesis to be thermodynamically favourable for methanogens (Lovley and Phillips 1987). Thus, availability of poorly crystalline Fe(III) as an alternative electron acceptor for microbial respiration can be a major factor suppressing methane production in lowland paddy cultivation.

5.3.10.7 Application of Urease Inhibitor and Nitrification Inhibitor

Urea, the dominant form of N fertilizer applied to rice in Asia, is subjected to various forms of loss and lead to a drop in redox potential of the soil and results in higher CH_4 production (Bharati et al. 2000). Urea is converted to ammonia by the enzyme urease. This ammonia is converted to nitrate via nitrite by a process called as nitrification. The application of urease and nitrification inhibitor prevents/delays the release of available forms of nitrogen in the soil and thus prevents its loss and suppresses the rapid release of methane from paddy fields (Xu et al. 2002). Application of a urease inhibitor (hydroquinone) and a nitrification inhibitor (dicyandiamide) together with urea reduced the total emission of N_2O and CH_4 respectively by about 1/3 and 1/2 as compared to that in control (only urea) (Xu et al. 2002).

Sodium azide, dicyandiamide, pyridine, aminopurine, ammonium thiosulfate, and thiourea are few nitrification inhibitors that are used in different parts of the world. The nitrification inhibitor-based inhibition of CH_4 production followed the order: sodium azide > dicyandiamide > pyridine > aminopurine > ammonium thiosulfate > Thiourea (Bharati et al. 2000). Nitrification inhibitors like calcium carbide and nitrapyrin have been reported to inhibit CH_4 emission from flooded soil planted to rice besides controlling N losses (Bronson and Mosier 1991; Keerthisinghe et al. 1993). Nitrification inhibitors inhibit the process of nitrification and oxidize most of the methane produced in flooded soils to CO_2 before being released into the atmosphere, and thus, it plays an important role in the biogeochemical cycling of methane. Although, several nitrification inhibitors regulate CH_4 production differentially in flooded soil, the exact mechanism of such inhibition on the selected microorganisms is still a matter of investigation.

5.4 Conclusion

A complex set of parameters including various climatic factors, physico-chemical properties of soil and cultural practices among many others influence the CH_4 production and emission from paddy fields. The immense variability of environmental factors affecting the 140 million hectares of annually harvested rice fields denies the use of blanket strategies to reduce methane emission. However, methane emission from water-logged paddy fields can be reduced by adoption of one or more mitigation strategies like managing organic inputs in soil, judicious use of nitrogen fertilizer, improved irrigation practice, use of improved crop cultivars etc. Reduction, early incorporation of straw in soil, or organic amendment of soil with rice straw manure derived from aerobic composting techniques significantly reduces emissions as compared to fresh straw (Corton et al. 2000). Application of sulfur coated urea, ammonium sulfate $(NH_4)_2SO_4$ and Gypsum ($CaSO_4$) increase competition between sulfate reducing and methanogenic bacteria and reduce methane emission from paddy fields. Gypsum application proves effective in neutralizing the pH of alkaline soils. Application of nitrification inhibitors like encapsulated calcium carbide and dicyandiamide can drastically reduce N losses, help enhance fertilizer N-use efficiency, and reduces the emission of N_2O, another greenhouse gas.

Water management practices like shallow flooding, midseason drainage and intermittent irrigation conserve water and increase yields and are considered one of the most promising mitigation option suitable for reducing methane emissions in irrigated rice fields. Midseason drainage decreases the net global warming potential (GWP) of paddy fields as long as nitrogen is applied in appropriate doses. However, high dose application of nitrogen fertilizer may offset the reduction of methane emission by encouraging emission of N_2O, another green house gas. Improvement of crop cultivars through breeding for specific traits like short duration, 'aerobic' rice varieties, and increased yield potential enhance plant performance and reduce methane emission. However, water management may prove to be the most promising mitigation option if irrigation water is available sufficiently and irrigation/drainage systems are well established (Wang et al. 2000).

Flooded rice paddies represent a microbiologically highly complex agroecosystem in which diverse groups of microorganisms like hydrolytic, cellulolytic, fermentative, homoacetogenic, syntrophic acetate and H_2 utilizing sulfate reducing bacteria and methanogenic archaea interact with each other. Detailed research need to be carried out to understand the effect of different factors mentioned above on adaptive mechanisms, physiological responses, and microbial community structural changes of the functional groups of microorganisms. Culture-independent approaches like analysis of DNA extracted directly from the environmental samples may be used for describing bacterial communities in such a complex agroecosystem. There is an urgent need of mitigating emissions from paddy fields at the national and international level using funding through 'Clean Development Mechanism' projects, introduced in Kyoto Protocol or from other resources. For mitigation programs to be successful, reduction of emissions should be concomitant with other benefits like higher yields, less fertilizer and water needs, and targeting both N_2O and CO_2 beside methane. However, further research is needed in order to combine geographic information, green house gas emission models, crop yield models, and socio-economic information to develop site-specific packages of mitigation technologies.

Acknowledgements Authors humbly acknowledge the assistance provided by the Vice Chancellor, S.D. Agricultural University (Gujarat, India) for preparation of this manuscript. This article does not attract any conflict of interest among the authors/institutions.

References

Adhya TK, Pattanaik P, Sathpathy SN, Kumaraswamy S, Sethunathan N (1998) Influence of phosphorus application on methane emission and production in flooded paddy soils. Soil Biol Biochem 30:177–181. https://doi.org/10.1016/S0038-0717(97)00104-1

Adhya TK, Bharati K, Mohanty SR, Ramakrishnan B, Rao VR, Sethunathan N, Wassmann R (2000) Methane emission from rice fields at Cuttack, India. Nutr Cycl Agroecosyst 58:95–105. https://doi.org/10.1023/A:1009886317629

Ali MA, Lee CH, Kim PJ (2008) Effect of silicate fertilizer on reducing methane emission during rice cultivation. Biol Fertil Soils 44:597–604. https://doi.org/10.1007/s00374-007-0243-5

Ali MA, Lee CH, Kim SY, Kim PJ (2009) Effect of industrial by-products containing electron acceptors on mitigating methane emission during rice cultivation. Waste Manag 29 (10):2759–2764. https://doi.org/10.1016/j.wasman.2009.05.018

Anastasi C, Dowding M, Simpson VJ (1992) Future CH_4 emissions from rice production. J Geophys Res 97:7521–7525

Asakawa S, Kimura M (2008) Comparison of bacterial community structures at main habitats in paddy field ecosystem based on DGGE analysis. Soil Biol Biochem 40:1322–1329. https://doi.org/10.1016/j.soilbio.2007.09.024

Asari N, Ishihara R, Nakajima Y, Kimura M, Asakawa S (2007) Succession and phylogenetic composition of eubacterial communities in rice straw during decomposition on the surface of paddy field soil. Soil Sci Plant Nutr 53(1):56–65. https://doi.org/10.1111/j.1747-0765.2007.00110.x

Aulakh MS, Bodenbender J, Wassmann R, Rennenberg H (2000) Methane transport capacity of rice plants. I. Influence of methane concentration and growth stage analyzed with an automated measuring system. Nutr Cycl Agroecosyst 58:357–366. https://doi.org/10.1023/A:1009831712602

Aulakh MS, Wassmann R, Rennenberg H (2001) Methane emissions from rice fields-quantification, mechanisms, role of management, and mitigation options. Adv Agron 70:193–260

Bharati K, Mohanty SR, Singh DP, Rao VR, Adhya TK (2000) Influence of incorporation or dual cropping of Azolla on methane emission from a flooded alluvial soil planted to rice in eastern India. Agric Ecosyst Environ 79:73–83

Bloom A, Swisher M (2010) Emissions from rice production. In: Cleveland CJ (ed) Encyclopedia of earth, Washington, DC. http://www.eoearth.org/article/Emissions from RiceProduction?topic=54486. Accessed on 11 July 2012

Bodelier PLE, Roslev P, Henckel T, Frenzel P (2000) Stimulation by ammonium-based fertilizers of methane oxidation in soil around rice roots. Nature 403:421–424. https://doi.org/10.1038/35000193

Bronson KF, Mosier AR (1991) Effect of encapsulated calcium carbide on dinitrogen, nitrous oxide, methane and carbon dioxide emissions from flooded rice. Biol Fertil Soils 11:116–120. https://doi.org/10.1007/BF00336375

Cai ZC, Xing GX, Yan XY, Xu H, Tsuruta H, Yagi K, Minami K (1997) Methane and nitrous oxide emissions from rice paddy fields as affected by nitrogen fertilizers and water management. Plant Soil 196:7–14

Cai ZC, Tsuruta H, Minami K (2000) Methane emission from rice fields in China: measurements and influencing factors. J Geophys Res 105:17231–17242

Cai ZC, Tsuruta H, Gao M, Xu H, Wei CF (2003) Options for mitigating methane emission from a permanently flooded rice field. Glob Chang Biol 9:37–45

Chen J, Xuan J, Du C, Xie J (1997) Effect of potassium nutrition of rice on rhizosphere redox status. Plant Soil 188:131–137

Chen HZ, Zhu DF, Lin XQ, Zhang YP (2007) Effects of soil permeability on root growth and nitrogen utilization in rice. Chin J Eco-Agric 15:34–37

Chen G, Zheng Z, Yang S, Fang C, Zou X, Zhang J (2010) Improving conversion of *Spartina alterniflora* into biogas by co-digestion with cow feces. Fuel Process Technol 91:1416–1421

Chidthaisong A, Conrad R (2000) Specificity of chloroform, 2-bromoethanesulfonate and fluoroacetate to inhibit methanogenesis and other anaerobic processes in anoxic rice field soil. Soil Biol Biochem 32:977–988

Chin KJ, Conrad R (1995) Intermediary metabolism in methanogenic paddy soil and the influence of temperature. FEMS Microbiol Ecol 18:85–102

Chin KJ, Lukow T, Conrad R (1999) Effect of Temperature on structure and function of the methanogenic archaeal community in an anoxic rice field, soil. Appl Environ Microbiol 65:2341–2349

Chin KJ, Lueders T, Friedrich MW, Klose M, Conrad R (2004) Archaeal community structure and pathway of methane formation on rice roots. Microb Ecol 47:59–67

Cicerone RJ, Shetter JD (1981) Sources of atmospheric methane: measurements in rice paddies and a discussion. J Geophys Res 86:7203–7209

Clegg CD, Lovell RDL, Hobbus PJ (2003) The impact of grassland management regime on the community structure of selected bacterial groups in soil. FEMS Microbiol Ecol 43:263–270

Conrad R (1999) Contribution of hydrogen to methane production and control of hydrogen concentrations in methanogenic soils and sediments (review). FEMS Microbiol Ecol 28:19–202

Conrad R (2002) Control of microbial methane production in wetland rice fields. Nutr Cycl Agroecosyst 64:59–69. https://doi.org/10.1023/A:1021178713988

Conrad R, Klose M (2005) Effect of potassium phosphate fertilization on production and emission of methane and its 13C-stable isotope composition in rice microcosms. Soil Biol Biochem 37:2099–2108

Conrad R, Klose M, Claus P (2000) Phosphate inhibits acetotrophic methanogenesis on rice roots. Appl Environ Microbiol 66:828–833

Corton TM, Bajita J, Grospe F, Pamplona R, Wassmann R, Lantin RS (2000) Methane emission from irrigated and intensively managed rice fields in Central Luzon (Philippines). In: Wassmann R, Lantin RS, Neue HU (eds) Methane emissions from major rice ecosystems in Asia, Developments in plant and soil sciences, vol 91. Springer, Dordrecht

Dannenberg S, Conrad R (1999) Effect of rice plants on methane production and rhizospheric metabolism in paddy soil. Biogeochemistry 45:53–71. https://doi.org/10.1007/BF00992873

Dubey SK (2001) Methane emission and rice agriculture. Curr Sci 81:345–346

FAOSTAT Database (2008) FAO, Rome. 22 September 2008. http://beta.irri.org/solutions/index.php?option=com_content&task=view&id=250

Feng JN, Hsieh YP (1998) Sulfate reduction in freshwater wetland soils and the effects of sulfate and substrate loading. J Environ Qual 27:968–972

Fey A, Conrad R (2000) Effect of temperature on carbon and electron flow and on the archaeal community in methanogenic rice field soil. Appl Environ Microbiol 66:4790–4797

Fey A, Chin KJ, Conrad R (2001) Thermophilic methanogens in rice field soil. Environ Microbiol 3:295–303

Fierer N, Jackson RB (2006) The diversity and biogeography of soil bacterial communities. Proc Natl Acad Sci U S A 103:626–631. https://doi.org/10.1073/pnas.0507535103

Frenzel P, Bosse U, Janssen PH (1999) Rice roots and methanogenesis in a paddy soil: ferric iron as an alternative electron acceptor in the rooted soil. Soil Biol Biochem 31:421–430

Furukawa Y, Inubushi K (2002) Feasible suppression technique of methane emission from paddy soil by iron amendment. Nutr Cycl Agroecosyst 64:193–201

Glissmann K, Conrad R (2000) Fermentation pattern of methanogenic degradation of rice straw in anoxic paddy soil. FEMS Microbiol Ecol 31:117–126

Hadas O, Pinkas R (1995) Sulfate reduction processes in sediments at different sites in Lake Kinneret, Israel. Microb Ecol 30:55–66

Harada N, Nishiyama M, Otsuka S, Matsumoto S (2005) Effects of inoculation of phototrophic bacteria on grain yield of rice and nitrogenase activity of paddy soil in a pot experiment. Soil Sci Plant Nutr 51:361–367

Hattori C, Ueki A, Seto T, Ueki K (2001) Seasonal variations in temperature dependence of methane production in paddy soil. Microbes Environ 16:227–233

He JZ, Zheng Y, Chen CR, He YQ, Zhang LM (2008) Microbial composition and diversity of an upland red soil under long-term fertilization treatments as revealed by culture-dependent and culture-independent approaches. J Soils Sediments 8:349–358

Holzapfel-Pschorn A, Conrad R, Seiler W (1986) Effects of vegetation on the emission of methane from submerged paddy soil. Plant Soil 92:223–233

Hua L, Wu W, Liu Y, McBride MB, Chen Y (2009) Reduction of nitrogen loss and Cu and Zn mobility during sludge composting with bamboo charcoal amendment. Environ Sci Pollut Res 16:1–9

Inubushi K, Muramatsu Y, Umebayashi M (1992) Influence of percolation on methane emission from flooded paddy soil. Japan J soil Sci Plant Nutr 63:184–189

Inubushi K, Sugii H, Nishono S, Nishino E (2001) Effect of aquatic weeds on methane emission from submerged paddy soil. Am J Bot 88:975–979

IPCC (1996) XII Summary for policy makers. In: Houghton IT, Meira F, Callander LG, Harris BA, Kattenberg A, Maskell K (eds) Climate change 1995: the scientific basis of climate. Cambridge University Press, Cambridge, p 572

IPCC (2007) Climate change 2007: couplings between changes in the climate system and biogeochemistry. http://www.ipcc.ch/pdf/assessment-report/ar4/wg1/ar4-wg1-chapter7pdf

IRRI (2006) http://www.irri.org/science/ricestat/pdfs/WRS2005-Table02.pdf

Islam R, Trivedi P, Madhaiyan M, Seshadri S, Lee G, Yang Y, Kim M, Han G, Singh Chauban P, Sa T (2010) Isolation, enumeration and characterization of diazotrophic bacteria from paddy soil sample under long-term fertilizer management experiment. Biol Fertil Soils 46:261–269

Jackel U, Schnell S, Conrad R (2001) Effect of moisture, texture and aggregate size of paddy soil on production and consumption of CH_4. Soil Biol Biochem 33:965–971

Jagadeesh Babu Y, Li C, Frolking S, Nayak DR, Adhya TK (2006) Field validation of DNDC model for methane and nitrous oxide emissions from rice-based production systems of India. Nutr Cycl Agroecosyst 74:157–174. https://doi.org/10.1007/s10705-005-6111-5

Jain MC, Kumar S, Wassman R, Mitra S, Singh SD, Sing JP, Singh R, Yadav AK, Gupta S (2000) Methane emissions from irrigated rice fields in northern India (New Delhi). Nutr Cycl Agroecosyst 58:75–83

Jia ZJ, Cai ZC, Xu H, Tsuruta H (2002) Effects of rice cultivars on methane fluxes in a paddy soil. Nutr Cycl Agroecosyst 64:87–94

Jiang CS, Wang YS, Zheng XH, Zhu B, Huang Y, Hao QJ (2006) Methane and nitrous oxide emissions from three paddy rice based cultivation systems in Southwest China. Adv Atmos Sci 23:415–424

Kang GD, Cai ZC, Feng XZ (2002) Importance of water regime during the non-rice growing period in winter in regional variation of CH_4 emissions from rice fields during following rice growing period in China. Nutr Cycl Agroecosyst 64:95–100

Keerthisinghe DG, Freney JR, Mosier AR (1993) Effect of wax-coated calcium carbide and nitrapyrin on nitrogen loss and methane emission from dry-seeded flooded rice. Biol Fertil Soils 16:71–75

Keppler F, Hamilton JT, Brass M, Röckmann T (2006) Methane emissions from terrestrial plants under aerobic conditions. Nature 439:187–191

Khalil MAK, Rasmussen RA (1991) Methane emission from the rice field in China. Environ Sci Technol 25:979–981

Kimura M (2000) Anaerobic microbiology in waterlogged rice fields. In: Bollag JM, Stotzky G (eds) Soil biochemistry. Marcel Dekker, New York, pp 35–138

Kimura M, Murase J, Lu YH (2004) Carbon cycling in rice field ecosystems in the context of input, decomposition and translocation of organic materials and the fates of their end products (CO_2 and CH_4). Soil Biol Biochem 36:1399–1416

Kirk GJD, Bajita JB (1995) Root induced iron oxidation, pH change and zinc solubilization in the rhizosphere of low land rice. New Phytol 131:129–137

Kluber HD, Conrad R (1998) Effects of nitrate, nitrite, NO and N_2O on methanogenesis and other redox processes in anoxic rice field soil. FEMS Microbiol Ecol 25:301–318

Kludze HK, DeLaune RD, Patrick WH Jr (1993) Aerenchyma formation and methane and oxygen exchange in rice. Soil Sci Soc Am J 57:386–391

Kristjansson JK, Scheonheit P, Thauer RK (1982) Different Ks values for hydrogen and methanogenic and sulphate reducing bacteria: an explanation for the apparent inhibition of methanogenesis by sulphate. Arch Microbiol 131:278–282

Kruger M, Frenzel P (2003) Effects of N-fertilisation on CH4 oxidation and production, and consequences for CH4 emissions from microcosms and rice fields. Glob Chang Biol 9:773–784

Kruger M, Frenzel P, Kemnitz D, Conrad R (2005) Activity, structure and dynamics of the methanogenic archaeal community in a flooded Italian rice field. FEMS Microbiol Ecol 51:323–331

Lee CH, Park KD, Jung KY, Ali MA, Lee D, Gutierrez J, Kim PJ (2010) Effect of Chinese milk vetch (*Astragalus sinicus* L.) as a green manure on rice productivity and methane emission in paddy soil. Agric Ecosyst Environ 138:343–347

Lee SY, Lee SH, Jang JK, Cho KS (2011) Comparison of methanotrophic community and methane oxidation between rhizospheric and non-rhizospheric soils. Geomicrobiol J 28(8):676–685. https://doi.org/10.1080/01490451.2010.511984

Li D, Liu M, Cheng Y, Wang D, Qin J, Jiao J, Li H, Hu F (2011) Methane emissions from double-rice cropping system under conventional and no tillage in southeast China. Soil Tillage Res 113 (2):77–81

Liesack W, Schnell S, Revsbech NP (2000) Microbiology of flooded rice paddies. FEMS Microbiol Rev 24:625–645

Liu CW, Wu CY (2004) Evaluation of methane emissions from Taiwanese paddies. Sci Total Environ 333:195–207

Liu DY, Ding WX, Jia ZJ, Cai ZC (2011) Relation between methanogenic archaea and methane production potential in selected natural wetland ecosystems across China. Biogeosciences 8:329–338. https://doi.org/10.5194/bg-8-329-2011

Lovley DR, Phillips EJP (1987) Rapid assay for microbially reducible ferric iron in aquatic sediments. Appl Environ Microbiol 53:1536–1540

Lu W, Liao Z, Zhang J, Cen C (1999) Effects of differnet rice-vegetable rotation systems on CH_4 emission from paddy soils. Agro Environ Prot 18(5):200–202

Lu WF, Chen W, Duan BW, Guo WM, Lu Y, Lantin RS, Wassmann R, Neue HU (2000) Methane emissions and mitigation options in irrigated rice fields in southeast China. Nutr Cycl Agroecosyst 58:65–73

Major J, Steiner C, Di Tommaso A, Falcao NPS, Lenmann J (2005) Weed composition and cover after three years of soil fertility management in the central Brazilian Amazon: compost, fertilizer, manure and charcoal applications. Weed Biol Manage 5:69–76

Masscheleyn PH, DeLaune RD, Pattrick WH Jr (1993) Methane and nitrous oxide emissions from laboratory measurements of rice soil suspensions: effect of soil oxidation reduction status. Chemosphere 26:251–260

Matthews RB, Wassmann R, Knox JK, Buendia LV (2000) Using a crop/soil simulation model and GIS techniques to assess methane emissions from rice fields in Asia. IV. Upscaling to national levels. Nutr Cycl Agroecosyst 58:201–217

Mingxing W, Jing L (2002) CH_4 emission and oxidation in Chinese rice paddies. Nutrt Cycl Agroecosyst 64:43–55

Nakayama N, Okabe A, Toyota K, Kimura M, Asakawa S (2006) Phylogenetic distribution of bacteria isolated from the flood water of a Japanese paddy field. Soil Sci Plant Nutr 52 (3):305–312

Naser HM, Nagata O, Tamura S, Hatano R (2007) Methane emissions from five paddy fields with different amounts of rice straw application in central Hokkaido, Japan. Soil Sci Plant Nutr 53:95–101

Neue HU, Roger PA (2000) Rice agriculture: factors controlling emissions. In: Khalil MAK (ed) Atmospheric methane. Its role in the global environment. Springer, Berlin, pp 134–169

Neue HU, Scharpenseel HW (1984) Gaseous products of the decomposition of organic matter in submerged soils. In: Organic matter and rice. International Rice Research Institute, Manila, pp 311–328

Neue HU, Latin RS, Wassmann R, Aduna JB, Alberto CR, Andales MJF (1992) Methane emission from rice soils of the Philippines. In: Minami K, Mosier A, Sass R (eds) CH_4 and N_2O global emissions and controls from rice fields and other agriculture and industrial sources, NIAES series 2. National Institute of Agro-Environmental Sciences, Tsukuba, pp 55–64

Nirmal Kumar JI, Viyol SV (2009) Short-term diurnal and temporal measurement of methane emission in relation to organic carbon, phosphate and sulphate content of two rice fields of central Gujarat, India. Paddy Water Environ 7:11–16. https://doi.org/10.1007/s10333-008-0147-5

Noll M, Matthies D, Frenzel P, Frenzel M, Liesack W (2005) Succession of bacterial community structure and diversity in a paddy soil oxygen gradient. Environ Microbiol 7:382–395

Okabe A, Toyota K, Kimura M (2000) Seasonal variations of phospholipid fatty acid composition in the flood water of a Japanese paddy field under a long-term fertilizer trial. Soil Sci Plant Nutr 46(1):177–188

Oude Elferink SJWH, Visser A, Hulshoff Pol LW, Stams AJM (1994) Sulphate reduction in methanogenic bioreactors. FEMS Microbiol Rev 15:119–136

Patrick WH Jr (1981) The role of inorganic redox systems in controlling reduction in paddy soils. In: Proceedings of symposium on paddy soil. Institute of Soil Science, Academia Sinica/ Springer, Beijing/Berlin, pp 107–117

Prinn RG (1995) Global atmospheric-biospheric chemistry. In: Prinn RG (ed) Global atmospheric-bioshperic chemistry. Plenum, New York, pp 1–18

Qin Z, Zhang JE, Luo SM, Xu HQ, Zhang J (2010) Estimation of ecological services value for the rice-duck farming system. Resour Sci 32(5):864–872

Qiu QF, Noll M, Abraham WR, Lu YH, Conrad R (2008) Applying stable isotope probing of phospholipid fatty acids and rRNA in a Chinese rice field to study activity and composition of the methanotrophic bacterial communities in situ. ISME J 2:602–614

Ramakrishnan B, Lueders T, Dunfield PF, Conrad R, Friedrich MW (2001) Archaeal community structures in rice soils from different geographical regions before and after initiation of methane production. FEMS Microbiol Ecol 37:175–186

Ratering S, Schnell S (2000) Localization of iron-reducing activity in paddy soil by profile studies. Bioegeochemistry 48:341–365

Renner R (2007) Rethinking biochar. Environ Sci Technol 41:5932–5933

Rondon M, Ramirez JA, Lehmann J (2005) Charcoal additions reduce net emissions of greenhouse gases to the atmosphere. In: Proceedings of the 3rd USDA symposium on greenhouse gases and carbon sequestration, Baltimore, USA, March 21–24, p 208

Rondon MA, Molina D, Hurtado M, Ramirez J, Lehmann J, Major J, Amezquita E (2006) Enhancing the productivity of crops and grasses while reducing greenhouse gas emissions through bio-char amendments to unfertile tropical soils. In: 18th world congress of soil science, July 9–15, Philadelphia, PA, http://crops.confex.com/crops/wc2006/techprogram/P16849.HTM. Accessed on Dec 2012

Rui J, Peng J, Lu Y (2009) Succession of Bacterial Populations during Plant Residue Decomposition in Rice Field Soil. Appl Environ Microbiol 75(14):4879–4886. https://doi.org/10.1128/AEM.00702-09

Saenjan P, Wada H 1990 Effects of salts on methane formation and sulfate reduction in submerged soil. In: Transactions of the 14th international congress of soil science vol 2, Commission 2, 12–18 August, Kyoto, Japan, pp 244–248

Sass RL, Fisher FM (1992) CH4 emission from paddy fields in the United States gulf coast area. In: Minami K, Mosier A, Sass R (eds) CH4 and N2O global emissions and controls from rice fields and other agriculture and industrial sources, NIAES series 2. National Institute of Agro-Environmental Sciences, Tsukuba, pp 65–78

Sass RL, Fisher FM Jr (1994) CH4 emission from paddy fields in the United States gulf coast area. In: Minami CK, Mosier A, Sass RL (eds) CH4 and N2O: global emissions and controls from rice fields and other agricultural and industrial sources, NIAES series 2. National Institute of Agro-Environmental Sciences, Tsukuba, pp 65–77

Sass RL, Fisher FM, Harcombe PA, Turner FT (1990) Methane production and emission in Texas rice fields. Glob Biogeochem Cycles 4:47–68

Satpathy SN, Mishra S, Adhya TK, Ramakrishnan B, Rao VR, Sethunathan N (1998) Cultivar variation in methane efflux from tropical rice. Plant Soil 202:223–229

Schimel J (2000) Global change: rice, microbes and methane. Nature 403:375–377. https://doi.org/10.1038/35000325

Schutz H, Holzapfel-Pschorn A, Conrad R, Rennenberg H, Seiler W (1989) A 3-year continuous record on the influence of daytime, season, and fertilizer treatment on methane emission rates from an Italian rice paddy. J Geophys Res 94:16405–16416. https://doi.org/10.1029/JD094iD13p16405

Sebacher DI, Harriss RC, Bartlett KB, Sebacher SM, Grice SS (1986) Atmospheric methane sources: alaskan tundra bogs, an alpine fen, and a subarctic boreal marsh. Tellus 38B:1–10

Singh NK, Dhar DW (2006) Sewage effluent: a potential nutrient source for microalgae. Proc Indian Natl Sci Acad 72:113–120

Singh NK, Dhar DW (2007) Nitrogen and phosphorous scavenging potential in microalgae. Indian J Biotechnol 6:52–56

Singh NK, Dhar DW (2011) Phylogenetic relatedness among *Spirulina* and related cyanobacterial genera. World J Microbiol Biotechnol 27:941–951. https://doi.org/10.1007/s11274-010-0537-x

Singh NK, Patel DB (2012) Microalgal remediation of distillery effluent: a review. In: Lichtfouse E (ed) Farming for food and water security, Sustainable agriculture reviews, vol 10. Springer, Dordrecht, pp 83–109. https://doi.org/10.1007/978-94-007-4500-1

Singh NK, Dhar DW, Tabassum R (2016a) Role of cyanobacteria in crop protection. Proc Natl Acad Sci India Sect B Biol Sci 86(1):1–8. https://doi.org/10.1007/s40011-014-0445-1 ISSN 0369-8211

Singh NK, Desai CK, Rathore BS, Chaudhari BG (2016b) Bio-efficacy of herbicides on performance of mustard, Brassica juncea (L.) and population dynamics of agriculturally important bacteria. Proc Natl Acad Sci India Sect B Biol Sci 86(3):743–748. https://doi.org/10.1007/s40011-015-0521-1 ISSN 0369-8211

Steiner C, Teixeira WG, Lehmann J, Nehls T, de Macedo JLV, Blum WEH, Zech W (2007) Long term effects of manure, charcoal and mineral fertilization on crop production and fertility on a highly weathered Central Amazonian upland soil. Plant Soil 291:275–290

Sugano A, Tsuchimoto H, Cho TC, Kimura M, Asakawa S (2005) Succession of methanogenic archaea in rice straw incorporated into a Japanese rice field: estimation by PCR-DGGE and sequence analyses. Archaea 1:391–397

Takahashi E, Ma JF, Miyake Y (1990) The possibility of silicon as an essential element for higher plants. Comments Agric Food Chem 2:99–122

Takahashi S, Uenosono S, Ono S (2003) Short- and long-term effects office straw application on nitrogen uptake by crops and nitrogen mineralization under flooded and upland conditions. Plant Soil 251:291–301. https://doi.org/10.1023/A:1023006304935

Takai Y, Kamura T (1966) The mechanism of reduction in waterlogged paddy soil. Folia Microbiol 11:304–313

Tanaka H, Kyaw K, Toyota K, Motobayashi T (2010) Influence of application of rice straw, farmyard manure, and municipal biowastes on nitrogen fixation, soil microbial biomass N, and mineral N in a model paddy microcosm. Biol Fertil Soils 42:501–505

Towprayoon S, Smakgahn K, Poonkqew S (2005) Mitigation of methane and nitrous oxide emissions from drained irrigated rice fields. Chemosphere 59:1547–1556

van der Gon HAD, van Bodegom PM, Wassmann R, Lantin RS, Metra-Corton T (2001) Assessing sulfate-containing soil amendments to reduce methane emissions from rice fields: mechanisms, effectiveness and costs. Mitig Adapt Strateg Glob Chang 6:69–87. https://doi.org/10.1023/A:1011380916490

van der Gon HAD, Kropff MJ, van Breemen N, Wassmann R, Lantin RS, Aduna E, Corton TM (2002) Optimizing grain yields reduces CH4 emissions from rice paddy fields. Proc Natl Acad Sci U S A 99:12021–12024. https://doi.org/10.1073/pnas.192276599

Wang MX (2001) Methane emission from chinese rice fields. Science Press, Beijing, p 223

Wang B, Adachi K (2000) Differences among rice cultivars in root exudation, methane oxidation, and populations of methanogenic and methanotrophic bacteria in relation to methane emission. Nutr Cycl Agroecosyst 58:349–356

Wang ZP, Delaune RD, Patrick WH Jr, Masscheleyn PH (1993) Soil redox and pH effects on methane production in a flooded rice soils. Soil Sci Soc Am J 57:382–385

Wang B, Neue HU, Samonte HP (1997) Effect of cultivar difference ('IR72', 'IR65598', and 'Dular') on methane emission. Agric Ecosyst Environ 62:31–40

Wang ZY, Xu YC, Li Z, Guo YX, Wassmann R, Neue HU, Lantin RS, Buendia LV, Ding YP, Wang ZZ (2000) A four-year record of methane emissions from irrigated rice fields in the Beijing region of China. Nutr Cycl Agroecosyst 58:55–63

Wang H, Lin K, Hou Z, Richardson B, Gan J (2010) Sorption of the herbicide terbuthylazine in two New Zealand forest soils amended with biosolids and biochars. J Soils Sediments 10:283–289

Watanabe T, Kimura M, Asakawa S (2006) Community structure of methanogenic archaea in paddy field soil under double cropping (rice-wheat). Soil Biol Biochem 38:1264–1274

Watanabe T, Kimura M, Asakawa S (2007) Dynamics of methanogenic archaeal communities based on rRNA analysis and their relation to methanogenic activity in Japanese paddy field soils. Soil Biol Biochem 39:2877–2887

Watanabe T, Hosen Y, Agbisit R, Llorca L, Fujita D, Asakawa S Kimura M (2010) Changes in community structure and transcriptional activity of methanogenic archaea in a paddy field soil brought about by a water-saving practice – Estimation by PCR-DGGE and qPCR of 16S rDNA and 16S rRNA. In: 19th World Congress Of Soil Science, Soil Solutions For A Changing World, 1–6 August 2010, Brisbane, Australia

Weber S, Lueders T, Friedrich MW, Conrad R (2001) Methanogenic populations involved in the degradation of rice straw in anoxic paddy soil. FEMS Microbiol Ecol 38:11–20

Win KT, Nonaka R, Toyota K, Motobayashi T, Hosomi M (2010) Effects of option mitigating ammonia volatilization on CH_4 and N_2O emissions from a paddy field fertilized with anaerobically digested cattle slurry. Biol Fertil Soils 46:589–595. https://doi.org/10.1007/s00374-010-0465-9

Wu XL, Chin KJ, Stubner S, Conrad R (2001) Functional patterns and temperature response of cellulose-fermenting microbial cultures containing different methanogenic communities. Appl Microbiol Biotechnol 56:212–219

Wu L, Ma K, Li Q, Ke X, Lu Y (2009) Composition of archaeal community in a paddy field as affected by rice cultivar and N fertilizer. Microbial Ecol 58:819–826

Wu M, Qin H, Chen Z, Wu J, Wei W (2011) Effect of long-term fertilization on bacterial composition in rice paddy soil. Biol Fertil Soils 47:397–405. https://doi.org/10.1007/s00374-010-0535-z

Xie B, Zheng X, Zhou Z, Gu J, Zhu B, Chen X, Shi Y, Wang Y, Zhao Z, Liu C, Yao Z, Zhu J (2010) Effects of nitrogen fertilizer on CH4 emission from rice fields: multi-site field observations. Plant Soil 326:393–401. https://doi.org/10.1007/s11104-009-0020-3

Xiong ZQ, Xing GX, Zhu ZL (2007) Nitrous oxide and methane emissions as affected by water, soil and nitrogen. Pedosphere 17:146–155

Xu Q (2001) Evolution of soil fertility in relation to soil quality in paddy fields of the Tai Lake area, Yangtze Basin. Res Environ 10(4):323–328

Xu H, Cai ZC, Li XP, Tsuruta H (2000) Effect of antecedent soil water regime and rice straw application time on CH_4 emission from rice cultivation. Aust J Soil Res 38:1–12

Xu H, Cai ZC, Jia ZJ (2002) Effect of soil water contents in the non-rice growth season on CH_4 emission during the following rice-growing period. Nutr Cycl Agroecosyst 64:101–110

Xu H, Cai ZC, Tsuruta H (2003) Soil moisture between rice growing season affects methane emission, production, and oxidation. Soil Sci Soc Am J 67:1147–1157

Yagi K, Minami K (1990) Effect of organic matter application on methane emission from some Japanese paddy fields. Soil Sci Plant Nutr 36:599–610

Yamane I, Sato K (1964) Decomposition of glucose and gas formation in flooded soils. Soil Sci Plant Nutr 10:127–133

Yan XY, Akiyama H, Yagi K, Akimoto H (2009) Global estimations of the inventory and mitigation potential of methane emissions from rice cultivation conducted using the 2006 intergovernmental panel on climate change guidelines. Global Biogeochem Cycles 23:1–15

Yanai Y, Toyota K, Okazaki M (2007) Effects of charcoal addition on N2O emissions from soil resulting from rewetting air-dried soil in short-term laboratory experiments. Soil Sci Plant Nutr 53:181–188

Yang SS, Chang HL (1998) Effect of environmental conditions on methane production and emission from paddy soil. Agric Ecosyst Environ 69:69–80

Yang SS, Chang HL (1999) Diurnal variation of methane emission from paddy fields at different growth stages of rice cultivation in Taiwan. Agric Ecosyst Environ 76:75–84

Yao H, Conrad R, Wassmann R, Neue HU (1999) Effect of soil characteristics on sequential reduction and methane production in sixteen rice paddy soils from China, the Philippines, and Italy. Biogeochemistry 47:269–295

Yue J, Shi Y, Liang W, Wu J, Wang C, Huang G (2005) Methane and nitrous oxide emissions from rice field and related microorganism in black soil, northeastern China. Nutr Cycl Agroecosyst 73:293–301. https://doi.org/10.1007/s10705-005-3815-5

Zhang A, Cui L, Pan G, Li L, Hussain Q, Zhang X, Zheng J, Crowley D (2010a) Effect of biochar amendment on yield and methane and nitrous oxide emissions from a rice paddy from Tai Lake plain. China Agric Ecosyst Environ 139:469–475. https://doi.org/10.1016/j.agee.2010.09.003

Zhang H, Lin K, Wang H, Gan J (2010b) Effect of *Pinus radiate* derived biochars on soil sorption and desorption of phenanthrene. Environ Pollut 158:2821–2825

Zhang G, Zhang X, Ma J, Xu H, Cai Z (2011) Effect of drainage in the fallow season on reduction of CH_4 production and emission from permanently flooded rice fields. Nutr Cycl Agroecosyst 89:81–91. https://doi.org/10.1007/s10705-010-9378-0

Zheng Y, Zhang LM, Zheng YM, Di HJ, He JZ (2008) Abundance and community composition of methanotrophs in a Chinese paddy soil under long-term fertilization practices. J Soils Sediments 8:406–414

Zou JW, Huang Y, Jiang JY, Zheng XH, Sass RL (2005) A 3-year field measurement of methane and nitrous oxide emissions from rice paddies in China: effects of water regime, crop residue, and fertilizer application. Global Biogeochem Cycles 19:GB2021. https://doi.org/10.1029/2004GB002401

Zou JW, Huang Y, Zheng XH, Wang Y (2007) Quantifying direct N_2O emissions in paddy fields during rice growing season in mainland China: dependence on water regime. Atmos Environ 41:8032–8042

Zwieten VL, Singh B, Joseph S, Kimber S, Cowie A, Chan YK (2009) Biochar and emissions of non-CO2 greenhouse gases from soil. In: Lehmann J, Joseph S (eds) Biochar for environmental management science and technology. Earthscan Press, London, pp 227–224

Chapter 6
Physical and Biological Processes Controlling Soil C Dynamics

Pratap Srivastava, Rishikesh Singh, Rahul Bhadouria, Pardeep Singh, Sachchidanand Tripathi, Hema Singh, A. S. Raghubanshi, and P. K. Mishra

Abstract Globally, land use change and management have declined soil organic carbon (SOC), thus emitting more CO_2 contributing to global warming. Here we review factors that control the fate of soil organic carbon. We found that dry tropical soils are considerably away from carbon saturation, and thus have the potential for high carbon sequestration, if managed properly. Integrated indicators have been set up, such as relative availability of inorganic nitrogen pools, carbon management index, macro-aggregate water stability and metabolic quotient. For example, the relative, rather than absolute, availability of inorganic

P. Srivastava (✉) · P. K. Mishra
Department of Chemical Engineering and Technology, Indian Institute of Technology, Banaras Hindu University, Varanasi, India
e-mail: pkmishra.che@itbhu.ac.in

R. Singh · A. S. Raghubanshi
Institute of Environment & Sustainable Development (IESD), Banaras Hindu University, Varanasi, India

R. Bhadouria
Department of Chemical Engineering and Technology, Indian Institute of Technology, Banaras Hindu University, Varanasi, India

Ecosystems Analysis Laboratory, Department of Botany, Institute of Science, Banaras Hindu University, Varanasi, India

P. Singh
Ecosystems Analysis Laboratory, Department of Botany, Institute of Science, Banaras Hindu University, Varanasi, India

Ecosystems Analysis Laboratory, Department of Environmental Studies, PGDAV College, University of Delhi, New Delhi, India
e-mail: psingh.rs.apc@itbhu.ac.in

S. Tripathi
Department of Environmental Studies, PGDAV College, University of Delhi, New Delhi, India

Department of Botany, Deen Dayal Upadhyay College, University of Delhi, New Delhi, India

H. Singh
Ecosystems Analysis Laboratory, Department of Botany, Institute of Science, Banaras Hindu University, Varanasi, India

E. Lichtfouse (ed.), *Sustainable Agriculture Reviews 33*, Sustainable Agriculture Reviews 33, https://doi.org/10.1007/978-3-319-99076-7_6

nitrogen pools has been found associated with resource conservation mechanisms in soils.

Keywords C accumulation · Dry tropical ecosystems · Organic amendments · Relative availability · Soil aggregates

6.1 Introduction

Soil is a structurally and functionally complex ecosystem, which plays a regulatory role in biogeochemical cycle and biosphere functioning. It generally encompasses a matrix of non-living material (made up of inorganic and organic components) along with the embedded air, water and web of interacting organisms. The multi-functionality and living nature of soil is a function of soil organic carbon (SOC) primarily, which closely relates with its physical, chemical and biological properties (Fig. 6.1) (Smith et al. 2000). Globally, soil holds the largest reservoir of carbon (about 1500 Pg, petagram), which comprises more than the combined carbon present in the vegetation (550 Pg) and atmosphere (750 Pg). SOC provides many ecosystem services such as plant nutrient retention and supply, reduction in soil erosion and improvement in aggregation, and water holding capacity (Tisdall and Oades 1982; Brady and Weil 2002). It is, therefore, considered as an important indicator of soil quality and fertility. Whereas, it's various components reflects the degree of soil viability (Vinther et al. 2004; Dominy et al. 2002; Lal 2006; Pan et al. 2009). However, global soil CO_2 emission represents a significant carbon flux (about 75 Pg C y^{-1}) which is identified as the second largest flux (contributes 20–38%) of carbon between soils and the atmosphere (Raich and Schlesinger 1992; Schlesinger and Andrews 2000). Therefore, climate change and the declining soil quality

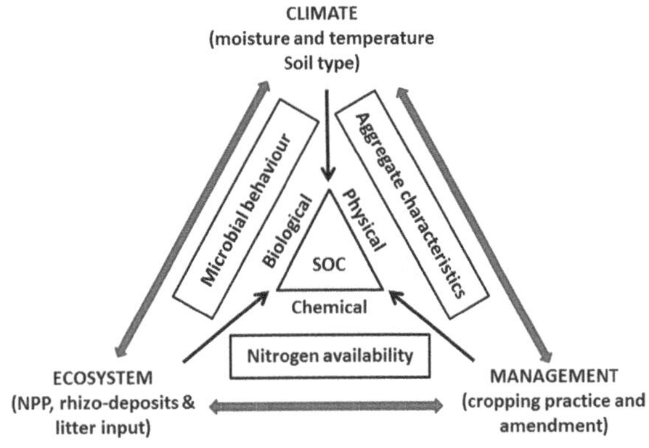

Fig. 6.1 Illustration showing integrative nature of soil organic carbon

has drawn global attention in recent years towards soil ecosystem, in general, and soil carbon dynamics, in particular, due to their evident recognition at the core of these interlinked problems. It has been observed that soil may act as both, source as well as sink of atmospheric CO_2 depending upon management. Therefore, delineation of various pools and fluxes of SOC along with factors affecting them is required to better understand the SOC dynamics for increased carbon accumulation in soils (Lal 2004a, b). It would indicate about the relative importance of various pools and processes in the soils, which govern soil's nature as source and sink of atmospheric CO_2. Moreover, how these pools and processes are affected under various agro-management systems and can be appropriately manipulated to achieve a balanced soil carbon dynamics for improved soil fertility and climate stability is an important emerging topic of research in recent times.

Cultivation decreases SOC, however, specific management systems, (such as reduced tillage, rotation and manure addition) enhance the same (Bronick and Lal 2005). SOC pools in the agricultural soil particularly in dry tropical ecosystem, is considered to be lower than their potential sink capacity (Lal et al. 2007). Stewart et al. (2007) argue that ecosystems which are quite away from carbon saturation inherently (due to low productivity under some abiotic constrains) or due to anthropogenic degradation are better suited for soil carbon sequestration. These dry tropical ecosystems, covering approximately 40% of the global land area (Murphy and Lugo 1986; FAO 2000, 2001) having low SOC saturation and severe anthropogenic influence, may have the greatest potential as carbon sink (Scurlock and Hall 1998; Rosenberg et al. 1999). The holistic understanding of soil carbon dynamics in these dry ecosystems might help in striking an appropriate balance between carbon mineralization and immobilization in an effective manner (Fig. 6.2). Here, we reviewed various SOC pools, its fluxes and the major processes governing SOC accumulation and release to holistically understand the possible mechanism of soil carbon dynamics to delineate the carbon accumulation approaches in dry tropical ecosystems.

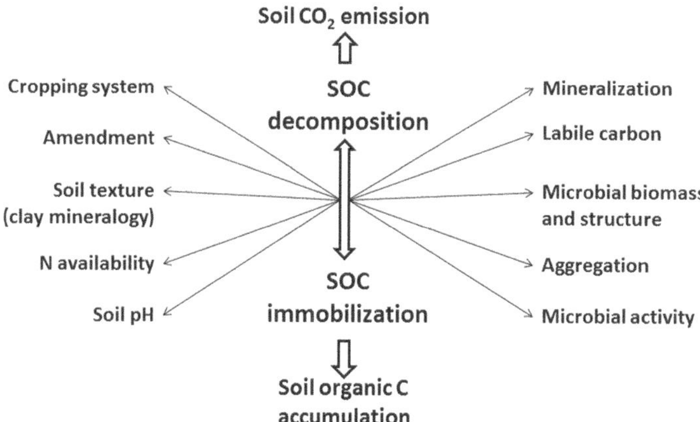

Fig. 6.2 Factors affecting the compromise between soil carbon decomposition and immobilization

6.2 Soil Carbon Pools

The world's soil holds the largest reservoir of carbon (about 1500 Pg, petagram) in terrestrial ecosystems, which comprises more than the combined carbon present in the vegetation and atmosphere (Fig. 6.3) (Post et al. 1982; Batjes and Sombroek 1997). Alternatively, majority of carbon (around 75%) in the terrestrial pool is stored as SOC (Batjes 1996), which holds substantial impact in biosphere functioning (Eswaran et al. 1993; Jobbagy and Jackson 2000). Tropical soil shares 32% of the total SOC stock (Eswaran et al. 1993). The contribution of carbon in the soil below 1 m is especially relevant in significantly deep tropical soils (Sombroek et al. 1993). It indicates that tropical soils store a considerable amount of SOC, higher than the temperate soils (Moraes et al. 1995). SOC consists of several pools with variable turnover rate, ranging from months to thousands of years (Parton et al. 1987; Silveira et al. 2008). Out of total SOC stock, roughly 80–160 Pg carbon resides in the surface detritus (Matthews 1997), 200–300 Pg C as soil organic matter (SOM), having turnover less than a century, and remainder as stable carbon, having turnover of centuries to millennia (Schimel 1995). SOC incorporates mainly of the plant residues, decomposition products and byproducts, microbial biomass, and humic materials.

Ecosystems also show great variation in the SOC content due to their distinct ecology. Generally, natural ecosystems (such as forest, savanna and grassland) have higher SOC and resource use efficiency than their modified ecosystems (such as, plantation, pastureland and cropland), due to greater net primary production (NPP) and carbon inputs as well as synchronized microbial processes/activity. Generally, grassland shows a lower or similar SOC as compared to forest ecosystem; however, savanna has been found to shows intermediate SOC between forest and cropland systems. The conversion of native vegetation to a cropland always leads to loss of about 30–75% SOC with time. Basically, such changes in land use and management shift the ecosystem towards an altogether "unique and new equilibrium", which is

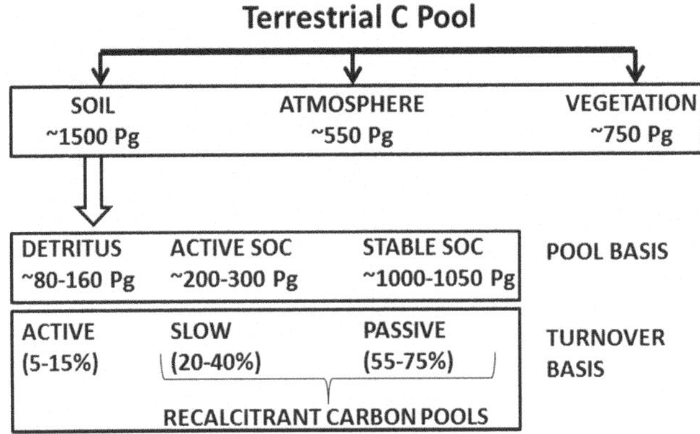

Fig. 6.3 Terrestrial carbon pools

considerably different from the native ecosystems. Such modified ecosystems are called as “alternative or novel ecosystem”. Additionally, various agro-management systems further affect this equilibrium variably affecting the soil properties, depending on the type of external physical (such as tillage) and chemical (such as chemical fertilization and organic amendments) disturbance. This altogether affects the SOC dynamics drastically in derived system as compared to their native counterpart.

6.2.1 Inorganic and Organic Carbon

World’s soil comprises two distinct carbon components: soil organic carbon (SOC) and soil inorganic carbon (SIC) pools, which is estimated to be around1576 Gt and 938 Gt, respectively, up to 1 meter depth (Post et al. 1982; Eswaran et al. 1993; Schlesinger 1995). Their regional distribution indicates that SOC pool is concentrated in arctic, boreal, and temperate soils, whereas the inorganic carbon pool is concentrated in the arid and semiarid soils. The organic carbon pool is mostly comprised of humus and the relatively passive charcoal carbon (Schnitzer 1991; Stevenson 1994; Singh et al. 2015); whereas the inorganic pool includes the elemental carbon and carbonate minerals, such as calcite, dolomite, and gypsum. Here, SIC has not been discussed further in relation to management of soil carbon dynamics in agro-ecosystems, as SOC holds crucial importance in this regard.

6.2.2 Active and Passive Carbon

The SOC consist of various fractions varying in degree of decomposition, recalcitrance, and turnover rates (Huang et al. 2008). On this basis, SOC pools are broadly divided into active (labile) and stable (non-labile) carbon pools (Fig. 6.3). The active/labile carbon pool has a greater turnover rate (or shorter mean residence time in soils) of several weeks to months or years compared with more recalcitrant or non-labile carbon pools (Paul et al. 2001). Thus, the relative size of labile carbon pool in soil is much smaller than non-labile carbon pools. Labile compounds (such as simple sugars and microbial biomass, etc.) comprise 5–15% of the total SOC pool. It includes microbial biomass carbon (MBC), dissolved organic carbon (DOC), water soluble carbon (WSC), particulate organic carbon (POC) and $KMnO_4$-oxidizable carbon etc., which turn over quickly and respond more rapidly to soil management than SOC (Blair et al. 1995; Ghani et al. 2003; Haynes 2005; Purakayastha et al. 2008; Gong et al. 2009). Hence, these fractions have been suggested as early and sensitive indicators of the effects of land use change and management on the soil quality (Gregorich et al. 1994; Blair et al. 1995; Yang et al. 2005; Rudrappa et al. 2006). The intermediate (slow) fraction comprises 20–40% of the total SOC pool, and its turnover rate spans over several decades. However, the stable or recalcitrant carbon fraction comprises the remaining 60–70% of the total SOC pool, with a turnover time of hundreds to thousands of years (Lichtfouse 1997; Rice 2002). It has been well identified that active and stable SOC play differential roles in SOM dynamics and nutrient cycling. Therefore, it is argued that the pool size of these two carbon fractions in the bulk soil as well as aggregate-size fractions may be of crucial importance in the assessment of soil management practices (Blair et al. 1995; Srivastava et al. 2016). However, none of the physical or chemical fractionation methods is so far able to satisfactorily separate the active SOC from stable SOC. Some of the most widely discussed labile carbon fractions have been discussed here due to their high sensitivity to land use change and management.

6.2.2.1 Dissolved Organic Carbon (DOC)

Dissolved organic carbon (DOC) can be considered as synonymous to the water soluble carbon (WSC) and water extractable organic carbon, as they seems to differ largely due to their method of extraction. It has been reported that carbon extracted with the water includes comparatively more labile carbon fractions as compared to other methods (Cook and Allan 1992). This soluble fraction of organic matter serves as the main energy substrate in soils required for microbial activity in the soil (Marschner and Kalbitz 2003). In spite of comparatively small amount in the soil, it governs important soil functions in multiple ways (Zsolnay 1996; Chantigny 2003), such as soil physical stability (Rilling and Steinberg 2002) and C sequestration (Guggenberger and Kaiser 2003). Moreover, the turnover of DOC is strongly linked to soil microbial activity (Chantigny 2003), and therefore, significantly contributes to nutrient availability and cycling (Haynes 2005). The knowledge about DOC fluxes and dynamics in agro-ecosystems is scarce and contradictory, particularly under different fertilization regime (Embacher et al. 2008). It has been reported that the influence of organic amendments is more consistent on WSC as compared to the contradictory impact of nitrogen fertilization (Chantigny 2003; Chantigny et al. 2002; Blair et al. 2006). It has been attributed to the variable quality of WSC (i.e. aromaticity), which depends on the multitude of biotic and abiotic factors such as temperature, preferential consumption of non-aromatic compounds, exudation of low molecular weight compounds and easily degradable carbon from microorganisms and plant roots.

6.2.2.2 Microbial Biomass Carbon

Microbial biomass represents a small, though the most important labile carbon pool, which is vital for SOC dynamics and nutrient cycling (Powlson et al. 1987; Singh et al. 1989). It represents a functional index of the soil redevelopment (Srivastava and Singh 1989). It is highly dynamic and fluctuates more over time than total SOC. The higher conversion of added carbon to microbial biomass suggests a better stability of SOC (Sparling 1992). Therefore, its measurement may show the effect of soil management practices on potential changes in SOM long before such effects can be detected by measuring the total SOC (Powlson et al. 1987). Long term studies have showed an increase in soil microbial biomass under fertilization (Yan et al. 2007; Šimon 2008; Wang et al. 2008). Microbial biomass as well as activity has been reported to be higher in organic amended soil than chemically fertilized soils (Tu et al. 2006). Application of organic manure with/without inorganic fertilizer has been stated to manipulate the soil microorganisms to improve the soil health and fertility (Chaparro et al. 2012). It is attributed to the strong influence of organic manure on the soil microbial biomass and community structure (Singh and Singh 1993; Carpenter-Boggs et al. 2000; Esperschütz et al. 2007) and activity (Goyal et al. 1999).

6.2.3 *Soil Aggregates and Soil Organic Carbon*

Recent understanding shows that not only the chemical recalcitrance of soil organic matter but biophysical processes, such as aggregate dynamics might be the major determinants of SOC turnover and sequestration (Schmidt et al. 2011). Tisdall and Oades (1982) demonstrated that SOC closely relates with soil aggregate formation and stability. The loss of SOC and aggregate stability represents the unsustainable soil management (Carter 2002). The aggregate represents the integrative effects of soil type, environment, plant species, and soil management (i.e. crop rotations, tillage and fertilizer management) practices (Martens and Frankenberger 1992; Nyamangara et al. 1999; Martens 2000). Therefore, it represents the integrative output of the interactions among physico-chemical and biological components inside the soil (Fig. 6.4). It suggest that the consideration of soil aggregate stability (both physical and chemical) as crucial integrative indicator or soil functional trait might be helpful in the improved soil management in a simple and cost effective manner. Therefore, soil aggregate organization and its behavior in response to land use change needs to be monitored in order to understand and manage soil carbon dynamics of an ecosystem.

6.2.3.1 Interrelationships of Soil Organic Carbon and Soil Aggregates

The SOC is a primary factor influencing soil structure and aggregate stability (Kay 1998), and in turn is influenced by the dynamics of soil structure (i.e. aggregation distribution and stability) (Elliott and Cambardella 1991). Tisdall and Oades (1982) proposed the aggregate hierarchy concept based upon the influence of SOM as a binding agent. It considered that binding agents act through three major mechanisms: temporary (mainly by polysaccharides), transient (by roots and fungal hyphae) and persistent (by humic substances and polyvalent metal cation complexes, oxides). In this model, it was proposed that fine particles (<20 μm) are bound together by persistent binding agents to form micro-aggregates (53–250 μm).

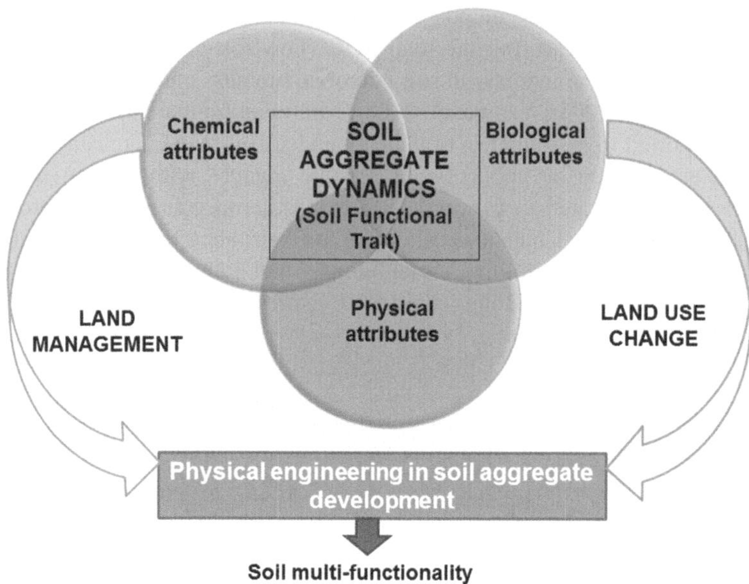

Fig. 6.4 Soil aggregate dynamics as an integrative functional trait of crucial importance

These micro-aggregates, in turn, are bound together by means of temporary and persistent binding agents into macro-aggregates (>250 μm). However, Six et al. (2000) reported an contradictory theory that micro-aggregates are formed within macro-aggregates, which is mediated by the encapsulation of POM by mineral particles and microbial by-products. In addition, the inorganic binding agents (such as oxides and/or carbonates) have also been found to play an important role in aggregate formation (Six et al. 2004). The dynamic relationship between soil organic carbon and aggregate development, factors affecting it and their role in soil multi-functionality is illustrated in Fig. 6.5.

The assessment of chemical characteristics of SOM is commonly used to infer its potential reactivity (Kögel-Knabner et al. 2008), which is responsible for carbon accumulation in the soils. Several researchers (Tisdall and Oades 1982; Saroa and Lal 2001; Kong et al. 2005; Mikha and Rice 2004; Marx et al. 2005; Lagomarsino et al. 2009) showed a higher concentration of carbon and nitrogen in macro-aggregate than micro-aggregate. It is due to the fact that micro-aggregates are bound together by SOM to form macroaggregate (Fig. 6.3). Puget et al. (1995) suggested that greater carbon content in macro-aggregate could be due to less decomposable SOM associated with these aggregates. It is proposed that increase of organic carbon concentration in the free mineral fraction is likely to play a key role in aggregation as well as carbon sequestration (Yu et al. 2012).

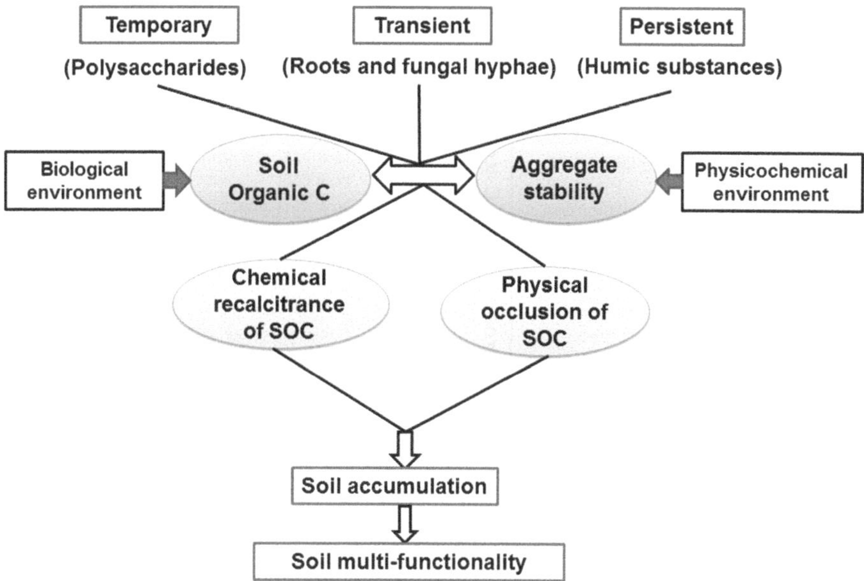

Fig. 6.5 Illustration showing importance of soil organic carbon-aggregate stability interaction in soil multi-functionality

6.2.3.2 Soil Microbe, Plant Roots and Aggregate

Soil microbes play a significant role in soil carbon management by affecting soil structural dynamics (Fig. 6.6). Differential microbial biomass distribution has been reported across aggregate size fractions, which further depends on management systems (Singh and Singh 1996). Similarly, the microbial community composition and function also differ considerably across different aggregate size classes (Gupta and Germida 1988; Hattori 1988; Mummey et al. 2006). Root exudates, which constitute around one third of the plant's photosynthetic production, stimulate the aggregate formation by providing substrate to the carbon-limited microorganisms. Root mucilage can cause short-term stabilization of aggregates via sticking the soil particles together (Morel et al. 1991). Similar biopolymers secreted by the soil microorganisms such as hydrophobins and glomalins (called as 'sticky string bag'; Miller and Jastrow 2000), helps to improve aggregation, supplementing hyphal enmeshment. These microbial secretions serve various purposes in soils such as attachment, nutrient capture and desiccation resistance (Rillig 2005). These fungal metabolic products are either secreted outside or contained in the hyphal wall, which have long been implicated as an important mechanism in soil aggregation (Tisdall and Oades 1982; Chenu 1989). However, despite seemingly higher fungal contribution in soil structural dynamics and C sequestration, it has not been studied extensively in such relations due to technical limitations.

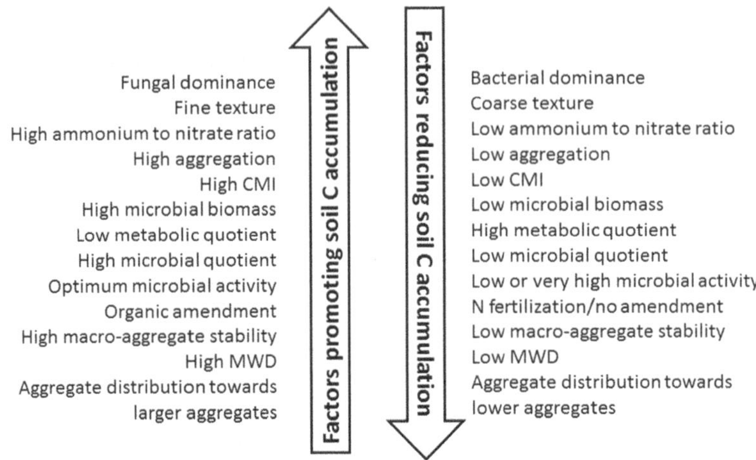

Fig. 6.6 Factors affecting soil carbon dynamics

6.3 Soil Carbon Processes

6.3.1 Litter Inputs and Organic Amendments

Litter-fall represents a major influx of the vegetative carbon to soil, and hence, change in litter input is likely to have wide-reaching consequences for soil carbon dynamics (Sayer et al. 2007). In agro-ecosystems, organic matter above the surface of the soil (termed as surface litter) is unaccounted in the assessment of SOC stocks. However, it may hold significant effect on SOC dynamics in agro-ecosystems. This litter pool represents a small fraction (i.e. 68–97 Pg C) of the global terrestrial carbon pool (Matthews 1997). Its global distribution across the vegetation types has been discussed in detail by Matthews (1997). Litter decomposition has been identified as a critical stage in SOM accumulation and nutrient mineralization (Austin and Ballaré 2010). It is found that the litter decomposition is controlled by climate (Hobbie 1996; Gholz et al. 2000; Hobbie et al. 2000; Zhang et al. 2008) and litter quality (such as, initial litter lignin content or lignin to N ratio (Melillo et al. 1982; Shaw and Harte 2001; Zhang et al. 2008). Litter provides the organic carbon that supports the heterotrophic activity in the soils. It contains biological macromolecules such as proteins, carbohydrates, cellulose, etc. which are highly favored for enzymatic attack, due to their relatively simple chemical structures. In contrast to litter, SOM lacks such a simple structure, and is a highly unfavorable substrate for enzymatic mineralization (Kemmitt et al. 2008). Therefore, litter dynamics differ from SOM dynamics as the mineralization of SOM proceeds at a much slower rate than the decomposition of organic residues, it arises from. The readily decomposable carbon in litter allows rapid growth of the soil microbial community (Agren and Bosatta 1996). Carbon entering into the soils is divided into extant soil carbon pools and soil

CO_2 efflux as decomposition returns most of the carbon added in litter into the atmosphere, except for a very small fraction stored as recalcitrant humus.

Organic amendment shows beneficial impact on SOC sequestration in the labile carbon pool and to a greater extent in recalcitrant carbon pool. The increase in recalcitrant carbon pool could be related to the biochemical resistance of organic carbon compounds contained either in the organic manure or the plant materials (McLauchlan et al. 2006; Yan et al. 2007). Studies had shown that farmyard manure application result in an increase in lignin and similar products, which are major components of the resistant carbon pool (Paul et al. 1997; Rovira and Vallejo 2002; Belay-Tedla et al. 2009). Moreover, combined amendment has also been found to increase both labile and recalcitrant SOC pool in soil than chemical fertilization (Ding et al. 2012). Its long term application has been reported to increase labile SOC fractions like WSC, MBC, POC and $KMnO_4$-oxidisable carbon as compared to the chemical fertilizer alone (Lou et al. 2011). As a result, organic amendment has also been found to increase the soil aggregate stability and structure (Reganold et al. 1987; Pulleman et al. 2003; Hati et al. 2008). Regular addition of organic material to soil is required to improve SOC pools, regulate nutrient fluxes, microbial biomass and activities, and soil physical properties (Marinari et al. 2000).

6.3.2 Carbon Mineralization

Carbon mineralization (also known as soil respiration) is a process in which organic carbon from soil is decomposed and released in the form of inorganic CO_2 into the atmosphere. “Soil respiration” or “soil CO_2 efflux”, which is usually defined as CO_2 released from soil to the atmosphere via the combined activity of roots (i.e. autotrophic or root respiration), and micro- and macro-organisms decomposing litter and organic matter present in the soils (i.e. heterotrophic respiration) (Högberg et al. 2009). Autotrophic respiration includes respiration by root and associated ecto-mycorrhiza and other microorganisms which are dependent on root secretions. However, heterotrophic respiration is reserved for the decomposition of more complex organic molecules present in the litter and other forms of SOM. Several biotic and abiotic factors influence soil CO_2 production: soil temperature and moisture (Epron et al. 1999; Yu et al. 2011), SOM quantity and quality (Couteaux et al. 1995), root and microbial biomass (Ryan et al. 1997), soil physico-chemical properties and site productivity (Subke et al. 2011). The reduction in soil disturbance decreases the rates of SOC mineralization and CO_2 emission (Dendoncker et al. 2004; Al-Kaisi and Yin 2005). Moreover, nitrogen application has been found to reduce both emission rate and season-long cumulative emission of CO_2-carbon from soil (Al-Kaisi et al. 2008). Root exudates and decomposition of litter (both above-ground and belowground), which provide carbon to soils also exert strong control of soil respiration or soil CO_2 efflux (Luyssaert et al. 2007). Globally, it represents a significant carbon flux (about 75 Pg C year^{-1}) which is identified as the second largest flux (contributes 20–38%) of carbon between soils and the atmosphere (Raich and Schlesinger 1992; Schlesinger and Andrews 2000). Annually, it once amounted to about ten time greater addition than that via fossil fuel burning (Mooney et al. 1987). Alternatively, it is also estimated to range from 64 to 72 Gt C year^{-1}, which accounts for 20–40% of the annual CO_2-C input to the atmosphere from terrestrial and marine sources (Houghton and Woodwell 1989; Raich and Schlesinger 1992). It is suggested that even a small shift in the soil CO_2 efflux may dramatically affect the soil carbon sequestration rates and atmospheric CO_2 concentration. Therefore, understanding the processes which affect soil CO_2 efflux and SOC is important for the management of future global climate.

The partitioning of soil respiration (R_s) in the autotrophic (R_a) and heterotrophic respiration (R_h) components is necessary for understanding the implications of environmental change on soil carbon cycling and sequestration (Tian et al. 2011). However, it is inherently difficult to make a precise separation of autotrophic and heterotrophic respiration from soils. The partitioning between these two components is highly variable spatiotemporally, and taxonomic autotrophs and heterotrophs may perform the function of the other group to some extent (Högberg et al. 2009). In reality, the complex situation is perhaps best described as a continuum from strict autotrophy to strict heterotrophy. As a result of this, and associated methodological problems, estimates of the contribution of autotrophic respiration to total soil respiration have been highly variable (Högberg et al. 2009). The observed variations in the sensitivity of soil respiration to temperature or moisture may be ascribed to the variations in the proportions of autotrophic and heterotrophic components of soil respiration (Gomez-Casanovas et al. 2012).

6.3.3 Decomposition of Soil Organic Matter

Carbon sequestration and its release as soil CO_2 efflux primarily depend on decomposition of SOM, which is a major biological process in biogeochemical transformation. Major differences between the tropical and temperate climate in the SOC pool in soils relates with the rate of decomposition, in addition to the land use change and management. Due to difference in the chemical composition of organic materials, the rate of decomposition in the tropics can be four times faster than that in temperate climates under high temperatures (Jenkinson and Ayanaba 1977). It indicates the differential temperature sensitivity of SOC majorly governs its accumulation in the soils. Further, tropical soils are reported to contain more humified, chemically recalcitrant and stable organic matter than that of temperate soils (Grisi et al. 1998), which hold importance in carbon sequestration due to its major share in SOC pool (Rosell et al. 2000). Additionally, the mechanism of protection of the SOC pool also differs between temperate and tropical conditions.

Decomposition is highly sensitive to change in land use and management, as it is a predominantly microbial process. Its differential behavior under varied management practices bears complementary relationship with the abiotic-biotic interaction (i.e. the biophysical interactions) happening inside the soils. Further, decomposition defines the characteristics, storage, turnover and transfer of carbon among various aggregate and particle size fractions and thus, indirectly defines SOC accumulation and soil CO_2 efflux. These SOC fractions with different stabilities and turnover rates are important variables, which are used to detect the influence of agricultural management practices on SOM quality (Silveira et al. 2008). However, these changes in the different SOC fractions under long term organic amendment is not well understood (Ding et al. 2012).

6.3.4 Soil Organic Carbon Immobilization

Organic carbon is immobilized in the soil via its incorporation in microorganisms. SOC immobilization has been widely discussed in relation to nitrogen input in the agro-ecosystems. The higher conversion of organic inputs into microbial biomass represents the stability of SOC. Generally, MBC accounted for 1–5% of total SOC (Jenkinson and Ladd 1981; Smith and Paul 1990). Its relative proportion (i.e. MBC/SOC), which is known as microbial quotient, is reported to range from 0.27 to 7.0 (Jinbo et al. 2007). It reflects the soil's potential to stabilize the organic matter after its addition to the soil (Pascual et al. 1997). The higher ratio represents higher tendency of the organic matter to stabilize. It is a reliable indicator of a turnover of SOC (Joergenson et al. 1994). In general, a situation favoring the accumulation of organic matter increases both the amount of MBC and its proportion to SOC (Collins et al. 1992). It is known that long term application of farmyard manure maintains the soil nutrient levels and stimulates the different aspect of soil fertility, because it ensures the constant presence and turnover of active microorganisms in the soil (Nardi et al. 2004). The carbon immobilization efficiency is found higher in fungi as compared to bacteria, as the former emits less carbon per g of carbon assimilated in the biomass as compared to later.

6.4 Determinants of Soil Carbon Pools and Fluxes

Soil carbon dynamics comprises the study of rate of transfer of SOC among its various soil pools (i.e. turnover) and its regulatory variables. Therefore, it includes the kinetics as well as the variables which define the temporal variations in SOC across its various compartments. The soil carbon dynamics and sequestration depends on various factors such as vegetation, soil texture, temperature, precipitation and management (Ladd et al. 1996; Lal 2004a, b; Alston et al. 2009; Van Wesemael et al. 2010; De Gryze et al. 2011). Factors affecting soil carbon accumulation variably are shown in Fig. 6.6. The important determinants are discussed in brief in the following subsections.

6.4.1 Soil Moisture and Temperature

Soil moisture and temperature conditions are the major regulators of SOC dynamics, which strongly affect the soil CO_2 efflux (Franzluebbers et al. 1995; Ren et al. 2013; Iqbal et al. 2008; Liu et al. 2008). Soil moisture availability promotes the microbial activity, and thus decomposition and mineralization of SOM. The optimal microbial activity occurs at near field capacity, when ~60% of the soil pores are water-filled (Rice 2002). Temperature strongly affects soil CO_2 efflux by modifying the soil physico-chemical and microbial properties (composition and activity) depending upon temperature range and ecosystem-type (Melling et al. 2005). It, thus, defines the rate of organic matter decomposition and nutrient mineralization (Pregitzer and King 2005). Appropriate moisture and temperature conditions in the soil cause a high rate of SOC decomposition, and thus, lower rate of SOC accumulation (Reichstein et al. 2002). Therefore, it is stated that soil moisture and temperature may define the spatial variations of SOC accumulation (Paustian et al. 1998; Freibauer et al. 2004). Moreover, their interaction has also been found to regulate the relative availability of soil nitrate-N and ammonium-N in dry tropical ecosystem, which has been identified as possible mechanism of nutrient conservation and climate change adaptation (Srivastava et al. 2015, 2016).

6.4.2 Soil Texture

Soil texture may affect the SOC sequestration rate (McLauchlan 2006). Gami et al. (2009) reported a positive relationship between silt and clay fractions, and SOC sequestration. Clay particles show higher protective effect on the biophysical and chemical processes of carbon stabilization (Christensen 1996), which may however differ depending upon the mineralogy (Laird et al. 2001). It is, therefore, that SOC decomposition rate decreases with increasing clay content (Hassink 1997; Kong et al. 2009). Clay and silt particle serve as a fixed capacity level (Hassink and Whitmore 1997), while the combination of micro-, meso- and macro-aggregate associated carbon provide an additional, though highly variable capacity. The former is soil-specific, while the later depends on the soil type and management. It is found that SOC content is generally determined by the amount of clay in most studies (Raghubanshi 1992), however, some studies in the tropical soils indicate either an opposite trend or no effect at all.

Tropical soils with predominantly low-activity clays (such as Kaolinite) have low ability to sequester carbon. These soils, mainly Oxisols and Ultisols, cover almost 70% of tropical soils where clay and silt fractions play important role in the amount of SOC pool (Lepsch et al. 1994). Kaolinite is the dominant clay-type mineral (Uehara 1982) in many Oxisols and Ultisols. Highly weathered conditions lead to the presence of sesquioxides (Fe and Al oxides and hydroxides) in tropical soils. In these highly weathered tropical soils, the rate of SOC loss by cultivation is more than that for temperate soils (Shang and Tiessen 1997), which is responsible for the rapid soil degradation in the former. The high clay content in fine textured soil provides the potential for stabilization of added organic residue by its protection within stable soil aggregates (Six et al. 2002). The decomposition of organic residue is reported to be faster in coarse than fine textured soil (Van veen and Kuikman 1990; Strong et al. 2004). It is found associated with the faster macroaggregate turnover and lower SOC and nitrogen in coarse textured soil (Bossuyt et al. 2001; Six et al. 2001).

6.4.3 Soil Aggregate Dynamics

Soil aggregate development is a hierarchical process (Tisdall and Oades 1982). It affects the SOC dynamics as is explained in macroaggregate-microaggregate conceptual model, which is predominantly dependent on the microbial activity (Elliott 1986). The mechanism of aggregate development and its association with soil carbon sequestration and efflux is shown in Fig. 6.7. Different management practices affect the SOC storage variably via its effect on aggregate size distribution and carbon and nitrogen present among them (Beare et al. 1994). Drastic decrease in

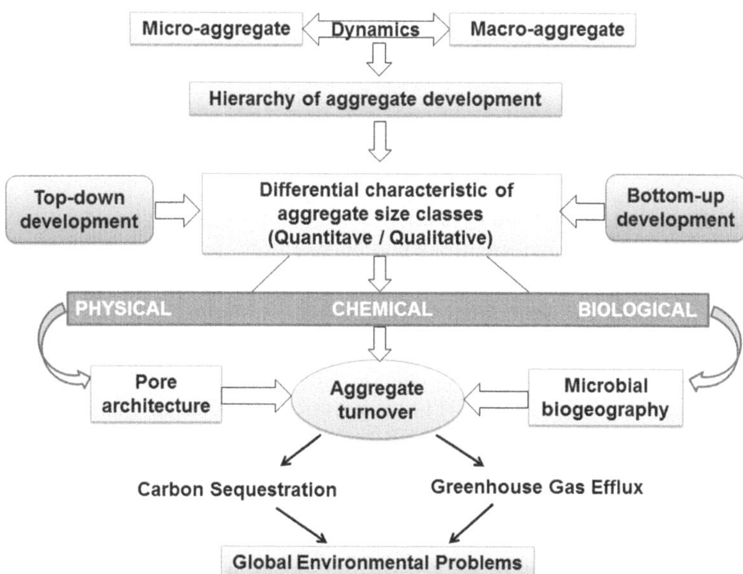

Fig. 6.7 Macroaggregate-microaggregate dynamics with global environmental problems

macro-aggregate water stability (WAS_{macro}) has been observed after cultivation of native soil (Hernández-Hernández and López-Hernández 2002). Mean weight diameter (MWD), which defines the overall structure of soil, has been suggested as sensitive indicator of SOM change (Sodhi et al. 2009). Labile SOC, which sensitively responds to the changes in soil management, has been found to correlate positively with MWD. Despite a history of a century of exploration, only few studies have considered the effect of organic based management practices on the quality of SOC across aggregate size fractions (Hernández-Hernández and López-Hernández 2002). Therefore, the effect of organic amendment on the aggregate physical (i.e. mass distribution and stability) and chemical characteristics (i.e. carbon quality and quantity) across aggregate size fractions is essential to understand the soil carbon dynamics.

Soil structural dynamics, represented by aggregate distribution and carbon turnover across various physical size fractions, represent a trade-off between SOM occlusion and decomposition. Soil aggregates not only protect carbon, but also regulate the soil microbial community structure and functions (Gupta and Germida 1988; Miller et al. 2009). On the contrary, soil microorganisms also help in soil aggregate formation by their secretions such as gum and mucilage (Tisdall and Oades 1982; Watt et al. 2005). The microbial access to SOC and its activity in the undisturbed natural ecosystems is highly constrained, because a major part of the SOM remains occluded within the aggregates pore network. However, human-managed ecosystems show higher SOM decomposition and mineralization due to inappropriate soil aggregate development. It is now widely accepted that the processes which release SOC are regulated by the physical protective capacity of aggregates, which limits the decomposition of SOM (Elliott 1986).

6.4.4 Soil pH

Soil pH affects the soil carbon dynamics by regulating soil enzymatic activity, and thus biogeochemical transformations and energy flow. Its changes are significant in temperate than tropical region, which is generally acidic in nature. It affects the microbial structural and functional attributes (i.e. community structure, dominance and activity) in the soil. Acidic soil pH favors the fungi (e.g. forest), while the basic soil pH favors bacteria (e.g. croplands). This shifts in microbial community dominance due to change in soil pH holds significance in human managed systems, such as cropland, due to chemical fertilization and irrigation. Fungi play a comparatively greater role in soil carbon accumulation than bacteria; however, it has been underestimated so far. Therefore, temporal change in soil pH may have a significant contribution to soil carbon dynamics in tropical croplands due to its significant effect on the microbial growth, composition and activity.

6.4.5 Net Primary Productivity of the Ecosystem

The natural ecosystems (e.g. forest and grassland) have a higher net primary productivity (NPP) than human managed systems (e.g. croplands). The former adds a considerable amount of carbon in soil as rhizo-deposits and litter (above and below the ground). Rhizo-deposit constitutes around 33% of its total photosynthetic production. Chemical fertilization in agro-ecosystems has been reported to shift the relative allocation of photosynthate towards the aboveground parts as compared to belowground. As the plant roots and rhizo-deposits add major carbon pool in the agro-ecosystems than the aboveground litter deposition, the effect of management systems on the carbon input in the agro-ecosystems need to be evaluated. Organic amendments have been found to increase both above and belowground carbon, which leads to increased carbon storage (Ryals and Silver 2013). In some studies, the increase in SOC under mineral fertilization has also been reported, which is ascribed to increased primary production due to nutrient enrichment in the soil.

6.4.6 Nutrient Availability

Soil nutrient availability affects the SOM cycling in nutrient-limited systems (Gärdenäs et al. 2011). The natural ecosystems (such as forests) are nutrient efficient due to their inherent and efficient nutrient scavenging mechanisms. Loss of nutrient from the soil through leaching and continuous removal of SOM is generally observed under agricultural systems. Agricultural management defines the SOC dynamics affecting soil processes (Ogle et al. 2005) including nitrification (Srivastava et al. 2015, 2016). In agro-ecosystems, enhanced nutrient availability through chemical fertilization though enhances the yield, but negatively affects the SOC with the time. More specifically, soil nitrogen availability is a crucial driver of SOM dynamics in agro-ecosystem. For example, the differential dynamics of soil NH_4^+-N and NO_3^--N is reported to have an important ecological significance as it may affect the important ecosystem properties (Bijlsma et al. 2000; Srivastava et al. 2015, 2016). Moreover, it may have important ecosystem consequences in different ways in the changing environment (Cruz et al. 2003; Srivastava et al. 2015, 2016). The consequent differential availability of soil NH_4^+-N and NO_3^--N may have implication on SOC turnover because of their contrasting effects (Currey et al. 2010; Srivastava et al. 2015). Literature findings (Min et al. 2011; Yang et al. 2014; Srivastava et al. 2015, 2016) suggests that the forms of inorganic nitrogen (i.e. soil NH_4^+-N or NO_3^--N) could differentially influence the rate of SOC cycling thus carbon mineralization via changes in soil chemical and biological attributes. Jha et al. (1996) reported that soil moisture content affects the soil nitrification more than ammonification, importantly at low water tension, through dehydration and substrate limitation. A significant relationship between gross nitrogen mineralization and soil CO_2 efflux is also reported in the literature (Flavel and Murphy 2006).

Compost based organic systems shows a direct linkage between nutrient availability and SOC dynamics (Buchanan and King 1992). Organic amendment has been found to increase nutrient availability (Su et al. 2006), though in a different way (Srivastava et al. 2015). It increases the relative availability of ammonium as compared to nitrate affecting microbial processes. The positive correlation of SOC with soil NH_4^+-N availability has also been reported previously (Trasar-Cepeda et al. 1998). Soil nitrogen input via deposition has also been reported to affect the global carbon cycle influencing microbial activities, particularly the carbon acquiring enzymes (Carreiro et al. 2000). It regulates the extracellular enzyme activity and WSC flux by controlling the oxidative enzyme production by soil microbial communities. The regulation of oxidative enzymes by different microbial communities in response to soil NO_3^--N deposition indicates that microbial community composition and function may directly control the ecosystem-level response to environmental changes (Waldrop and Firestone 2006).

6.4.7 Microbial Behavior

Variations in biological and biochemical properties of the soil are closely related with SOC and SON (Trasar-Cepeda et al. 1998). The microbial contribution to soil C sequestration is governed by the interactions between the microbial attributes (such as biomass, community structure, synthesized by-products) and other soil physical properties (such as texture, clay mineralogy, pore-size distribution, and aggregate dynamics) (Six et al. 2006). Furthermore, it could be related to the diverse soil processes, including decomposition of organic residues, nutrient cycling and mobilization of recalcitrant organic matter. The microbial behavior has not been dealt rationally so far despite its crucial role in soil structural dynamics and SOC accumulation, especially the role of fungi as compared to bacteria (Lynch and Bragg 1985). Fungi possess higher growth efficiency than bacteria and produce more recalcitrant organic compounds due to its unique metabolic pathways (Six et al. 2006; Lehmann et al. 2011). Moreover, it helps to associate small aggregates into larger aggregate with the help of various secretions (*i.e.* polysaccharides, organic acids and glycoprotein) in association with extensive hyphal matrix (Tisdall 1991) and plant roots (Gale et al. 2000). Therefore, fungi help to build soil structure necessary for improved soil quality, mediating organic matter accumulation (Tisdall and Oades 1982).

Soil microbes are typically carbon-limited (Smith and Paul 1990) and act as a source of plant nutrient in dry tropical ecosystems (Singh et al. 1989). The changes in SOM are highly related to soil microorganisms due to their primary role in nutrient cycling and SOM turnover (Paul and Clark 1989; Fließbach and Mäder 2000; Chu et al. 2007; Ryan et al. 2009). Soil greenhouse gases (GHGs) production and emission, which are the crucial elements in soil carbon dynamics, is mediated by several microbial processes (Conrad 1996). Recent studies have shown that biological control predominates than chemical recalcitrance (i.e. molecular structure) in SOM stability (Schmidt et al. 2011). Most researches showed that application of organic manure with/without inorganic fertilizers influences SOC (Goyal et al. 1999), soil microbial biomass and community structure (Singh and Singh 1993; Carpenter-Boggs et al. 2000; Peacock et al. 2001; Jimenez et al. 2002; Esperschütz et al. 2007) and activity (Goyal et al. 1999). Thus, it manipulates the soil microorganisms to improve soil health and fertility (Chaparro et al. 2012). Other long term studies have also showed an increase in microbial biomass under fertilization (Yan et al. 2007; Šimon 2008; Wang et al. 2008). The type of amendment (organic or mineral) had an effect on microbial biomass size and activity (Stark et al. 2007). It has been reported that microbial biomass and activity shows higher value in organically managed soils than conventional ones (Tu et al. 2006). Lucas et al. (2014), Larkin et al. (2011) and Saison et al. (2006) observed that compost additions increases fungal biomass and dominance in the soil. Fungal population has been reported higher in acidic soils and in treatments under continuous inorganic fertilization treatments, whereas higher number of bacteria is found in integrated fertilization (Vineela et al. 2012).

Microbial enzymatic activities have been found related with SOC (Jimenez et al. 2002). Their activities represent the limitation of nutrients in soil, for which microorganisms produces the respective enzymes. Strong correlations of enzymatic activities with MBC have been reported in the literature (Ekenler and Tabatabai 2004) except some contrasting studies (Bohme et al. 2005). Soil dehydrogenase (DHA), β glucosidase (β GLU) and alkaline phosphomonoesterase (PME) enzymes play a regulatory role in the SOM decomposition and, thus have been majorly studied in relation to SOC dynamics. The DHA is a membrane-bound enzyme which is related to the living microbial activity. Its increased activity under organic amendment signifies the enhanced biological activity due to carbon input. β GLU is an abiontic and rate limiting enzyme in the carbohydrate decomposition, which determines the availability of carbon in the soil ecosystems. Its activity has also been identified as most sensitive indicator of soil health and management (Caldwell et al. 1999). However, alkaline phosphatase is a microbial-origin enzyme, which determines inorganic phosphorus availability in the soil by acting at the ester bond present in SOM, especially in the nucleic acid and phospholipids.

Long term application of manure has been found to enhance the soil microbiological activities (Parham et al. 2002). Under organic amendment, β GLU activity has been reported to approach to that in the forest (Moeskops et al. 2010). Organic amendment has also been reported to increase the microbial biomass, and DHA and PME activity (Malik et al. 2013; Hueso et al. 2012). In tropical forest soils, biological activity is often limited by P availability (Nottingham et al. 2012). Also, SOC have been found positively correlated with PME activity (Trasar-Cepeda et al. 1998). Therefore, β GLU and PME activity has been reported as the suitable, short and long term indicator of soil quality, respectively (Wick et al. 2002; de la Horra et al. 2003).

6.5 Parameters for Monitoring Soil Ecosystem Health and Quality

6.5.1 *Microbial Eco-physiological Properties*

Soil eco-physiological traits indicate about the ecological well-being of the soil ecosystems. In this respect, soil microbial quotient, basal respiration and metabolic quotient (qCO_2) are the three well known soil eco-physiological traits, which are used to monitor the ecosystem health. Microbial quotient, which is a proportion of MBC to total C (MBC/SOC), denotes (1) C stability and turnover, (2) capability of soil to support microbial growth, (3) efficiency of C utilization by the microorganism, and (4) soil quality (Insam and Domsch 1988). However, basal soil respiration denotes the microbial activity, and the maintenance energy requirement of the soil ecosystem. Alternatively, it also refers to the in vitro soil respiration measured under standard set of conditions. qCO_2, which represents the ratio of basal soil respiration to the MBC (Insam and Haselwandter 1989), demonstrates the efficiency of microbial communities in C substrate utilization in the soil (Insam 1990). Higher the qCO_2 value, lower is the efficiency of the soil ecosystem, and vice-versa. Soils under high stress are expected to show higher qCO_2 than less-stressed soils (Insam and Haselwandter 1989). For example, no-tillage shows lower qCO_2 (Kaschuk et al. 2009) as it does not shatter aggregates and hyphal networks, and therefore, supports a higher ratio of fungi to bacteria (Beare 1997; Frey et al. 1999; Bailey et al. 2002). The reason being, that the fungi have a lower energy requirement for maintenance than the bacteria, and thus, transform the substrate- carbon into microbial- carbon more efficiently (Alvarez et al. 1995; Haynes 1999). It has been reported that the organic systems show higher qCO_2 than conventional systems (Araujo et al. 2008).

6.5.2 *Carbon Management Index*

It is now recognized that it is essential to maintain an appropriate balance between the labile carbon (which is required for nutrient cycling and soil structure) and non-labile carbon (which has potential to get sequestered in soil as reserve carbon due to longer half life) in the soils (Blair et al. 2006). It is because, both the quantity as well as the quality of SOC represents the important indicators of soil health and ecosystem productivity (Liu et al. 2005). Carbon management index (CMI) has been found as an integrative variable in this regard, which integrates the changes in both, labile and non-labile carbon in the soils. It has been found to be a useful variable for monitoring the rate of SOC change in agricultural systems, as compared to precursor reference system. Higher CMI value indicates toward a better carbon sequestration in the soil. The observations on changes in CMI across aggregate size fractions and its relationship with other soil properties in long term management experiments are lacking, which could be helpful in understanding the mechanism of soil carbon dynamics.

The positive relationship between MBC, carbon mineralization and CMI (Xu et al. 2011) suggests that microbial behavior may possibly influence SOC sequestration by differentially affecting the labile and non-labile carbon in the soil as well as across aggregate size fractions. He et al. (2008) reported that CMI of macro-aggregate may act as a promising variable for soil quality assessment in subtropical land use and management. Moreover, Srivastava et al. (2016) reported that SOC dynamics across land use change may be linked with the physical (water stability) and chemical stability (CMI) of macro-aggregate, which could be linked with the change in relative availability of nitrate-N and ammonium-N in the soil. The dynamics of aggregate total, labile and non-labile carbon across age series of organic management and their interaction with biotic (microbial biomass, composition and activity) and abiotic (notably, physico-chemical attributes) is mechanistically not well understood for their impact on SOC and soil CO_2 emission. Though, organic amendment has been found to increase the CMI (Lou et al. 2011; Verma et al. 2013); however, its differential impact on CMI value of aggregate size fractions with age has not been well explored so far.

6.6 Conclusion

The present article clearly emphasizes upon the importance of carbon pools and fluxes in the dry tropical soil ecosystems. Also, it highlights the fact that proper management of soil carbon stocks in dry tropical agricultural soils would provide a dual benefit: enhanced soil fertility with climate change mitigation via carbon sequestration. Therefore, an in-depth mechanistic study of the SOC dynamics is required in these ecosystems to identify the integrative determinants of soil carbon and CO_2 efflux, considering all the important factors together. In recent years, the relative availability of soil inorganic nitrogen pools and carbon management index has shown interesting association with SOC dynamics and other soil properties in dry tropical agro-ecosystems. Such finding needs more exploration and validation across the tropical climate and ecosystems for its identification as an important ecological indicator for enhanced soil carbon sequestration. We advocate for multi-factorial experimentation to study soil carbon pools and fluxes in relation to physico-chemical and biological attributes of integrative nature under various ecosystems, climate and management for a better understanding of SOC dynamics. It would help in the improved soil management for sustainable soil fertility and mitigation of climate change in a simple and cost effective manner.

Acknowledgement We are thankful to the book editors and reviewers for their constructive comments, which helped us in improving the chapter. Also, we acknowledge the financial support from DST-SERB (PDF/2016/003503) and University Grant Commission (UGC), New Delhi, India. PS and RS extend their thanks to Shikha Singh, Institute of Environment & Sustainable Development (IESD), Banaras Hindu University, Varanasi, for helping in drafting the figures of this chapter.

References

Agren GI, Bosatta E (1996) Quality: a bridge between theory and experiment in soil organic matter studies. Oikos 76:522–528

Al-Kaisi MM, Yin X (2005) Tillage and crop residue effects on soil carbon and carbon dioxide emission in corn–soybean rotations. J Environ Qual 34:437–445

Al-Kaisi MM, Kruse ML, Sawyer JE (2008) Effect of nitrogen fertilizer application on growing season soil carbon dioxide emission in a corn–soybean rotation. J Environ Qual 37:325–332

Alston JM, Beddow JM, Pardey PG (2009) Agricultural research, productivity, and food prices in the long run. Science 325:1209–1210

Alvarez R, Dıaz RA, Barbero N, Santanatoglia OJ, Blotta L (1995) Soil organic carbon, microbial biomass and CO_2-C production from three tillage systems. Soil Tillage Res 33:17–28

Araujo ASF, Santos VB, Monteiro RTR (2008) Responses of soil microbial biomass and activity for practices of organic and conventional farming systems in Piauı state, Brazil. Eur J Soil Biol 44:225–230

Austin AT, Ballaré CL (2010) Dual role of lignin in litter decomposition in terrestrial ecosystems. Proc Natl Acad Sci 107:4618–4612

Bailey VL, Smith JL, Bolton H Jr (2002) Fungal-to-bacterial ratios in soils investigated for enhanced C sequestration. Soil Biol Biochem 34:997–1007

Batjes NH (1996) Total carbon and nitrogen in the soils of the world. Eur J Soil Sci 47:151–163

Batjes NH, Sombroek WG (1997) Possibilities for carbon sequestration in tropical and subtropical soils. Glob Chang Biol 3:161–173

Beare MH (1997) Fungal and bacterial pathways of organic matter decomposition and nitrogen mineralization in arable soils. In: Brussaard L, Ferrera-Cerrato R (eds) Soil ecology in sustainable agricultural systems. CRC/Lewis Publishers, Boca Raton, pp 37–70

Beare MH, Hendrix PF, Coleman DC (1994) Water-stable aggregates and organic matter fractions in conventional-tillage and no-tillage soils. Soil Sci Soc Am J 58:777–786

Belay-Tedla A, Zhou XH, Su B, Wan SQ, Luo YQ (2009) Labile, recalcitrant and microbial carbon and nitrogen pools of a tall grass prairie soil in the US Great Plains subjected to experimental warming and clipping. Soil Biol Biochem 41:110–116

Bijlsma RJ, Lambers H, Kooijman SALM (2000) A dynamic whole-plant model of integrated metabolism of nitrogen and carbon. 1. Comparative ecological implications of ammonium-nitrate interactions. Plant Soil 220:49–69

Blair GJ, Lefroy RDB, Lisle L (1995) Soil carbon fractions based on their degree of oxidation, and the development of a carbon management index for agricultural systems. Aust J Agric Res 46:1459–1466

Blair N, Faulkner R, Till A, Poulton P (2006) Long-term management impacts on soil C, N and physical fertility: part I: Broadbalk experiment. Soil Tillage Res 91:30–38

Bohme L, Langer U, Bohme F (2005) Microbial biomass, enzyme activities and microbial community structure in two European long-term field experiments. Agric Ecosyst Environ 109:141–152

Bossuyt H, Denef K, Six J, Frey SD, Merckx R, Paustian K (2001) Influence of microbial populations and residue quality on aggregate stability. Appl Soil Ecol 16:195–208

Brady NC, Weil RR (2002) The nature and properties of soils. Pearson Education Inc, Upper Saddle River

Bronick CJ, Lal R (2005) Soil structure and management: a review. Geoderma 124:3–22

Buchanan M, King LD (1992) Seasonal fluctuations in soil microbial biomass carbon, phosphorus, and activity in no-till and reduced-chemical-input maize agroecosystems. Biol Fertil Soils 13:211–217

Caldwell BA, Griffiths RP, Sollins P (1999) Soil enzyme response to vegetation disturbance in two lowland Costa Rican soils. Soil Biol Biochem 31:1603–1608

Carpenter-Boggs L, Kennedy AC, Reganold JP (2000) Organic and biodynamic management: effects on soil biology. Soil Sci Soc Am J 64:1651–1659

Carreiro MM, Sinsabaugh RL, Repert DA, Parkhurst DF (2000) Microbial enzyme shifts explain litter decay responses to simulated nitrogen deposition. Ecology 81:2359–2365

Carter MR (2002) Soil quality sustainable land management: organic matter and aggregation interactions that maintain soil functions. Agron J 94:38–47

Chantigny MH (2003) Dissolved and water-extractable organic matter in soils: a review on the influence of land use and management practices. Geoderma 113:357–380

Chantigny MH, Angers DA, Rochette P (2002) Fate of carbon and nitrogen from animal manure and crop residues in wet and cold soils. Soil Biol Biochem 34:509–517

Chaparro JM, Sheflin AM, Manter DK, Vivanco JM (2012) Manipulating the soil microbiome to increase soil health and plant fertility. Biol Fertil Soils 48:489–499

Chenu C (1989) Influence of a fungal polysaccharide, scleroglucan, on clay microstructures. Soil Biol Biochem 21:299–305

Christensen BT (1996) Carbon in primary and secondary organomineral complexes. In: Structure and organic matter storage in agricultural soils. CRC Press, Boca Raton, pp 97–165

Chu H, Lin XG, Fujii T, Morimoto S, Yagi K, Hu J, Zhang J (2007) Soil microbial biomass, dehydrogenase activity, bacterial community structure in response to long-term fertilizer management. Soil Biol Biochem 39:2971–2976

Collins HP, Rasmussen PE, Douglas CL (1992) Crop rotation and residue management effects on soil carbon and microbial dynamic. Soil Sci Soc Am J 56:783–788

Conrad R (1996) Soil microorganisms as controllers of atmospheric trace gases (H_2, CO, CH_4, OCS, N_2O, and NO). Microbiol Rev 60:609–640

Cook BD, Allan DL (1992) Dissolved organic carbon in old field soils: compositional changes during the biodegradation of soil organic matter. Soil Biol Biochem 24:595–600

Couteaux MM, Bottner P, Berg B (1995) Litter decomposition, climate and liter quality. Trends Ecol Evol 10:63–66

Cruz C, Lips H, Martins-Loução MA (2003) Nitrogen use efficiency by a slow-growing species as affected by CO 2 levels, root temperature, N source and availability. J Plant Physiol 160:1421–1428

Currey PM, Johnson D, Sheppard LJ, Leith ID, Toberman H, René VDW, Artz RR (2010) Turnover of labile and recalcitrant soil carbon differ in response to nitrate and ammonium deposition in an ombrotrophic peatland. Glob Chang Biol 16:2307–2321

De Gryze S, Lee J, Ogle S, Paustian K, Six J (2011) Assessing the potential for greenhouse gas mitigation in intensively managed annual cropping systems at the regional scale. Agric Ecosyst Environ 144:150–158

de la Horra AM, Conti ME, Palma RM (2003) Beta-glucosidase and proteases activities as affected by long-term management practices in a typic argiudoll soil. Commun Soil Sci Plant Anal 34:2395–2404

Dendoncker N, Van Wesemael B, Rounsevell MD, Roelandt C, Lettens S (2004) Belgium's CO_2 mitigation potential under improved cropland management. Agric Ecosyst Environ 103:101–116

Ding X, Han X, Liang Y, Qiao Y, Li L, Li N (2012) Changes in soil organic carbon pools after 10 years of continuous manuring combined with chemical fertilizer in a Mollisol in China. Soil Tillage Res 122:36–41

Dominy CS, Haynes RJ, van Antwerpen R (2002) Loss of soil organic matter and related soil properties under long-term sugarcane production on two contrasting soils. Biol Fertil Soils 36:350–356

Ekenler M, Tabatabai MA (2004) Arylamidase and amidohydrolases in soils as affected by liming and tillage systems. Soil Tillage Res 77:157–168

Elliott ET (1986) Aggregate structure and carbon, nitrogen and phosphorus in native and cultivated soils. Soil Sc Soc Am J 50:627–633

Elliott ET, Cambardella CA (1991) Physical separation of soil organic matter. Agric Ecosyst Environ 34:407–419

Embacher A, Zsolnay A, Gattinger A, Munch JC (2008) The dynamics of water extractable organic matter (WEOM) in common arable top soils: II. Influence of mineral and combined mineral and manure fertilization in a Haplic Chernozem. Geoderma 148:63–69

Epron D, Farque L, Lucot É, Badot P-M (1999) Soil CO efflux in a beech forest: dependence on soil temperature and soil water content. Ann For Sci 56(3):221–226

Esperschütz J, Gattinger A, Mäder P, Schloter M, Fließbach A (2007) Response of soil microbial biomass and community structures to conventional and organic farming systems under identical crop rotations. FEMS Microb Ecol 61:26–37

Eswaran H, Vandenberg E, Reich P (1993) Organic carbon in soils of the world. Soil Sci Soc Am J 57:192–194

FAO (2000) Land resource potential and constraints at regional and country levels, World Soil Resources Report 90. Food and Agriculture Organisation of the United Nations, Rome

FAO (2001) Soil carbon sequestration for improved land management, World Soil Resources Reports 96, FAO, Rome

Flavel TC, Murphy DV (2006) Carbon and nitrogen mineralization rates after application of organic amendments to soil. J Environ Qual 35:183–193

Fließbach A, Mäder P (2000) Microbial biomass and size-density fractions differ between soils of organic and conventional agricultural systems. Soil Biol Biochem 32:757–768

Franzluebbers AJ, Hons FM, Zuberer DA (1995) Tillage-induced seasonal changes in soil physical properties affecting soil CO2 evolution under intensive cropping. Soil Tillage Res 34:41–60

Freibauer A, Rounsevell MDA, Smith P, Verhagen J (2004) Carbon sequestration in the agricultural soils of Europe. Geoderma 122:1–23

Frey SD, Elliott ET, Paustian K (1999) Bacterial and fungal abundance and biomass in conventional and no-tillage agroecosystems along two climatic gradients. Soil Biol Biochem 31:573–585

Gale WJ, Cambardella CA, Bailey TB (2000) Surface residue and root-derived carbon in stable and unstable aggregates. Soil Sci Soc Am J 64:196–201

Gami SK, Lauren JG, Duxbury JM (2009) Influence of soil texture and cultivation on carbon and nitrogen levels in soils of the eastern indo-gangetic plains. Geoderma 153:304–311

Gärdenäs AI, Ågren GI, Bird JA, Clarholm M, Hallin S, Ineson P, Kätterer T, Knicker H, Nilsson SI, Näsholm T, Ogle S (2011) Knowledge gaps in soil carbon and nitrogen interactions–from molecular to global scale. Soil Biol Biochem 43:702–717

Ghani A, Dexter M, Perrott KW (2003) Hot-water extractable carbon in soils: a sensitive measurement for determining impacts of fertilisation, grazing and cultivation. Soil Biol Biochem 35:1231–1243

Gholz HL, Wedin DA, Smitherman SM, Harmon ME, Parton WJ (2000) Long-term dynamics of pine and hardwood litter in contrasting environments: toward a global model of decomposition. Glob Chang Biol 6:751–765

Gomez-Casanovas N, Matamala R, Cook DR, Gonzalez-Meler MA (2012) Net ecosystem exchange modifies the relationship between the autotrophic and heterotrophic components of soil respiration with abiotic factors in prairie grasslands. Glob Chang Biol 18:2532–2545

Gong W, Yan X, Wang J, Hu T, Gong Y (2009) Long-term manure and fertilizer effects on soil organic matter fractions and microbes under a wheat-maize cropping system in northern China. Geoderma 149:318–324

Goyal S, Chander K, Mundra MC, Kapoor KK (1999) Influence of inorganic fertilizers and organic amendments on soil organic matter and soil microbial properties under tropical conditions. Biol Fertil Soils 29:196–200

Gregorich EG, Carter MR, Angers DA, Monreall CM, Ellert BH (1994) Towards a minimum data set to assess soil organic matter quality in agricultural soils. Can J Soil Sci 74:367–385

Grisi B, Grace C, Brookes PC, Benedetti A, Dell'Abate MT (1998) Temperature effects on organic matter and microbial biomass dynamics in temperate and tropical soils. Soil Biol Biochem 30:1309–1315

Guggenberger G, Kaiser K (2003) Dissolved organic matter in soil: challenging the paradigm of sorptive preservation. Geoderma 113:293–210

Gupta VVSR, Germida JJ (1988) Distribution of microbial biomass and its activity in different soil aggregate size classes as affected by cultivation. Soil Biol Biochem 20:777–786

Hassink J (1997) The capacity of soils to preserve organic C and N by their association with clay and silt particles. Plant Soil 191:77–87

Hassink J, Whitmore AP (1997) A model of the physical protection of organic matter in soils. Soil Sci Soc Am J 61:131–139

Hati KM, Swarup A, Mishra B, Manna MC, Wanjari RH, Mandal KG, Misra AK (2008) Impact of long-term application of fertilizer, manure and lime under intensive cropping on physical properties and organic carbon content of an Alfisol. Geoderma 148:173–179

Hattori T (1988) Soil aggregates as microhabitats of microorganisms. Rep Inst Agric Res Tohoku Univ 37:23–36

Haynes RJ (1999) Size and activity of the soil microbial biomass under grass and arable management. Biol Fertil Soils 30:210–216

Haynes RJ (2005) Labile organic matter fractions as central components of the quality of agricultural soils: an overview. Adv Agron 85:221–268

He Y, Xu Z, Chen C, Burton J, Ma Q, Ge Y, Xu J (2008) Using light fraction and macroaggregate associated organic matters as early indicators for management-induced changes in soil chemical and biological properties in adjacent native and plantation forests of subtropical Australia. Geoderma 147:116–125

Hernández-Hernández R, López-Hernández D (2002) Microbial biomass, mineral nitrogen and carbon content in savanna soil aggregates under conventional and no-tillage. Soil Biol Biochem 34:1563–1570

Hobbie SE (1996) Temperature and plant species control over litter decomposition in Alaskan tundra. Ecol Monogr 66:503–522

Hobbie SE, Schimel JP, Trumbore SE, Randerson JR (2000) Controls over carbon storage and turnover in high-latitude soils. Glob Chang Biol 6:196–210

Högberg P, Löfvenius MO, Nordgren A (2009) Partitioning of soil respiration into its autotrophic and heterotrophic components by means of tree-girdling in old boreal spruce forest. For Ecol Manag 257:1764–1767

Houghton RA, Woodwell GM (1989) Global climatic change. Sci Am U S 260(4):36–44

Huang ZQ, Xu ZH, Chen CR (2008) Effect of mulching on labile soil organic matter pools, microbial community functional diversity and nitrogen transformations in two hardwood plantations of subtropical Australia. Appl Soil Ecol 40:229–239

Hueso S, Garcia C, Hernandez T (2012) Severe drought conditions modify the microbial community structure, size and activity in amended and unamended soils. Soil Biol Biochem 50:167–173

Insam H (1990) Are the soil microbial biomass and basal respiration governed by the climatic regime? Soil Biol Biochem 22:525–532

Insam H, Domsch KH (1988) Relationship between soil organic carbon and microbial biomass on chronosequences of reclamation sites. Microb Ecol 15:177–188

Insam H, Haselwandter K (1989) Metabolic quotient of the soil microflora in relation to plant succession. Oecologia 79:174–178

Iqbal J, Ronggui H, Lijun D, Lan L, Shan L, Tao C, Leilei R (2008) Differences in soil CO2 flux between different land use types in mid-subtropical China. Soil Biol Biochem 40:2324–2333

Jenkinson DS, Ayanaba A (1977) Decomposition of carbon-14 labeled plant material under tropical conditions. Soil Sci Soc Am J 41:912–915

Jenkinson DS, Ladd JN (1981) Microbial biomass in soil: measurement and turnover. In: Paul EA, Ladd JN (eds) Soil biochemistry, vol 5. Mercel Decker, Inc, New York, pp 415–471

Jha PB, Singh JS, Kashyap AK (1996) Dynamics of viable nitrifier community and nutrient availability in dry tropical forest habitat as affected by cultivation and soil texture. Plant Soil 180:277–285

Jimenez M, Horra AM, Pruzzo L, Palma RM (2002) Soil quality: a new index based on microbiological and biochemical parameters. Biol Fertil Soil 35:302–306

Jinbo Z, Changchun S, Wenyan Y (2007) Effects of cultivation on soil microbiological properties in a freshwater marsh soil in Northeast China. Soil Tillage Res 93:231–235

Jobbagy EG, Jackson RB (2000) The vertical distribution of soil organic carbon and its relation to climate and vegetation. Ecol Appl 2:423–436

Joergenson RG, Meyer B, Mueller T (1994) Time course of soil microbial biomass under wheat: a one year field study. Soil Biol Biochem 26:987–994

Kaschuk G, Alberton O, Hungria M (2009) Three decades of soil microbial biomass studies in Brazilian ecosystems: lessons learned about soil quality and indications for improving sustainability. Soil Biol Biochem 42:1–13

Kay BD (1998) Soil structure and organic carbon: a review. In: Lal R, Kimble JM, Follett RF, Stewart BA (eds) Soil processes and the carbon cycle. CRC Press, Boca Raton, pp 169–197

Kemmitt SJ, Lanyon CV, Waite IS, Wen Q, Addiscott TM, Bird NRA, Brookes PC (2008) Mineralization of native soil organic matter is not regulated by the size, activity or composition of the soil microbial biomass—a new perspective. Soil Biol Biochem 40:61–73

Kögel-Knabner I, Ekschmitt K, Flessa H, Guggenberger G, Matzner E, Marschner B (2008) An integrative approach of organic matter stabilization in temperate soils: linking chemistry, physics and biology. J Plant Nutr Soil Sci 171:5–13

Kong AYY, Six J, Bryant DC, Denison RF, Kessel C (2005) The relationship between carbon input, aggregation, and soil organic carbon stabilization in sustainable cropping systems. Soil Sci Soc Am J 69:1078–1085

Kong X, Dao TH, Qin J, Qin H, Li C, Zhang F (2009) Effects of soil texture and land use interactions on organic carbon in soils in North China cities' urban fringe. Geoderma 154:86–92

Ladd JN, Foster RC, Nannipieri P, Oades JM (1996) Soil structure and biological activity. In: Stotzky G, Bollag J (eds) Soil biochemistry. Marcel Dekker, New York, pp 23–78

Lagomarsino A, Grego S, Marhan S, Moscatelli MC, Kandeler E (2009) Soil management modifies micro-scale abundance and function of soil microorganisms in a Mediterranean ecosystem. Eur J Soil Sci 60:2–12

Laird DA, Martens DA, Kingery WL (2001) Nature of clay-humic complexes in an agricultural soil. Soil Sci Soc Am J 65:1413–1418

Lal R (2004a) Carbon sequestration in dryland ecosystems. Environ Manag 33:528–544

Lal R (2004b) Soil carbon sequestration to mitigate climate change. Geoderma 123:1–22

Lal R (2006) Enhancing crop yields in the developing countries through restoration of the soil organic carbon pool in agricultural lands. Land Degrad Dev 17:197–209

Lal R, Follett RF, Stewart BA, Kimble JM (2007) Soil carbon sequestration to mitigate climate change and advance food security. Soil Sci 172:943–956

Larkin RP, Honeycutt CW, Griffin TS, Olanya OM, Halloran JM, He Z (2011) Effects of different potato cropping system approaches and water management on soilborne diseases and soil microbial communities. Phytopathology 101:58–67

Lehmann J, Rillig MC, Thies J, Masiello CA, Hockaday WC, Crowley D (2011) Biochar effects on soil biota–a review. Soil Biol Biochem 43:1812–1836

Lepsch IF, Menk JRF, Oliveira JD (1994) Carbon storage and other properties of soils under agriculture and natural vegetation in Sao Paulo State, Brazil. Soil Use Manag 10:34–42

Lichtfouse E (1997) Heterogeneous turnover of molecular organic substances from crop soils as revealed by 13C labeling at natural abundance with Zea mays. Naturwissenschaften 84:23–25

Liu X, Herbert SJ, Hashemi AM, Zhang X, Ding G (2005) Effects of agricultural management on soil organic matter and carbon transformation – a review. Plant Soil Environ 52:531–543

Hui Liu, Ping Zhao, Ping Lu, Yue-Si Wang, Yong-Biao Lin, Xing-Quan Rao (2008) Greenhouse gas fluxes from soils of different land-use types in a hilly area of South China. Agric Ecosyst Environ 124(1–2):125–135

Lou Y, Wang J, Liang W (2011) Impacts of 22-year organic and inorganic N managements on soil organic C fractions in a maize field, Northeast China. Catena 87:386–390

Lucas ST, D'Angeloa EM, Williams MA (2014) Improving soil structure by promoting fungal abundance with organic soil amendments. Appl Soil Ecol 75:13–23

Luyssaert S, Inglima I, Jung M, Richardson AD, Reichstein M, Papale D, Piao SL, Schulze ED, Wingate L, Matteucci G, Aragao LE (2007) CO_2 balance of boreal, temperate, and tropical forests derived from a global database. Glob Chang Biol 13:2509–2537

Lynch JM, Bragg E (1985) Microorganisms and aggregate stability. Adv Soil Sci 2:133–171

Malik MA, Khan KS, Marschner P, Ali S (2013) Organic amendments differ in their effect on microbial biomass and activity and on P pools in alkaline soils. Biol Fertil Soils 49:415–425

Marinari S, Mascinadaro G, Ceccanti B, Grego S (2000) Influence of organic and mineral fertilizers on soil biological and physical properties. Bioresour Technol 72:9–17

Marschner AD, Kalbitz K (2003) Controls of bioavailability and biodegradability of dissolved organic matter in soils. Geoderma 113:211–235

Martens DA (2000) Management and crop residue influence soil aggregate stability. J Environ Qual 29:723–727

Martens DA, Frankenberger WT Jr (1992) Modification of infiltration rates in an organic-amended irrigated soil. J Agron 84:707–717

Marx MC, Kandeler E, Wood M, Wermbter N, Jarvis SC (2005) Exploring the enzymatic landscape: distribution and kinetics of hydrolytic enzymes in soil particle-size fractions. Soil Biol Biochem 37:35–48

Matthews E (1997) Global litter production, pools, and turnover times: estimates from measurement data and regression models. J Geophys Res 102:18771–18800

McLauchlan KK (2006) Effects of soil texture on soil carbon and nitrogen dynamics after cessation of agriculture. Geoderma 136:289–299

McLauchlan KK, Hobbie SE, Post WM (2006) Conversion from agriculture to grassland builds soil organic matter on decadal timescales. Ecol Appl 16:143–153

Melillo JM, Aber JD, Muratore JF (1982) Nitrogen and lignin control of hardwood leaf litter decomposition dynamics. Ecology 63:621–626

Melling L, Hatano R, Goh KJ (2005) Soil CO_2 flux from three ecosystems in tropical peatland of Sarawak, Malaysia. Tellus B 57:1–11

Mikha MM, Rice CW (2004) Tillage and manure effects on soil and aggregate-associated carbon and nitrogen. Soil Sci Soc Am J 68:809–816

Miller RM, Jastrow JD (2000) Mycorrhizal fungi influence soil structure. In: Kapulnik Y, Douds DD (eds) Arbuscular mycorrhizae: molecular biology and physiology. Kluwer Academic Publishers, Dordrecht, pp 3–18

Miller MN, Zebarth BJ, Dandie CE, Burton DL, Goyer C, Trevors JT (2009) Denitrifier community dynamics in soil aggregates under permanent grassland and arable cropping systems. Soil Sci Soc Am J 73:1843–1851

Min K, Kang H, Lee D (2011) Effects of ammonium and nitrate additions on carbon mineralization in wetland soils. Soil Biol Biochem 43:2461–2469

Moeskops B, Sukristiyonubowo Buchan D, Sleutel S, Herawaty L, Husen E, Saraswati R, Setyorini D, De Neve S (2010) Soil microbial communities and activities under intensive organic and conventional vegetable farming in West Java, Indonesia. Appl Soil Ecol 45:112–120

Mooney HA, Vitousek PM, Matson PA (1987) Exchange of materials between terrestrial ecosystems and the atmosphere. Science 238:926–932

Moraes JL, Cerri CC, Melillo JM, Kicklighter D, Neill C, Steudler PA, Skole DL (1995) Soil carbon stocks of the Brazilian Amazon basin. Soil Sci Soc Am J 59:244–247

Morel JL, Habib L, Plantureux S, Guckert A (1991) Influence of maize root mucilage on soil aggregate stability. Plant Soil 136:111–119

Mummey DL, Rillig MC, Six J (2006) Endogeic earthworms differentially influence bacterial communities associated with different soil aggregate size fractions. Soil Biol Biochem 38:1608–1614

Murphy PG, Lugo AE (1986) Ecology of tropical dry forest. Annu Rev Ecol Syst 17:67–88

Nardi S, Morari F, Berti A, Tosoni M, Giardini L (2004) Soil organic matter properties after 40 years of different use of organic and mineral fertilizers. Eur J Agron 21:357–367

Nottingham AT, Turner BL, Chamberlain PM, Stott AW, Tanner EVJ (2012) Priming and microbial nutrient limitation in lowland tropical forest soils of contrasting fertility. Biogeochemistry 111:219–237

Nyamangara J, Piha MI, Kirchmann H (1999) Interactions of aerobically decomposed cattle manure and nitrogen fertilizer applied to soil. Nutr Cycl Agroecosyst 54:183–188

Ogle SM, Breidt FJ, Paustian K (2005) Agricultural management impacts on soil organic carbon storage under moist and dry climatic conditions of temperate and tropical regions. Biogeochemistry 72:87–121

Pan G, Smith P, Pan W (2009) The role of soil organic matter in maintaining the productivity and yield stability of cereals in China. Agric Ecosyst Environ 129:344–348

Parham JA, Deng SP, Raun WR, Johnson GV (2002) Long-term cattle manure application in soil – I. Effect on soil phosphorus levels, microbial biomass C, and dehydrogenase and phosphatase activities. Biol Fertil Soils 35:328–337

Parton WJ, Schimel DS, Cole CV, Ojima DS (1987) Analysis of factors controlling soil organic levels of grasslands in the Great Plains. Soil Sci Soc Am J 51:1173–1179

Pascual JA, Pascual G, Garcıa JA (1997) Changes in the microbial activity of an arid soil amended with urban organic wastes. Biol Fertil Soils 24:429–434

Paul EA, Clark FE (1989) Soil microbiology and biochemistry, 2nd edn. Academic, San Diego, pp 131–146

Paul EA, Follett RF, Leavitt SW, Halvorson A, Peterson GA, Lyon DJ (1997) Radiocarbon dating for determination of soil organic matter pool sizes and dynamics. Soil Sci Soc Am J 61:1058–1067

Paul EA, Collins HP, Leavitt SW (2001) Dynamics of resistant soil carbon of Midwestern agricultural soils measured by naturally occurring ^{14}C abundance. Geoderma 104:239–256

Paustian K, Cole CV, Sauerbeck D, Sampson N (1998) CO_2 mitigation by agriculture: an overview. Clim Chang 40:135–162

Peacock AD, Mullen MD, Ringelberg DB, Tyler DD, Hedrick DB, Gale PM, White DC (2001) Soil microbial community responses to dairy manure or ammonium nitrate applications. Soil Biol Biochem 33:1011–1019

Post WM, Emanuel WR, Zinke PJ, Stangenberger AG (1982) Soil carbon pools and world life zones. Nature 298:156–159

Powlson DS, Brookes PC, Christensen BT (1987) Measurement of soil microbial biomass provides an early indication of changes in total soil organic matter due to straw incorporation. Soil Biol Biochem 19:159–164

Pregitzer KS, King JS (2005) Effects of soil temperature on nutrient uptake. In: Nutrient acquisition by plants. Springer, Berlin/Heidelberg, pp 277–310

Puget P, Chenu C, Balesdent J (1995) Total and young organic matter distributions in aggregates of silty cultivated soils. Eur J Soil Sci 46:449–459

Pulleman M, Jongmans A, Marinissen J, Bouma J (2003) Effects of organic versus conventional arable farming on soil structure and organic matter dynamics in a marine loam in the Netherlands. Soil Use Manag 19:157–165

Purakayastha TJ, Rudrappa L, Singh D, Swarup A, Bhadraray S (2008) Long-term impact of fertilizers on soil organic carbon pools and sequestration rates in maize- wheat.-cowpea cropping system. Geoderma 144:370–378

Raghubanshi AS (1992) Effect of topography on selected soil properties and nitrogen mineralization in a dry tropical forest. Soil Biol Biochem 24:145–150

Raich JW, Schlesinger WH (1992) The global carbon dioxide flux in soil respiration and its relationship to vegetation and climate. Tellus B 44:81–99

Reganold JP, Elliott LF, Unger YL (1987) Long-term effects of organic and conventional farming on soil erosion. Nature 330:370–372

Reichstein M, Tenhunen JD, Roupsard O, Ourcival J, Rambal S, Miglietta F, Peressotti A, Pecchiari M, Tirone G, Valentini R (2002) Severe drought effects on ecosystem CO_2 and H_2O fluxes at three Mediterranean evergreen sites: revision of current hypotheses? Glob Chang Biol 10:999–1017

Ren LY, Xiang L, Rong SQ, Chun XUV (2013) Enzyme activity in water stable soil aggregates as affected by long term application of organic manure and chemical fertilizer. Pedosphere 23:111–119

Rice CW (2002) Storing carbon in soil: why and how? Geotimes 47:14–17

Rillig MC (2005) A connection between fungal hydrophobins and soil water repellency? Pedobiologia 49:395–399

Rilling MC, Steinberg PD (2002) Glomalin production by an arbuscularmycorrhizal fungus: a mechanism of habitat modification? Soil Biol Biochem 34:1371–1374

Rosell RA, Galantini JA, Suñer LG (2000) Long-term crop rotation effects on organic carbon, nitrogen, and phosphorus in haplustoll soil fractions. Arid Soil Res Rehabil 14:309–315

Rosenberg NJ, Izaurralde RC, Malone EL (1999) Carbon sequestration in soils: monitoring and beyond. Battelle Press, Columbus

Rovira P, Vallejo VR (2002) Labile and recalcitrant pools of carbon and nitrogen in organic matter decomposing at different depths in soil: an acid hydrolysis approach. Geoderma 107:109–141

Rudrappa L, Purakayastha T, Singh D, Bhadraray S (2006) Long-term manuring and fertilization effects on soil organic carbon pools in a Typic Haplustept of semi-arid sub-tropical India. Soil Tillage Res 88:180–192

Ryals R, Silver WL (2013) Effects of organic matter amendments on net primary productivity and greenhouse gas emissions in annual grasslands. Ecol Appl 23:46–69

Ryan MG, Lavigne MB, Gower ST (1997) Annual carbon cost of autotrophic respiration in boreal forest ecosystems in relation to species and climate. J Geophys Res Atmos 102:28871–28883

Ryan J, Masri S, Singh M (2009) Seasonal changes in soil organic matter and biomass and labile forms of carbon as influenced by crop rotations. Commun Soil Sci Plant Anal 40:188–199

Saison C, Degrange V, Oliver R, Millard P, Commeaux C, Montange D, Le Roux X (2006) Alteration and resilience of the soil microbial community following compost amendment: effects of compost level and compost-borne microbial community. Environ Microbiol 8:247–257

Saroa GS, Lal R (2001) Mulching effect on aggregation and carbon sequestration in a Miamian soil in Central Ohio. Land Degrad Dev 14:481–493

Sayer JA, Maginnis S, Laurie M (2007) Forests in landscapes: ecosystem approaches to sustainability. Routledge, London

Schimel DS (1995) Terrestrial ecosystems and the carbon cycle. Glob Chang Biol 1:77–91

Schlesinger WH (1995) An overview of the carbon cycle. In: Soils and global change. CRC Press, Boca Raton, pp 9–25

Schlesinger WH, Andrews JW (2000) Soil respiration and the global carbon cycle. Biogeochemistry 48:7–20

Schmidt MWI, Torn MS, Abiven S, Dittmar T, Guggenberger G, Jannssens IA, Kleber M, Kögel-Knabner I, Lehmann J, Manning DAC, Nannipieri P, Rasse DP, Weiner S, Trumbore SE (2011) Persistence of soil organic matter as an ecosystem property. Nature 478:49–56

Schnitzer M (1991) Soil organic matter-the next 75 years. Soil Sci 151:41–58

Scurlock JMO, Hall DO (1998) The global carbon sink: a grassland perspective. Glob Chang Biol 4:229–233

Shang C, Tiessen H (1997) Organic matter lability in a tropical oxisol: evidence from shifting cultivation, chemical oxidation, particle size, density, and magnetic fractionations. Soil Sci 162:795–807

Shaw MR, Harte J (2001) Control of litter decomposition in a subalpine meadow-sagebrush steppe ecotone under climate change. Ecol Appl 11:1206–1223

Silveira ML, Comerford NB, Reddy KR, Cooper WT, El-Rifai H (2008) Characterization of soil organic carbon pools by acid hydrolysis. Geoderma 144:405–414

Šimon T (2008) The influence of long-term organic and mineral fertilization on soil organic matter. Soil Water Res 3:41–51

Singh H, Singh KP (1993) Effect of residue placement and chemical fertilizer on soil microbial biomass under tropical dryland cultivation. Biol Fertil Soils 16:275–281

Singh S, Singh JS (1996) Water-stable aggregates and associated organic matter in forest, savanna, and cropland soils of a seasonally dry tropical region, India. Biol Fertil Soils 22:76–82

Singh JS, Raghubanshi AS, Singh RS, Srivastava SC (1989) Microbial biomass acts as a source of plant nutrients in dry tropical forests and savanna. Nature 338:499–500

Singh R, Babu JN, Kumar R, Srivastava P, Singh P, Raghubanshi AS (2015) Multifaceted application of crop residue biochar as a tool for sustainable agriculture: an ecological perspective. Ecol Eng 77:324–347

Six J, Elliott ET, Paustian K (2000) Soil macroaggregate turnover and microaggregate formation: a mechanism for C-sequestration under no-tillage agriculture. Soil Biol Biochem 32:2099–2103

Six J, Carpentier A, van Kessel C, Merckx R, Harris D, Horwath WR, Luscher A (2001) Impact of elevated CO_2 on soil organic matter dynamics as related to changes in aggregate turnover and residue quality. Plant Soil 234:27–36

Six J, Conant RT, Paul EA, Paustian K (2002) Stabilization mechanisms of soil organic matter: implications for C-saturation of soils. Plant Soil 241:155–176

Six J, Bossuyt H, Degryze S, Denef K (2004) A history of research on the link between (micro) aggregates, soil biota, and soil organic matter dynamics. Soil Tillage Res 79:7–31

Six J, Frey SD, Thiet RK, Batten KM (2006) Bacterial and fungal contributions to carbon sequestration in agroecosystems. Soil Sci Soc Am J 70:555–569

Smith JL, Paul EA (1990) The significance of soil microbial biomass estimations. In: Bollag JM, Stotzky G (eds) Soil biochemistry, vol 6. Marcel Dekker, Inc, New York, pp 357–396

Smith OH, Petersen GW, Needelman BA (2000) Environmental indicators of agroecosystems. Adv Agron 69:75–97

Sodhi GPS, Beri V, Benbi DK (2009) Using carbon management index to assess the impact of compost application on changes in soil carbon after ten years of rice-wheat cropping. Commun Soil Sci Plant Anal 40:3491–3502

Sombroek WG, Nachtergaele FO, Hebel A (1993) Amounts, dynam-ics and sequestering of carbon in tropical and subtropical soils. Ambio 22:417–425

Sparling GP (1992) Ratio of microbial biomass carbon to soil organic carbon as a sensitive indicator of changes in soil organic matter. Aust J Soil Res 30:195–207

Srivastava SC, Singh JS (1989) Effect of cultivation on microbial biomass C and N of dry tropical forest soil. Biol Fertil Soils 8:343–348

Srivastava P, Raghubanshi AS, Singh R, Tripathi SN (2015) Soil carbon efflux and sequestration as a function of relative availability of inorganic N pools in dry tropical agroecosystem. Appl Soil Ecol 96:1–6

Srivastava P, Singh PK, Singh R, Bhadouria R, Singh DK, Singh S, Afreen T, Tripathi SN, Singh P, Singh H, Raghubanshi AS (2016) Relative availability of inorganic N-pools shifts under land use change: an unexplored variable in soil carbon dynamics. Ecol Indic 64:228–236

Stark C, Condron LM, Stewart A, Di HJ, O'Callaghan M (2007) Influence of organic and mineral amendments on microbial soil properties and processes. Appl Soil Ecol 35:79–93

Stevenson FJ (ed) (1994) Humus chemistry. Wiley, New York

Stewart CE, Paustian K, Conant RT, Plante AF, Six J (2007) Soil carbon saturation: concept, evidence and evaluation. Biogeochemistry 86:19–31

Strong DT, Wever HD, Merckx R, Recous S (2004) Spatial location of carbon decomposition in the soil pore system. Eur J Soil Sci 55:739–750

Su YZ, Wang F, Suo DR, Zhang ZH, Du MW (2006) Long- term effect of fertilizer and manure application on soil-carbon sequestration and soil fertility under the wheat-wheat-maize cropping system in Northwest China. Nutr Cycl Agroecosyst 75:285–295

Subke J-A, Voke NR, Leronni V, Garnett MH Phil Ineson(2011) Dynamics and pathways of autotrophic and heterotrophic soil CO2 efflux revealed by forest girdling. J Ecol 99(1):186–193

Tian H, Xu X, Lu C, Liu M, Ren W, Chen G et al (2011) Net exchanges of CO2, CH4, and N2O between China's terrestrial ecosystems and the atmosphere and their contributions to global climate warming. J Geophys Res Biogeo 116(G2)

Tisdall JM (1991) Fungal hyphae and structural stability of soil. Aust J Soil Res 29:729–743

Tisdall JM, Oades JM (1982) Organic matter and water-stable aggregates in soils. J Soil Sci 33:141–163

Trasar-Cepeda C, Leiros C, Gil-Sotres F, Seoane S (1998) Towards a biochemical quality index for soils: an expression relating several biological and biochemical properties. Biol Fertil Soils 26:100–106

Tu C, Ristaino JB, Hu S (2006) Soil microbial biomass and activity in organic tomato farming systems: effects of organic inputs and straw mulching. Soil Biol Biochem 38:247–255

Uehara G (1982) Soil science for the tropics. In: Proceedings of the ASCE geotechnical engineering division specialty conference (ed) Engineering and construction in tropical and residual soils, Honolulu, pp 13–26

Van Veen JA, Kuikman PJ (1990) Soil structural aspects of decomposition of organic matter by micro-organisms. Biodegradation 11:213–233

Van Wesemael B, Paustian K, Meersmans J, Goidts E, Barancikova G, Easter M (2010) Agricultural management explains historic changes in regional soil carbon stocks. Proc Natl Acad Sci 107:14926–14930

Verma BC, Datta SP, Rattan RK, Singh AK (2013) Labile and stabilised fractions of soil organic carbon in some intensively cultivated alluvial soils. J Environ Biol 34:1069–1075

Vinther FP, Hansen EM, Olesen JE (2004) Effects of plant residues on crop performance, N mineralisation and microbial activity including field CO_2 and N_2O fluxes in unfertilised crop rotations. Nutr Cycl Agroecosyst 70:189–199

Vineela C, Wani SP, Srinivasarao C, Padmaja B, Vittal KPR (2008) Microbial properties of soils as affected by cropping and nutrient management practices in several long-term manurial experiments in the semi-arid tropics of India. Appl Soil Ecol 40(1):165–173

Waldrop MP, Firestone MK (2006) Response of microbial community composition and function to soil climate change. Microb Ecol 52:716–724

Wang QL, Bai YH, Gao HW, He J, Chen H, Chesney RC, Kuhn NJ, Li HW (2008) Soil chemical properties and microbial biomass after 16 years of no-tillage fanning on the Loess Plateau, China. Geoderma 144:502–508

Watt M, Kirkegaard JA, Rebetzke GJ (2005) A wheat genotype developed for rapid leaf growth copes well with the physical and biological constraints of unploughed soil. Funct Plant Biol 32:695–706

Wick B, Kuhne RF, Vielhauer K, Vlek PLG (2002) Temporal variability of selected soil microbiological and biochemical indicators under different soil quality conditions in South-Western Nigeria. Biol Fertil Soils 35:155–167

Xu M, Louy SX, Wang W, Baniyamuddin M, Zhao K (2011) Soil organic carbon active fractions as early indicators for total carbon change under straw incorporation. Biol Fertil Soils 47:745–752

Yan D, Wang D, Yang L (2007) Long-term effect of chemical fertilizer, straw, and manure on labile organic matter fractions in a paddy soil. Biol Fertil Soils 44:93–101

Yang C, Yang L, Ouyang Z (2005) Organic carbon and its fractions in paddy soil as affected by different nutrient and water regimes. Geoderma 124:133–142

Yang K, Zhu J, Xu S (2014) Influences of various forms of nitrogen additions on carbon mineralization in natural secondary forests and adjacent larch plantations in Northeast China. Can J For Res 44:441–448

Yu X, Zha T, Pang Z, Wu B, Wang X, Chen G, Li C, Cao J, Jia G, Li X, Wu H (2011) Response of soil respiration to soil temperature and moisture in a 50-year-old oriental arborvitae plantation in China. PLoS One 6:e28397

Yu H, Ding W, Luo J, Genga R, Cai Z (2012) Long-term application of organic manure and mineral fertilizers on aggregation and aggregate-associated carbon in a sandy loam soil. Soil Tillage Res 124:170–177

Zhang D, Hui D, Luo Y, Zhou G (2008) Rates of litter decomposition in terrestrial ecosystems: global patterns and controlling factors. J Plant Ecol 1:85–93

Zsolnay A (1996) Dissolved humus in soil waters. In: Piccolo A (ed) Humic substances in terrestrial ecosystems. Elsevier, Amsterdam, pp 171–223

Chapter 7
Halophilic Microbial Ecology for Agricultural Production in Salt Affected Lands

Sanjay Arora and Meghna J. Vanza

Abstract Halophiles microbes are present in hypersaline environments. Several alkaliphilic *Bacillus* species isolated from soils show halophilic characteristics. Genera that include halophilic species isolated from soil samples are *Halobacillus, Filobacillus, Tenuibacillus, Lentibacillus,* and *Thalassobacillus.* Species from *Filobacillus, Thalassobacillus, Lentibacillus* and *Tenuibacillus* genera are moderately halophile. The family *Nocardiopsaceae* predominate in saline or alkaline soils. Many Gram-negative, moderately halophilic, or halotolerant species are included in the family *Halomonadaceae*. Microorganisms from the genus *Streptomonospora,* which are Gram-positive, aerobic organisms with branching hyphae, are found to grow upto 15% NaCl.

Mycorrhizal fungi can increase the growth of plants growing in salinity. Vesicular arbuscular mycorrhizal fungi have the ability to protect plants from salt stress. Compatible solute strategy is employed by the majority of moderately halophilic and halotolerant bacteria. All halophilic microorganisms contain potent transport mechanisms, generally based on Na^+/H^+ antiporters, to expel sodium ions from the interior of the cell. Also, some halophiles express aminocyclopropane-1-carboxylic acid (ACC) deaminase activity that removes stress, ethylene from the rhizosphere and some produce auxins that promote root growth. Plant growth-promoting rhizobacteria induces plants salt stress tolerance. Inoculation of halophilic plant growth-promoting bacterial strains reduces sodium by 19% in soil. Also, with such method, the yield of wheat and *Zea mays* can be increased by 10–12% under salinity stress. Liquid bioformulations of efficient halophilic plant growth promoters improvs crop yields under salt stress.

Keywords Halophiles · Expremophiles · Salt tolerance · Hypersaline environment · Bio-remediation · Saline soil

S. Arora (✉)
ICAR-Central Soil Salinity Research Institute, Regional Research Station, Lucknow, Uttar Pradesh, India

M. J. Vanza
V.N. South Gujarat University, Surat, Gujarat, India

E. Lichtfouse (ed.), *Sustainable Agriculture Reviews 33*, Sustainable Agriculture Reviews 33, https://doi.org/10.1007/978-3-319-99076-7_7

7.1 Introduction

Life exists over the whole range of salt concentrations encountered in natural habitats: from freshwater environments to hypersaline lakes such as the Dead Sea, and other places saturated with respect to sodium chloride i.e. salt affected soils. Everything is everywhere but environment selects (Rutger and Bouvier 2006). In many cases the soil properties are the most definitory of the limitations for the ecosystem functioning and, quite always, the soil is the component of the ecosystem more resilient to changes. The influence of the high salt concentrations masks other soil forming processes or soil properties and environmental conditions, often altering them. Microorganisms play an important role in the maintenance and sustainability of any ecosystem as they are more capable of rapid adjustment towards environmental changes and deterioration. Microorganisms are considered to be the first life forms to have evolved; they are versatile and adaptive to various challenging environmental conditions. Microorganisms are omnipresent and they impact the entire biosphere. They play a major role in regulating biogeochemical cycles in the extreme environmental conditions (Seigle-Murandi et al. 1996). Microorganisms are responsible for carbon mineralization, nitrogen fixation, methane metabolism, and sulphur metabolism, thus controlling the global biogeochemical cycling (Das et al. 2006). They produce diverse metabolic enzymes that can be employed for the safe removal of contaminants, which can be achieved either by direct destruction of the chemical or through transformation of the contaminants to a safer or lesser toxic intermediate (Dash and Das 2012). Due to their versatility, microorganisms have provided a useful platform to be used for an enhanced model of bioremediation. There are many characteristics features of bacteria which make them suitable for application in bioremediation practices.

7.2 Salt Affected Soils

The soil that contains excess salts which impairs its productivity is called salt-affected. Salt accumulation in soil is characterized by saline soil, contains high amount of soluble salts Ca^{2+}, Mg^{2+}, K^+ & Na^+ salt of Cl^-, NO_3^-, SO_4^{2-} & CO_3^{3-} etc.; Sodic soil, dominated by Na^+ salt & saline-sodic soil that have high salt of Ca^{2+}, Mg^{2+} & K^+ as well as Na^+. Salt-affected soils occupy an estimated 952.2 million ha of land in the world that constitutes to nearly 7% of the total land area and nearly 33% of the area of potential arable land. In India, the salt affected soils account for 6.727 million ha i.e. 2.1% of geographical area of the country. A build-up of soluble salts in the soil may influence its behaviour for crop production through changes in the proportions of exchangeable cations, soil reaction, physical properties and the effects of osmotic and specific ion toxicity (Arora and Sharma 2017). A practical grouping of salt affected soils can be done considering their biogeochemical behavior (Szabolcs 1989). Salt

Table 7.1 Groups of salt affected soils

Group of Salt affected soils	Main electrolyte(s) causing salinity, alkalinity or acidity	Usual pH range	Natural origin
Saline soils	Sodium chloride	5.5–8.5	Occurrence in most lagoon ecosystems mainly in cases of permanent contact with sea water.
Alkali or sodic soils	Sodium ion capable of alkaline hydrolysis Na_2CO_3, $NaHCO_3$, Na_2SiO_3	5.5–12	Developing in cases of stagnant water jointly with reduction processes often following desiccation
Magnesium Salted soils	Magnesium ions	5–9	Concurrently with sodium chloride in case of high amount of dissolved or absorbed Mg ions
Gypsiferous soils	Calcium ions ($CaSO_4$)	4–8	Development from evaporitic rocks containing gypsum
Acid sulphate soils	Ferric & aluminium ions (mainly sulphates)	1.5–4.5	In heavy, sulphur containing sediments as a result of their emerging to the surface & oxidation

affected soils can be grouped into saline soils, alkali soils, magnesium salted soil, gypsiferous soils, acid sulphate soils (Table 7.1). Due to sodium ions the saline soils i.e. coastal soils occurred by having permanent contact with the sea water in most of the lagoon ecosystems.

7.3 Halophilic Microorganisms

The existence of high osmotic pressure, ion toxicity, unfavourable soil physical conditions and/or soil flooding, are serious constraints to many organisms and therefore salt-affected ecosystems are specialised ecotones. The organisms found over there have developed mechanisms to survive in such adverse media, and many endemisms. The halophilic microorganisms or "salt-loving" microorganisms live in environments with high salt concentration that would kill most other microbes. Halotolerant and halophilic microorganisms can grow in hypersaline environments, but only halophiles specifically require at least 0.2 M of salt for their growth. Halotolerant microorganisms can only tolerate media containing less than 0.2 M of salt. Distinctions between different kinds of halophilic microorganisms are made on the basis of their level of salt requirement and salt tolerance.

According to Kushner (1993) classification, of microbes response to salt in which they grow best, and they are grouped as non-halophilic (less than 0.2 M salt), slight halophiles (0.2–0.5 M salt), moderate halophiles (0.5–2.5 M salt), borderline extreme halophiles (1.5–4.0 M salt) and extreme halophiles (2.5–5.2 M salt). The halotolerant microbes grow best in media containing <0.2 M ($\sim 1\%$) salt and also can tolerate high salt concentrations. This definition is widely referred to in many reports (Arahal and Ventosa 2002; Ventosa et al. 1998; Yoon et al. 2003).

7.4 Microbial Ecology of Salt Affected Soils

Microbial community in the soil are not distributed at random. Factors such as soil composition, organic matter, pH, water and oxygen availability, along with the host plant, play major role in the selection of the natural flora (Ross et al. 2000). The soil gains importance, especially in saline agricultural soils, where high salinity results from irrigation practices and application of chemical fertilizer. This effect is always more pronounced in the rhizosphere as a result of increased water uptake by the plants due to transpiration. Hence, the rhizobacteria form a group of the best adapted microorganisms (Tripathi et al. 1998).

Saline or hypersaline soils have yielded many Gram-positive species, and these have been characterized taxonomically. The microbiota of hypersaline soils is more similar to those of non-saline soils than to the microbiota from hypersaline waters. This suggests that general features of the environments are more important in determining the microbiota in a particular habitat than are individual factors such as high salinity (Quesada et al. 1983).

Industrial biocatalysis has found in the halophilic micro-organisms a source of enzymes with novel properties of high interest. Over the years, different enzymes of halotolerant and halophilic micro-organisms isolated from saline soils have been described and a number of new possibilities for industrial processes have emerged due to their overall inherent stability at high salt concentrations. These enzymes could be used in harsh industrial processes such as food processing, biosynthetic processes, and washing (Ventosa et al. 2005). Halopholic enzymes are active and stable at high salt concentrations, showing specific molecular properties that allow them to cope with osmotic stress. Mevarech et al. (2000) showed that these enzymes present an excess of acidic residues over basic residues and a low content of hydrophobic residues at their surface.

Compatible solutes are low-molecular weight organic compounds such as polyols, amino acids, sugars, and betaines that the halophilic and halotolerant bacteria accumulate intracellularly to achieve osmotic balance (Brown 1976). Also, halophilic bacteria tolerant to heavy metals could be used as bioassay indicator organisms in saline-polluted environments. Several halotolerant and halophilic bacteria isolated from hypersaline soils tolerate high concentrations of different metals, such as Co, Ni, Cd, or Cr (Nieto et al. 1989; Rios et al. 1998).

Industrial processes such as the production of pesticides, herbicides, and pharmaceutical products, in addition to paper mills and petrochemical industries generate wastewaters containing toxic compounds and with different levels of salinities. In the biological treatment, the micro-organisms conventionally used show only poor degradative efficacy due to the highly saline conditions. The potential of halophilic organisms in effluent treatment offers the promise of innovative research. Other than that halophilic also used to recover saline soil by directly supporting the growth of vegetation, thus indirectly increasing crop yields in saline soil.

7.5 Diversity of Halophilic Microorganisms

The phylogenetic diversity of microorganisms living at high salt concentrations is surprising. Halophiles are found in each of the three domains: Archaea, Bacteria, and Eucarya. The metabolic diversity of halophiles is great as well: they include oxygenic and anoxygenic phototrophs, aerobic heterotrophs, fermenters, denitrifiers, sulfate reducers, and methanogens. The diversity of metabolic types encountered decreases with salinity. The upper salinity limit at which each dissimilatory process takes place is correlated with the amount of energy generated and the energetic cost of osmotic adaptation. The understanding of the biodiversity in salt-saturated environments has increased greatly in recent years.

The soil is an important habitat for bacteria. Soil bacteria can be found as single cells or as microcolonies, embedded in a matrix of polysaccharides. Bacteria inhabiting soil play a role in conservation and restoration biology of higher organisms. The domain Bacteria contains many types of halophilic and halotolerant microorganisms, spread over a large number of phylogenetic groups (Ventosa et al. 1998). The different branches of the Proteobacteria contain halophilic representatives often having close relatives that are non-halophilic. Halophiles are also found among the cyanobacteria, the Flavobacterium – Cytophaga branch, the Spirochetes and the Actinomycetes (Oren 2002). Within the lineages of gram positive bacteria (*Firmicutes*), halophiles are found both within the aerobic branches (*Bacillus* and related organisms) as also within the anaerobic branches. It may be stated that generally most halophiles within the domain bacteria are moderate rather than extreme halophiles. However, there are a few types that resemble the Archaeal halophiles of the family *Halobacteriaceae* in their salt requirements and tolerance. Rodriguez-Valera (1988) stated that there was an abundance of halophilic bacteria in saline soil and that the dominant types encountered in saline soil belong to genera of *Alcaligenes*, *Bacillus*, *Micrococcus,* and *Pseudomonas*. Garabito et al. (1998) isolated and studied 71 halotolerant Gram-positive endospore forming rods from saline soils and sediments of salterns located in different areas of Spain. These isolates were tentatively assigned to the genus *Bacillus*, and the majority of them were classified as extremely halotolerent microorganisms, being able to grow in most cases in up to 20 or 25% salts.

7.5.1 Moderately Halophilic Bacteria

Several alkaliphilic *Bacillus* species have been isolated from soils and that showed halophilic characteristics (Table 7.2). *Bacillus krulwichiae,* a facultatively anaerobic was isolated in Tsukuba, Japan, as reported by Yumoto et al. (2003). These are straight rod with peritrichous flagella that produces ellipsoidal spores, having ability to utilize benzoate or *m*-hydroxybenzoate as the sole carbon source. *Bacillus patagoniensis* (Olivera et al. 2005) was isolated from the rhizosphere of the perennial shrub *Atriplex lampa* from north-eastern Patagonia. Another is *Bacillus oshimensis* (Yumoto et al. 2005). It is a halophilic nonmotile, facultatively alkaliphilic species. Another example is the genus *Virgibacillus.* This genus was first proposed by Heyndrickx et al. (1998) based on polyphasic data and it was later described by Heyrman et al. (2003). Members of the genus *Virgibacillus* produce oval to ellipsoidal endospores, are Gram-positive motile rods (Heyrman et al. 2003). This genus comprises eight species, two of which are moderately halophilic and have been isolated from soil samples: *Virgibacillus salexigens* (Garabito et al. 1997; Heyrman et al. 2003) and *Virgibacillus koreensis* (Lee et al. 2006).

Several other aerobic or facultatively anaerobic, moderately halophilic, endospore-forming, Gram-positive bacteria have been classified within genera related to *Bacillus.* Genera that include halophilic species isolated from soil samples are *Halobacillus*, *Filobacillus*, *Tenuibacillus*, *Lentibacillus*, and *Thalassobacillus*. Species from *Filobacillus*, *Thalassobacillus* & *Tenuibacillus* genera are borderline halophile. The genus *Halobacillus* is clearly differentiated from other related genera on the basis of its cell-wall peptidoglycan type as the members of this genus have peptidoglycan (Spring et al. 1996; Schlesner et al. 2001). Within these genera, the halophilic species isolated from soils are: *Halobacillus halophilus* (Spring et al. 1996), *Halobacillus karajensis* (Amoozegar et al. 2003).

With respect to the genus *Lentibacillus*, two halophilic soil species are identified. A *Lentibacillus salicampi* isolated from a salt field in Korea (Yoon et al. 2002), and A *Lentibacillus salarius* from a saline sediment in China (Jeon et al. 2005a). The family *Nocardiopsaceae* contains three genera, namely *Nocardiopsis* (Meyer 1976), *Thermobifida* (Zhang et al. 1998), and *Streptomonospora* (Cui et al. 2001). At present, the genus *Nocardiopsis* comprises 19 validly published species names (Li et al. 2003a, 2004; Al-Zarban et al. 2002). These species comprise aerobic, Gram-positive, non-acid-fast, and nonmotile organisms. Originally, members of this genus had been isolated from mildewed grain (Brocq-Rousseau 1904), but the natural habitat of *Nocardiopsis* is soil. It has been reported to predominate in saline or alkaline soils (Tang et al. 2003) and several recognized species have been isolated from such sources (Al-Tai and Ruan 1994; Chun et al. 2000; Al-Zarban et al. 2002; Li et al. 2003a, 2004). Some examples of moderately halophilic species of the genus *Nocardiopsis* isolated from soil samples were: *Nocardiopsis gilva*, *Nocardiopsis rosea*, *Nocardiopsis rhodophaea*, *Nocardiopsis chromatogenes*, and *Nocardiopsis baichengensis* (Li et al. 2006). These all are isolated from saline sediment from Xinjiang Province, China.

Table 7.2 Moderate halophiles (3–15%)

Sr. No	Species	Gram nature	Isolation source	References
1	*Bacillus krulwichiae*	P	Soil from Tsukuba, Ibaraki, Japan	Yumoto et al. (2003)
2	*Bacillus haloalkaliphilus*	P	Showa, Saitama	Echigo et al. (2005)
3	*Bacillus oshimensis*	P	Soil from Oshymanbe, Oshima, Hokkaido, Japan	Yumoto et al. (2003)
4	*Bacillus patagoniensis*	P	Rhizosphere of the perennial shrub Atriplex lampa in north-eastern Patagonia, Argentina	Olivera et al. (2005)
5	*Gracilibacillus halotolerans*	P	Shiki, Saitama	Echigo et al. (2005)
6	*Halobacillus halophilus*	P	Salt marsh and saline soils	Spring et al. (1996), Ventosa et al. (1983)
7	*Halobacillus karajensis*	P	Saline soil of the Karaj region, Iran	Amoozegar et al. (2003)
8	*Halomonas anticariensis*	N	Soil from Fuente de Piedra. Málaga, Spain	Martinez-Canovas et al. (2004a)
9	*Halomonas boliviensis*	N	Soil around the lake Laguna Colorada, Bolivia	Quillaguaman et al. (2004)
10	*Halomonas maura*	N	Soil from a solar saltern at Asilah, Morocco	Bouchotroch et al. (2001)
11	*Halomonas organivorans*	N	Saline soil from Isla Cristina, Huelva, Spain	Garcia et al. (2004)
12	*Lentibacillus salaries*	P	Saline sediment of Xinjiang Province, China	Jeon et al. (2005a)
13	*Lentibacillus salicampi*	P	Salt field in Korea	Yoon et al. (2002)
14	*Marinobacter excellens*	N	Sediment collected from Chazhman Bay, Sea of Japan	Gorshkova et al. (2003)
15	*Marinobacter koreensis*	N	Sea sand in Pohang, Korea	Kim et al. (2006)
16	*Marinobacter lipolyiticus*	N	Saline soil from Cadiz, Spain	Martin et al. (2003)
17	*Marinobacter sediminum*	N	Marine coastal sediment from Peter the Great Bay, Sea of Japan	Romanenko et al. (2005)
18	*Microbacterium halotolerans*	P	Soil sediment of Qinghai Province, China	Li et al. (2005a)
19	*Natranobacterium sp-1*	N	Salt pan of Kovalam	Murugan et al. (2011)
20	*Nocardiopsis baichengensis*	P	Saline sediment from Xinjiang Province, China	Li et al. (2006)
21	*Nocardiopsis chromatogenes*	P	Saline sediment from Xinjiang Province, China	Li et al. (2006)

(continued)

Table 7.2 (continued)

Sr. No	Species	Gram nature	Isolation source	References
22	*Nocardiopsis gilva*	P	Saline sediment from Xinjiang Province, China	Li et al. (2006)
23	*Nocardiopsis rhodophaea*	P	Saline sediment from Xinjiang Province, China	Li et al. (2006)
24	*Nocardiopsis rosea*	P	Saline sediment from Xinjiang Province, China	Li et al. (2006)
25	*Palleronia marisminoris*	N	Hypersaline soil bordering a solar saltern in Murcia, Spain	Martinez-Checa et al. (2005)
26	*Salipiger mucosus*	N	Hypersaline soil from a solar saltern in Calblanche, Murcia, Spain	Martinez-Canovas et al. (2004b)
27	*Virgibacillus halodenitrificans*	P	Ranzan, Saitama	Echigo et al. (2005)
28	*Virgibacillus koreensis*	P	Salt field near Taean-Gun on the Yellow Sea in Korea	Lee et al. (2006)

From salt pans of Kovalam in Kanyakumari district of Kerala, India, gram negative moderately halophilic bacteria like *Natranobacterium sp-1.* identified in the study of the diversity over period of time (Murugan et al. 2011). Many Gram-negative, moderately halophilic, or halotolerant species are currently included in the family *Halomonadaceae*, which belongs to the Gammaproteobacteria (Arahal and Ventosa 2005). This family includes three genera with halophilic species: *Halomonas*, *Chromohalobacter*, and *Cobetia*, plus two genera of nonhalophilic bacteria, *Zymobacter* and *Carnimonas* (Arahal et al. 2002; Dobson and Franzmann 1996; Garriga et al. 1998; Okamoto et al. 1993). Among the genera that comprise this family, *Halomonas* covers the greatest number of species (more than 40) showing heterogeneous features.

During an extensive search on different hypersaline habitats in Spain and Morocco focused on the screening of new exopolysacharide (EPS)-producing bacteria, several strains were isolated from saline soils and described as new species belonging to the genus *Halomonas*: *Halomonas maura* (Bouchotroch et al. 2001) and *Halomonas anticariensis* (Martinez-Canovas et al. 2004a). Other halophilic exopolysacharide -producing species were also isolated in these studies: *Salipiger mucosus*, that was the first moderately halophilic exopolysacharide -producing micro-organism belonging to the Alphaproteobacteria (Martinez-Canovas et al. 2004b). The genus *Marinobacter*, with the type species *Marinobacter hydrocarbonoclasticus*, was created in 1992 to accommodate Gram-negative, moderately halophilic, aerobic Gammaproteobacteria that utilize a variety of hydrocarbons as the sole source of carbon and energy (Gauthier et al. 1992). It also accommodate moderately halophilic *Marinococcus halophilus* and *Marinococcus albus* (Hao et al. 1984). Another species,

Marinococcus halotolerans that is extremely halophilic was also reported (Li et al. 2005a, b). It was motile cocci that grew over a wide range of salt concentrations up to 20% NaCl.

A phylogenetic and chemotaxonomic reanalysis of the genus *Micrococcus* resulted in the proposal of the genus *Nesterenkonia* (Stackebrandt et al. 1995) constituted by coccoid or short Gram-positive rods, non-spore-forming, chemo-organotrophic with strictly respiratory metabolism. Species of this genus are aerobic, catalase-positive, and moderately halophilic or halotolerant (Stackebrandt et al. 1995; Collins et al. 2002; Li et al. 2005b). Two species isolated from soil habitats are *Nesterenkonia halotolerans* and *Nesterenkonia xinjiangensis* (Li et al. 2004). The genus *Marinobacter* comprises 13 species, some of which being moderately halophilic bacteria isolated from soil samples: *Marinobacter lipolyticus*, that shows lipolytic activity with potential industrial applications (Martin et al. 2003), *Marinobacter excellens* (Gorshkova et al. 2003), *Marinobacter sediminum* (Romanenko et al. 2005), and the *Marinobacter koreensis* (Kim et al. 2006).

7.5.2 Borderline Halophiles

Some moderately halophilic, spore-forming, Gram-positive rods were originally assigned to the genus *Bacillus*, but have been reclassified within new genera by the application of molecular methods and improved phenotypic approaches (Heyndrickx et al. 1998; Yoon et al. 2001, 2004) (Table 7.3). Indeed, 16S rRNA sequence and chemotaxonomic analyses revealed the existence of several phylogenetically distinct lineages within the genus *Bacillus* (Ash et al. 1991; Nielsen et al. 1994). One example of such independent lineage is a group comprising the species *Alkalibacillus haloalkaliphilus* (formerly *Bacillus haloalkaliphilus*; Fritze 1996; Jeon et al. 2005b) isolated from alkaline, high saline mud from Wadi Natrun, Egypt, and *Alkalibacillus salilacus*, isolated from soil sediment of a salt lake in China (Jeon et al. 2005b).

The genera *Tenuibacillus* and *Thalassobacillus* are comprised of only one species, *Tenuibacillus multivorans* isolated from a saline soil in Xin-Jiang, China (Ren and Zhou 2005) and *Thalassobacillus devorans*, isolated from a saline soil in South Spain (García et al. 2005). All these genera belong to the family Bacillaceae, included in the phylogenetic group of the low GC Gram-positive bacteria, and are closely related. The genus *Saccharomonospora* (Nonomura and Ohara 1971) includes actinomycetes producing predominantly single spores on aerial hyphae. The studies on halophilic actinomycetes carried out by Jiang and coworkers in hypersaline soils of the Xinjiang Province, China, led to the isolation of the novel species *Saccharomonospora paurometabolica* (Li et al. 2003b). Cui et al. (2001) isolated the microorganisms from the genus *Streptomonospora,* which is constituted by only two species, which are Gram-positive, aerobic organisms with branching hyphae that grow optimally in media with 15% NaCl: *Streptomonospora alba* and *Streptomonospora salina*, both of them isolated from the same sampling site in China.

Table 7.3 Borderline halophiles (9–23%)

Sr No	Species	Gram nature	Isolation source	References
1	*Actinopolyspora iraqiensis*	P	Soil sample in Iraq	Ruan et al. (1994)
2	*Actinopolyspora mortivallis*	P	Soil sample obtained from Death Valley, CA, USA	Yoshida et al. (1991)
3	*Alkalibacillus haloalkaliphilus*	P	Alkaline, highly saline mud from Wadi Natrun, Egypt	Jeon et al. (2005b)
4	*Alkalibacillus salilacus*	P	Soil sediment from a salt lake in Xinjiang Province, China	Jeon et al. (2005b)
5	*Bacillus megaterium*	P	Kasukabe, Saitama	Echigo et al. (2005)
6	*Filobacillus milosensis*	P	Beach sediment from Palaeochori Bay, Milos, Greece	Schlesner et al. (2001)
7	*Filobacillus milosensis*	P	Okabe, Saitama	Echigo et al. (2005)
8	*Halobacillus karajensis*	P	Katsushika, Tokyo	Echigo et al. (2005)
9	*Halobacillus litoralis*	P	Okegawa, Saitama	Echigo et al. (2005)
10	*Halobacillus salinus*	P	Salt pan of Kovalam	Murugan et al. (2011)
11	*Halobacillus tueperi*	P	Omiya, Saitama	Echigo et al. (2005)
12	*Halobacterium salinarum*	N	Salt pan of Kovalam	Murugan et al. (2011)
13	*Halococcus salifodinae*	N	Salt pan of Kovalam	Murugan et al. (2011)
14	*Lentibacillus salicampi*	P	Iwatsuki, Saitama	Echigo et al. (2005)
15	*Saccharomonospora paurometabolica*	P	Saline sediment of Xinjiang Province, China	Li et al. (2003b)
16	*Staphylococcus citreus*	P	Salt pan of Kovalam	Murugan et al. (2011)
17	*Staphylococcus epidermidis*	P	Salt pan of Kovalam	Murugan et al. (2011)
18	*Staphylococcus intermedius*	P	Salt pan of Kovalam	Murugan et al. (2011)
19	*Tenuibacillus multivorans*	P	Soil from Xinjiang Province, China	Ren and Zhou (2005)
20	*Thalassobacillus devorans*	P	Saline soil in South Spain	García et al. (2005)
21	*Vibrio fischeri*	N	Salt pan of Kovalam	Murugan et al. (2011)

Table 7.4 Extreme halophiles (15–32%)

Sr. No	Species	Gram Nature	Isolation source	References
1	*Bacillus subtilis*	P	Salt pan of Kovalam	Murugan et al. (2011)
2	*Marinococcus halophilus*	P	Saline soil from Alicante and Cadiz, Spain	Hao et al. (1984), Marquez et al. (1992)
3	*Marinococcus halotolerans*	P	Saline soil in Qinghai, north-west China	Li et al. (2005a)

From salt pans of Kovalam in Kanyakumari district of Kerala, India, gram positive bacteria like *Halobacillus salinus*, *Staphylococcus epidermidis*, *Staphylococcus intermedius*, *Staphlococcus citreus* obtained. While gram negative organisms like *Vibrio fischeri*, *Halobacterium salinarum*, *Halobacterium salifodinae* organisms identified in the study of the diversity over period of time (Murugan et al. 2011).

7.5.3 Extremely Halophiles

A very low proportion of extremely halophilic archaea (1%) were reported to be isolated from the soil (Table 7.4). They were isolated from salt pan or saline soils. They assigned to the genus *Halobacterium* and their presence in soil suggested the existence of local microsites with sufficiently high salt concentrations to allow the growth of halophilic *Archaea*. The genus *Marinobacter*, with the type species *Marinobacter hydrocarbonoclasticus*, was created in 1992 to accommodate Gram-negative, moderately halophilic, aerobic Gammaproteobacteria that utilize a variety of hydrocarbons as the sole source of carbon and energy (Gauthier et al. 1992). It also accommodates moderately halophilic *Marinococcus halophilus* and *Marinococcus albus* (Hao et al. 1984). Li et al. (2005a, b) described a third species, *Marinococcus halotolerans* that is extremely halophilic. They are motile cocci that grow over a wide range of salt concentrations and up to 20% NaCl.

Obligately anaerobic bacteria that exist in extremely hypersaline lake ecosystems are quite numerous, because in these environments oxygen availability is low due to poor solubility, and organic carbon availability is high because substrate from primary production is not degraded by secondary consumers (plants, animals). The anoxic sediments of hypersaline environments are often characterized by a large number of halophilic anaerobic organisms belonging to the domain Bacteria. Bacteria inhabiting soil produce potentially important biotechnology products and are critically important sources of knowledge about the strategies and limit of life.

7.6 Vesicular Arbuscular Mycorrhiza

Vesicular Arbuscular mycorrhizal fungi occur naturally in saline environment (Khan 1974; Khan and Belik 1994). Several researchers investigated the relationship between soil salinity and occurrence of mycorrhizae on halophytes. They reported that the number of Vesicular Arbuscular Mycorrhiza spores or infectivity of Vesicular Arbuscular Mycorrhizal fungi changed with change in salt concentration (Juniper and Abbott 1993). The stresses due to saline soils effect the growth of plants, fungus or both. Vesicular Arbuscular Mycorrhizal fungi most commonly observed in saline soils are *Glomus spp.* this suggest that this may be adapted to grow in saline conditions, but ecological specificity has not been demonstrated (Juniper and Abbott 1993). There is evidence that Vesicular Arbuscular Mycorrhiza species distribution is markedly changed with increased salinity (Stahl and Williams 1986).

The most predominant species of arbuscular mycorrhizal fungi in the severely saline soils of the Tabriz plains were *Glomus intraradices*, *G. versiform* and *G. etunicatum* (Aliasgharzadeh et al. 2001). It was also found that the number of Arbuscular Mycorrhizal Fungi spores did not significantly decrease with soil salinity and reported a relatively high spore number (mean of 100/10 g soil). The higher fungal spore density in saline soils may be due to the fact that sporulation is stimulated under salt stress (Tressner and Hayes 1971) which means that Arbuscular Mycorrhizal Fungi may produce spores at low root-colonization levels in severe saline conditions (Aliasgharzadeh et al. 2001). It was reported that abundant occurrence of Arbuscular Mycorrhizal Fungi spores in extremely alkaline soils of pH values up to 11, independently of the soil type and irrespective of NaCl, Na_2CO_3, Na_2SO_4 or $CaSO_4$ salt types, though the degree of colonization varied from one individual to the next (Landwehr et al. 2002). In most of the earlier studies identification of the Arbuscular Mycorrhizal Fungi spores was based mainly on the morphological criteria. Complementary to morphology based identification methods, use of molecular techniques such as polymerase chain reaction and restriction fragment length polymorphism for identification of Arbuscular Mycorrhizal Fungi has been on the rise. Many authors have employed molecular techniques for identification of Arbuscular Mycorrhizal Fungi spores (Landwehr et al. 2002; Regvar et al. 2003; Wilde et al. 2009). The molecular identification techniques can overcome some of the pitfalls in morphology based identification.

There are few studies indicating that mycorrhizal fungi can increase growth of plants growing in saline habitats (Ojala et al. 1983; Pond et al. 1984). Vesicular Arbuscular -mycorrhizal fungi may have the ability to protect plants from salt stress (Hirrel and Gerdemann 1980; Rosendahl and Rosendahl 1991), but the mechanism is not fully understood (Yadav et al. 2017). The few data available at present suggest that fungi do have a potential to enhance plant growth by increasing the uptake of the nutrients. To test the efficacy of three species of Arbuscular Mycorrhizal Fungi – *Glomus mosseae*, *G. intraradices* and *G. claroideum* – to alleviate salt stress in olive trees under nursery conditions, study was conduted by Porras-Soriano et al. (2009) who observed that *G. mosseae* was the most efficient fungus in terms of olive tree performance and particularly in the protection offered against the detrimental effects of salinity. These findings suggest that the capability of Arbuscular Mycorrhizal Fungi in protecting plants from the detrimental effects of salt stress may depend on the behaviour of each species.

7.7 Mechanisms for Halotolerance

Halotolerance is the adaptation of living organisms to conditions of high salinity. High osmolarity in hypersaline conditions can be deleterious to cells since water is lost to the external medium until osmotic equilibrium is achieved. Many microorganisms respond to increase in osmolarity by accumulating osmotica in their cytosol, which protects them from cytoplasmic dehydration (Yancey et al. 1982). As biological membranes are permeable to water, all microorganisms have to keep their cytoplasm at least isoosmotic with their environment to prevent water loss of cellular water; when a turgor pressure is to be maintained, the cytoplasm should even be slightly hyperosmotic. Adaptation to conditions of high salinity has an evolutionary significance. The concentration of brines during prebiotic evolution suggests haloadaptation at earliest evolutionary times (Dundas 1998). Osmophily is related to the osmotic aspects of life at high salt concentrations, especially turgor pressure, cellular dehydration and desiccation. Halophily refers to the ionic requirements for life at high salt concentrations.

Halophilic microorganisms usually adopt either of the two strategies of survival in saline environments: ‘compatible solute’ strategy and ‘salt-in’ strategy (Ventosa et al. 1998). When an isoosmotic balance with the medium is achieved, cell volume is maintained. Compatible solute strategy is employed by the majority of moderately halophilic and halotolerant bacteria, some yeasts, algae and fungi. In this strategy cells maintain low concentrations of salt in their cytoplasm by balancing osmotic potential through the synthesis or uptake of organic compatible solutes and exclusion of salts from cytoplasm as much as possible. The compatible solutes or osmolytes, small organic molecules that are soluble in water to molar concentrations, that accumulate in halophiles are available in great spectrum and used in all three domains of life. These are assigned in two classes of chemicals i.e. (1) the amino acids and their derivatives, such as glycine betaine, glutamine, glutamate, proline, ectoine or N-acetyl-β-lysine and (2) polyols e.g. glycine betaine, ectoine, sucrose, trehalose and glycerol, which do not disrupt metabolic processes and have no net charge at physiological pH. The accumulation can be accomplished either by uptake from the medium or by *de novo* synthesis (Shivanand and Mugeraya 2011).

The salt-in strategy is employed by true halophiles, including halophilic archaea and extremely halophilic bacteria. These the microorganisms that are adapted to high salt concentrations and cannot survive when the salinity of the medium is lowered. They generally do not synthesize organic solutes to maintain the osmotic equilibrium. In this adaptation the intracellular K^+ concentration is generally higher than that of outside, the intracellular Na^+ concentration is generally lower than that in the medium, the intracellular K^+ concentration increases with increasing external NaCl concentration in a non-linear pattern. All halophilic microorganisms contain potent transport mechanisms, generally based on Na^+/H^+ antiporters (Oren 1999). The electrical potential ($\Delta\psi$) that drives the uptake of potassium ions in these organisms, results from the concerted action of the membrane bound proton-pump bacteriorhodopsin and the "proton gradient-consuming" proteins ATP synthase and Na^+/H^+ antiporter (Wagner et al. 1978). Transport of Cl^- is accomplished *via* the light driven chloride pump halorhodopsin (Schobert and Lanyi 1982). Thus, ultimately it accumulate KCl into the cytoplasm to counterbalance the external salinity.

All enzymes and structural cell components must be adapted to high salt concentrations for proper cell function. Proteins of halophilic microorganisms contain an excess ratio of acidic to basic amino acids and are resistant to high salt concentration than the non-halophilic microbes (Oren and Mana 2002). These proteins always need a quite high intracellular salt concentration for correct protein folding and activity. Surface negative charges prevent denaturation and precipitation of proteins at high salt concentrations (DasSarma and Arora 2001).

Halobacillus is the first chloride-dependent bacterium reported, and several cellular functions depend on Cl^- for maximal activities, the most important being the activation of solute accumulation. *Halobacillus* switches its osmolyte strategy with the salinity in its environment by the production of different compatible solutes. Glutamate and glutamine dominate at intermediate salinities, and proline and ectoine dominate at high salinities. Chloride stimulates expression of the glutamine synthetase and activates the enzyme. The product glutamate then turns on the biosynthesis of proline by inducing the expression of the proline biosynthetic genes (Saum and Muller 2008). *Halobacillus dabanensis* is used by Su-Sheng Yang and his colleagues (Beijing, China) as a model organism to study the genes involved in halotolerance, including genes encoding Na^+/H^+ antiporters, enzymes involved in osmotic solute metabolism, and stress proteins (Yang et al. 2006).

7.8 Applications of Halophilic Bacteria

Halophilic bacteria provide a high potential for biotechnological applications for at least two reasons: (1) their activities in natural environments with regard to their participation in biogeochemical processes of C, N, S, and P, the formation and dissolution of carbonates, the immobilization of phosphate, and the production of growth factors and nutrients (Rodriguez-Valera 1993); and (2) their nutritional requirements are simple. The majority can use a large range of compounds as their sole carbon and energy source. Most of them can grow at high salt concentrations, minimizing the risk of contamination. Moreover, several genetic tools developed for the nonhalophilic bacteria can be applied to the halophiles, and hence their genetic manipulation seems feasible (Ventosa et al. 1998). The application of halophilic bacteria in environmental biotechnology is possible for (1) the recovery of saline soil, (2) the decontamination of saline or alkaline industrial wastewater, and (3) the degradation of toxic compounds in hypersaline environments.

The use of halophilic bacteria in the recovery of saline soils is covered by the following hypotheses. The first hypothesis is that microbial activities in saline soil may favor the growth of plants resistant to soil salinity. The second hypothesis is based on the utilization of these bacteria as bio-indicators in saline wells. Indicator microorganisms can be selected by their abilities to grow at different salt concentrations. Another hypothesis is the application of halophilic bacterium genes using a genetic manipulation technique to assist wild type plants to adapt to grow in saline soil by giving them the genes for crucial enzymes that are taken from halophiles. The production of genetically modified plants has however been controversial.

7.9 Halophilic Bacteria for Bioremediation and Agricultural Production

Both physical and chemical methods of their reclamation are not cost-effective and also the availability of mineral gypsum or other chemical amendments is a problem. The applications of halophilic bacteria include recovery of salt affected soils by directly supporting the growth of vegetation thus indirectly increasing crop yields in salt affected soils (Arora et al. 2016). All halophilic microorganisms contain potent transport mechanisms, generally based on Na^+/H^+ antiporters, to expel sodium ions from the interior of the cell (Oren 2002). Also, some halophiles express ACC deaminase activity that removes stress, ethylene from the rhizosphere and some produce auxins that promote root growth. Halophilic microbes are also found to remove salt from saline soils (Bhuva et al. 2013). There are reports that potential salt tolerant bacteria isolated from soil or plant tissues and having plant growth promotion trait, helps to alleviate salt stress by promoting seedling growth and increased biomass of crop plants grown under salinity stress (Chakraborty et al. 2011; Arora et al. 2013; Arora et al. 2014b).

Although the salt shows negligible effects on seed germination and seedling growth but salt sensitivity of many crops is well documented on plant dry weight and biomass as the major energy of the plant is utilized to maintain the osmotic balance under salt stress (Jamal et al. 2011; Saqib et al. 2012). Plant growth promoting rhizobacteria -induced plants salt stress tolerance has been well studied and is considered to be the cost-effective solution to the problem. PGPR isolated from saline soils improve the plant growth at high salt (Mayak et al. 2004; Yildirim and Taylor 2005; Barassi et al. 2006).

These halophilic plant growth promoting rhizobacteria tolerate wide range of salt stress and enable plants to withstand salinity by hydraulic conductance, osmotic accumulation, sequestering toxic Na^+ ions, maintaining the higher osmotic conductance and photosynthetic activities (Dodd and Perez-Alfocea 2012). The inocluation with halophilic strains of plant growth promoting rhizobacteria will help to improve the plants tolerance in stress environment especially salinity and promote their growth particularly in food crops which is a essentially required to meet the food demands of the country.

It was reported that aminocyclopropane-1-carboxylic acid (ACC)-deaminase containing halotolerant strain SAL-15 (*Planococcus* sp.) which is also an indole acetic acid (IAA) producing strain increased root and shoot growth and plant biomass under salt stress in the presence of ACC. Inoculated plants showed 71% increase in plant weight, 94% in root length and 183% in shoot length than uninoculated control plants. In the presence of salt, bacteria showing IAA-activity without ACC-deaminase activity inhibit root growth rather than root elongation showing the importance of and higher synthesis of ACC under stress (Cheng et al. 2007).

Plant growth promoting rhizobacteria assist in diminishing the accumulation of ethylene levels and re-establish a healthy root system needed to cope with environmental stress. The primary mechanism includes the destruction of ethylene via enzyme ACC deaminase. There are number of publications (Ghosh et al. 2003; Govindasamy et al. 2008; Duan et al. 2009) mentioning rhizosphere bacteria such as *Achromobacter, Azospirillum, Bacillus, Enterobacter, Pseudomonas and Rhizobium* with ACC deaminase activity. Most of the studies have demonstrated the production of ACC deaminase gene in the plants treated with Plant growth promoting rhizobacteria under environmental stress. Grichko and Glick (2001) inoculated tomato seeds with *Enterobacter cloacae* and *Pseudomonas putida* expressing ACC deaminase activity and registered an increase in plant resistance. Ghosh et al. (2003) recorded ACC deaminase activity in three *Bacillus* species namely, *Bacillus circulans* DUC1, *Bacillus firmus* DUC2 and *Bacillus globisporus* DUC3 that stimulated root elongation in *Brassica campestris*. Mayak et al. (2004) observed tomato plants inoculated with the bacterium *Achromobacter piechaudii* under water and saline stress conditions and reported a significant increase in fresh and dry weight of inoculated plants. Many rhizobia isolated from Acacia, such as *Sinorhizobium arboris*, turned out to be moderately salt tolerant, capable of growing in 0.3–0.5 M (2–3%) NaCl (Zahran et al. 1994). Various Plant growth promoting rhizobacteria including Rhizobium, *Pseudomonas*, *Acetobacter*, *Bacillus*, and *Flavobacterium* and several *Azospirillum* can maintain their plant growth promotion ability even at high saline conditions. Halophilic bacteria strain (CSSRO2 *Planococcus maritimus*) and CSSRY1 (*Nesterenkonia alba*) having plant growth promotion properties were isolated from rhizosphere of dominant halophytes from coastal ecosystem (Arora et al. 2012). Salt tolerant *rhizobium* species were isolated from coastal saline soils (Trivedi and Arora 2013).

Researchers have demonstrated the feasibility of *Azospirillum* inoculation to mitigate negative effects of NaCl on plant growth parameters. This beneficial effect of *Azospirillum* inoculation was observed in wheat seeds, where a mitigating effect of salt stress was also evident (Creus et al. 1997). *Azospirillum* inoculated wheat (*T. aestivum*) seedlings subjected to osmotic stress developed significant higher coleoptiles, with higher fresh weight and better water status than non-inoculated seedlings (Creus et al. 1998).

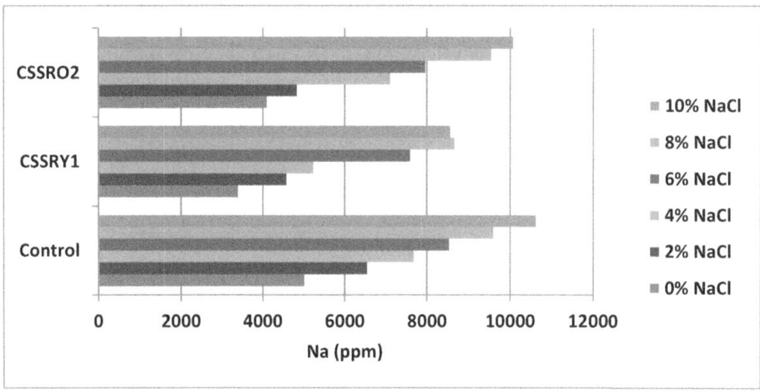

Fig. 7.1 Sodium removal by halophilic bacterial inoculates in saline soil

The biotic approach 'plant-microbe interaction' to overcome salinity problems has recently received considerable attention throughout the world. Plant-microbe interactions are beneficial associations between plants and microorganisms and also a more efficient method for reclamation of saline soils. Two promising halophilic bacterial strains that showed positive for plant growth promotion were selected and tested for salt removal efficiency. Halophilic bacteria strain (CSSRO2) was more efficient in reducing sodium concentration from 1,12,230 ppm in supernatant to 1,00,190 ppm at 24 h while strain CSSRY1 reduced Na concentration to 92,730 ppm at 48 h in halophilic broth with 15% NaCl. This shows that inoculation of strains in liquid media resulted in removal of 12,040 and 19,500 ppm of Na by halophilic bacterial strains CSSRO2 and CSSRY1 respectively (Fig. 7.1). The halophilic bacteria strains CSSRY1 and CSSRO2 were also shown to have high potential for removal of sodium ions from soil. CSSRY1 efficiently removed sodium at higher (6%, 8%, 10% NaCl) salt concentration in comparison of CSSRO2 and association of both organisms (CSSRY1 and CSSRO2) (Arora et al. 2012). This was also confirmed by reduction of electrical conductivity or total dissolved salts (TDS). It is hypothesized that once the sodium ion concentration is reduced in rhizosphere, plants are able to resume nutrient and water uptake.

To confirm about the sodium removal efficacy of these halophilic bacterial strains from soil, CSSRY1 and CSSRO2 were inoculated in sterile soil to test their efficacy for sodium removal from the soil containing different concentrations of NaCl (0–10% NaCl). It was observed that inoculation of strain CSSRY1 decreased soluble sodium content up to 31% at 4% NaCl concentration while at 10% NaCl concentration, it reduced only 19% sodium from soil. These selected cultures were further studied in greenhouse pot experiments for plant growth promotion. Results showed there was increase in plant growth parameters and yield of wheat when halophilic bacteria were inoculated with seeds and saline water irrigation was applied. It was observed that there was 10–12% increase in yield attributes and yield of wheat at 6% NaCl as compared to 2% NaCl. In the 5% NaCl treated soil, only the growth of the *Zea mays* was observed. Plants inoculated

Table 7.5 Plant growth promotion of halophilic bacteria on inoculation with maize

Halophilic Culture	Treatment (Conc. of NaCl)	Fresh weight (g/pot)	Dry weight (g/pot)	Shoot length(cm)	Root length (cm)
MB55	5%	1.945	0.660	14.50	12.00
MB66	5%	2.920	0.505	25.00	16.80
MB90	5%	2.900	0.665	18.36	17.16
MB94	5%	2.825	0.855	11.15	13.80
Consortium	5%	5.595	0.975	27.07	17.80
Consortium	10%	2.075	0.700	8.90	11.26
Control	5%	1.900	0.490	11.7	10.40

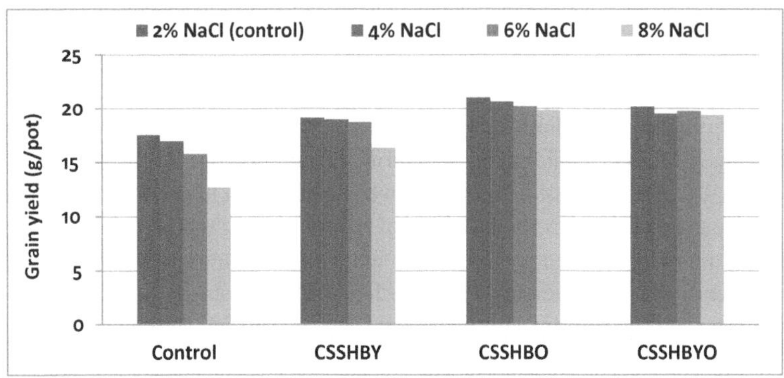

Fig. 7.2 Wheat performance in salt stress with halophilic bacteria inoculation

with a consortium of halophilic bacteria also showed growth at 10% NaCl, whereas inoculation with single isolates did not promote plant growth at this salt concentration (Table 7.5). The maximum fresh weight, dry weight, shoot length and root length of plant were found in the case of "Consortium 5% NaCl" treated pot, 194.5% increase in fresh weight, 98.97% increase in dry weight, 15.37 cm increase in shoot length and 7.4 cm increase in root length as compared to the uninoculated control plants (Arora et al. 2013). The results show that inoculation with these bacterial isolates can promote the growth of plants in salt affected soils due to production of hormone auxin and thus enhanced root growth. Another very likely mechanism may be alleviation of salinity stress via plant growth promoting rhizobacteria that express ACC deaminase activity. This enzyme removes stress ethylene from the rhizosphere. Also, the halophilic/halotolerant bacteria remove sodium from the surrounding soil and thus useful in plant growth promotion in salt affected soils (Arora and Vanza 2017).

Halophilic microbes were found to have the ability to remediate the saline soil and can be use for glycophytes/crop plants for optimum growth under saline condition (Fig. 7.2).

7.9.1 *Halophilic Endophytes for Bioremediation*

Endophytes are the microorganisms that thrive inside the plants. They face less competition for nutrients and are more protected from adverse changes in the environment than bacteria in the rhizosphere and phyllosphere as they interact closely with host plant (Weyens et al. 2009). They can help in degradation of the pollutants taken by the plants, thus lowering the phytotoxicity. From the leaves of dominant halophyte plant species dominant in coastal ecosystem of west coast of India, halophilic endophytic bacteria were isolated and assessed their plant growth promotion (Arora et al. 2014a). Recent evidence indicates that endophytes can contribute to phytoremediation of recalcitrant organic compounds and heavy metals (Thijs et al. 2014). Endophytic bacteria can positively enhance plant growth either: (1) directly through production of phytohormones (auxins and cytokinins) or by increasing the amounts of available nutrients by number of biochemical processes (e.g. N_2-fixation, phosphate solubilisation, siderophore release increasing Fe availability); or (2) indirectly through the suppression of ethylene production by 1-aminocyclopropane-1-carboxylic acid deaminase (ACCD), through chemical induction of plant defense mechanisms, or by the degradation of harmful contaminants (Thijs et al. 2014; Weyens et al. 2009). These properties of endophytic microorganisms make them suitable candidate for application in phytoremediation in salt and drought stress as well as organic pollutants in soil and enhancing the phyto-uptake of heavy metals.

7.10 Bioformulations of Halophilic Plant Growth Promoting Rhizobacteria

Bioformulations are best defined as biologically active products containing one or more beneficial microbial strains in easy to use and economical carrier materials. Usually, the term bioformulation refers to preparations of microorganism(s) that may be a partial or complete substitute for chemical fertilizers/pesticides (Arora et al. 2010). By the end of the nineteenth century, the practice of mixing "naturally inoculated" soil with seeds became a recommended method of legume inoculation in the USA (Smith 1992). The erratic performances of bioinoculants under field conditions have raised concerns about the practical potential offered by microbial releases into soil. Liquid bioformulations 'Halo-Azo', 'Halo-PSB', 'Halo-Zinc' and 'Halo-Mix' comprising of halophilic plant growth promoting bacteria isolated from salt affected soils were developed for application. These bioformulations were tested at research farm and farmers fields in salt affected areas and found to be effective. It was observed that growth and yield of rice, wheat, mustard, spinach and fodder crops enhanced under salt stress when the seeds were incoulated with bioinoculants (Arora and Singh 2018).

Although much is known about the survival of bacteria within the protective environment of an inoculant carrier, little is known about the stresses that bacteria must endure upon transfer to the competitive and often harsh soil environment (Heijnen et al. 1992). Inoculants have to be designed to provide a dependable source of beneficial bacteria that survive in the soil and become available to the plant. Mixed bacterial inoculants surviving in stress condition have to be developed so that these formulations encapsulate the living cells, protect the microorganisms against many environmental stresses, release them to the soil, and ultimately enhance crop yield. Use of stress-tolerating strains of Plant growth promoting rhizobacteria biopreparations either as aqueous suspensions or in bioformulations of sawdust, rice husk, tea waste, and talc-based bioformulants promoted growth in agricultural crops even under saline conditions (Ross et al. 2000). A successful plant growth promoting agent must be an aggressive colonizer with better competence and storage conditions in its formulation and use. Encapsulation enables slow and controlled cell release from the immobilization matrix of the alginate gel bead upon inoculation into soil, facilitates in establishing the stable plant growth promoting rhizobacterial population, and minimizes the possibilities of decline in population over time. The versatile nature of humic acid in the soil environment also extends the prospects of this encapsulation technique to the bioremediation of contaminated soil.

7.11 Conclusion

One of the recent focuses of research involves implication of Plant growth promoting rhizobacteria to combat salt stress. The development of biological products based on beneficial microorganisms can extend the range of options for maintaining the healthy yield of crops in saline habitat. In recent years, a new approach has been developed to alleviate salt stress in plants, by treating crop seeds and seedlings with Plant growth promoting rhizobacteria. The great opportunity for salt tolerance research now is its ability to be combined with halophilic Plant growth promoting rhizobacteria. The bottom line of every inoculation technology is its successful application under agricultural and industrial conditions. The microbial formulation and application technology are crucial for the development of commercial salt-tolerant bioformulation effective under salt stress conditions. Bioformulations offer an environmentally sustainable approach to increase crop production and health, contributing substantially in making the twenty-first century the age of biotechnology. Apart from bioformulation, reclamation and improving fertility of stressed sites is another aim to be focused on. The promising approach toward tackling the problem of soil salinity utilizing beneficial microorganism(s) including Plant growth promoting rhizobacteria will make the greatest contribution to the agricultural economy, if inexpensive and easy to use stress-tolerant strain formulation(s) could be developed.

References

Aliasgharzadeh N, Saleh Rastin N, Towfighi H, Alizadeh A (2001) Occurrence of arbuscular mycorrhizal fungi in saline soils of the Tabriz plain of Iran in relation to some physical and chemical properties of soil. Mycorrhiza 11:119–122

Al-Tai AM, Ruan JS (1994) *Nocardiopsis halophila* sp. nov., a new halophilic actinomycete isolated from soil. Int J Syst Bacteriol 44:474–478

Al-Zarban SS, Abbas I, Al-Musallam AA, Steiner U, Stackebrandt E, Kroppenstedt RM (2002) *Nocardiopsis halotolerans* sp. nov., isolated from salt marsh soil in Kuwait. Int J Syst Evol Microbiol 52:525–529

Amoozegar MA, Malekzadeh F, Malik KA, Schumann P, Sproer C (2003) *Halobacillus karajensis* sp. nov., a novel moderate halophile. Int J Syst Evol Microbiol 53:059–1063

Arahal DR, Ventosa A (2002) Moderately halophilic and halotolerant species of *Bacillus* and related genera. In: Berkeley R, Heyndrickx M, Logan N, De Vos P (eds) Applications and systematics of *Bacillus* and relatives. Blackwell, Oxford, pp 83–99

Arahal DR, Ventosa A (2005) The family *Halomonadaceae*. In: Dworkin M, Falkow S, Rosenberg E, Schleifer KH, Stackebrandt E (eds) The prokaryotes: an evolving electronic resource for the microbial community, release 3.20. Springer, New York http://141.150.157.117:8080/prokPUB/index.htm

Arahal DR, Castillo AM, Ludwig W, Schleifer KH, Ventosa A (2002) Proposal of *Cobetia marina* gen. nov., comb. nov., within the family *Halomonadaceae*, to include the species *Halomonas marina*. Syst Appl Microbiol 25:207–211

Arora NK, Khare E, Maheshwari DK (2010) Plant growth promoting rhizobacteria: constraints in bioformulation, commercialization, and future strategies. In: Maheshwari DK (ed) Plant growth and health promoting bacteria, vol 18. Springer, Berlin/Heidelberg, pp 97–116

Arora SV (2017) Reclamation and management of salt affected soils for safeguarding agricultural productivity. Journal of Safe Agriculture 01(1):1–10

Arora S, Singh YP, Vanza M, Sahni D (2016) Bioremediation of saline and sodic soils through halophilic bacteria to enhance agricultural production. J Soil Water Conserv, India 15(4):302–305

Arora S, Singh YP (2018) Bioremediation of salt affected soils of Uttar Pradesh through halophilic microbes to promote organic farming. In: Sharma PC, Singh A (eds) Annual report, 2017–18. ICAR-Central Soil Salinity Research Institute, Karnal, Haryana, India, pp 112–118

Arora S, Trivedi R, Rao GG (2012) Bioremediation of coastal and inland salt affected soils using halophilic soil microbes. Salinity News 18(2):3

Arora Sanjay, Trivedi R, Rao GG (2013) Bioremediation of coastal and inland salt affected soils using halophyte plants and halophilic soil microbes. CSSRI annual report 2012–13, CSSRI, Karnal, India, pp 94–100

Arora S, Patel P, Vanza M, Rao GG (2014a) Isolation and characterization of endophytic bacteria colonizing halophyte and other salt tolerant plant species from coastal Gujarat. Afr J Microbiol Res 8(17):1779–1788

Arora S, Vanza M, Mehta R, Bhuva C, Patel P (2014b) Halophilic microbes for bio-remediation of salt affected soils. Afr J Microbiol Res 8(33):3070–3078

Arora S, Vanza M (2017) Microbial approach for bioremediation of saline and sodic soils. In: Sanjay A, Singh AK, Singh YP (eds) Bioremediation of salt affected soils: an Indian perspective. Springer, Switzerland, pp 87–100

Ash C, Farrow JAE, Wallbanks S, Collins MD (1991) Phylogenetic heterogeneity of genus *Bacillus* as revealed by comparative analysis of small-subunit-ribosomal RNA sequence. Lett Appl Microbiol 13:202–206

Barassi CA, Ayrault G, Creus CM, Sueldo RJ, Sobrero MT (2006) Seed inoculation with *Azospirillum* mitigates NaCl effects on lettuce. Sci Hortic 109(1):8–14

Bhuva C, Arora S, Rao GG (2013) Efficacy of halophilic microbes for salt removal from coastal saline soils. In: National seminar with the theme "Microbes and Human Welfare", Bharathidasan University, Tiruchirappalli, India

Bouchotroch S, Quesada E, del Moral A, Llamas I, Bejar V (2001) *Halomonas maura* sp. nov., a novel moderately halophilic, exopolysaccharide-producing bacterium. Int J Syst Evol Microbiol 51:1625–1632

Brocq-Rousseau D (1904) Sur un Streptothrix. Ref Gen Botanique 16:20–26

Brown AD (1976) Microbial water stress. Bacteriol Rev 40:803–846

Chakraborty U, Roy S, Chakraborty AP, Dey P, Chakraborty B (2011) Plant growth promotion and amelioration of salinity stress in crop plants by a salt-tolerant bacterium. Recent Res Sci Technol 3(11):61–70

Cheng Z, Park E, Glick BR (2007) 1-Aminocyclopropane- 1-carboxylate deaminase from *Pseudomonas putida* UW4 facilitates the growth of canola in the presence of salt. Can J Microbiol 53:912–918

Chun J, Bae KS, Moon EY, Jung SO, Lee HK, Kim SJ (2000) *Nocardiopsis kunsanensis* sp. nov., a moderately halophilic actinomycete isolated from a saltern. Int J Syst Evol Microbiol 50:1909–1913

Collins MD, Lawson PA, Labrenz M, Tindall BJ, Weiss N, Hirsch P (2002) *Nesterenkonia lacusekhoensis* sp. Nov., isolated from hypersaline Ekho Lake, east Antarctica and emended description of the genus *Nesterenkonia*. Int J Syst Evol Microbiol 52:1145–1150

Creus CM, Sueldo RJ, Barassi CA (1997) Shoot growth and water status in *Azospirillum*-inoculated wheat seedlings grown under osmotic and salt stresses. Plant Physiol Biochem 35:939–944

Creus CM, Sueldo RJ, Barassi CA (1998) Water relations in *Azospirillum* inoculated wheat seedlings under osmotic stress. Can J Bot 76:238–244

Cui XL, Mao PH, Zeng M, Li WJ, Zhang LP, Xu LH, Jiang CL (2001) *Streptimonospora salina* gen. nov., sp. nov., a new member of the family *Nocardiopsaceae*. Int J Syst Evol Microbiol 51:357–363

Das S, Lyla PS, Khan SA (2006) Marine microbial diversity and ecology: present status and future perspectives. Curr Sci 90(10):1325–1335

Dash HR, Das S (2012) Bioremediation of mercury and importance of bacterial mer genes. Int Biodeterior Biodegrad 75:207–213

DasSarma S, Arora P (2001) Halophiles. Enc Life Sci. https://doi.org/10.1038/npg.els.0004356

de Rutger W, Bouvier T (2006) Environmental. Microbiology 8(4):755–758. https://doi.org/10.1111/j.1462-2920.2006.01017.x

Dobson SJ, Franzmann PD (1996) Unification of the genera *Deleya* (Baumann et al. 1983), *Halomonas* (Vreeland et al. 1980) and *Halovibrio* (Fendrich, 1988) and the species *Paracoccus halodenitrificans* (Robinson and Gibbons, 1952) into a single genus, *Halomonas* and placement of the genus *Zymobacter* in the family *Halomonadaceae*. Int J Syst Bacteriol 46:550–558

Dodd IC, Perez-Alfocea F (2012) Microbial alleviation of crop salinity. J Exp Bot 63:3415–3428

Duan J, Muller KM, Charles TC, Vesely S, Glick BR (2009) 1-Aminocyclopropane-1-carboxylate (ACC) deaminase genes in rhizobia from Southern Saskatchewan. Microb Ecol 57:423–436

Dundas I (1998) Was the environment for primordial life hypersaline? Extremophiles 2:375–377

Echigo A, Hino M, Fukushima T, Mizuki T, Kamekura M, Usami R (2005) Endospores of halophilic bacteria of the family *Bacillaceae* isolated from non-saline Japanese soil may be transported by Kosa event (Asian dust storm). Saline Syst 1:8

Fritze D (1996) *Bacillus haloalkaliphilus* sp. nov. Int J Syst Bacteriol 46:98–101

Garabito MJ, Arahal DR, Mellado E, Marquez MC, Ventosa A (1997) *Bacillus salexigens* sp. nov., a new moderately halophilic *Bacillus species*. Int J Syst Bacteriol 47:735–741

Garabito MJ, Marquez MC, Ventosa A (1998) Halotolerant *Bacillus* diversity in hypersaline environments. Can J Microbiol 44:95–102

Garcia MT, Mellado E, Ostos JC, Ventosa A (2004) *Halomonas organivorans* sp. nov., a moderate halophile able to degrade aromatic compounds. Int J Syst Evol Microbiol 54:1723–1728

García MT, Gallego V, Ventosa A, Mellado E (2005) *Thalassobacillus devorans* gen. nov., sp. nov., a moderately halophilic, phenol-degrading, Gram-positive bacterium. Int J Syst Evol Microbiol 55:1789–1795

Garriga M, Ehrmann MA, Arnau J, Hugas M, Vogel RF (1998) *Carnimonas nigrificans* gen. nov., sp. nov., a bacterial causative agent for black spot formation on cured meat products. Int J Syst Bacteriol 48:677–686

Gauthier MJ, Lafay B, Christen R, Fernandez L, Acquaviva M, Bonin P, Bertrand JC (1992) *Marinobacter hydrocarbonoclasticus* gen. nov., sp. nov., a new, extremely halotolerant, hydrocarbon- degrading marine bacterium. Int J Syst Bacteriol 42:568–576

Ghosh S, Penterman JN, Little RD, Chavez R, Glick BR (2003) Three newly isolated plant growth-promoting bacilli facilitate the seedling growth of canola, *Brassica campestris*. Plant Physiol Biochem 41:277–281

Gorshkova NM, Ivanova EP, Sergeev AF, Zhukova NV, Alexeeva Y, Wrighy JP, Nicolau DV, Mikhailov VV, Christen R (2003) *Marinobacter excellens* sp. nov., isolated from sediments of the Sea of Japan. Int J Syst Evol Microbiol 53:2073–2078

Govindasamy V, Senthilkumar M, Gaikwad K, Annapurna K (2008) Isolation and characterization of ACC deaminase gene from two plant growth-promoting rhizobacteria. Curr Microbiol 57 (4):312–317

Grichko VP, Glick BR (2001) Amelioration of flooding stress by ACC-deaminase containing plant growth promoting bacteria. Plant Physiol Biochem 39:11–17

Hao MV, Kocur M, Komagata K (1984) *Marinococcus* gen. nov., a new genus for motile cocci with meso-diaminopimelic acid in the cell wall; and *Marinococcus albus* sp. nov. and *Marinococcus halophilus* (Novitsky and Kushner) comb. nov. J Gen Appl Microbiol 30:449–459

Heijnen CE, Hok-A-Hin CH, van Veen JA (1992) Improvements to the use of bentonite clay as a protective agent, increasing survival levels of bacteria introduced into soil. Soil Biol Biochem 24:533–538

Heyndrickx M, Lebbe L, Kersters K, De Vos P, Forsyth C, Logan NA (1998) *Virgibacillus*: A new genus to accommodate *Bacillus pantothenticus* (Proom and Knight 1950). Emended description of *Virgibacillus pantothenticus*. Int J Syst Bacteriol 48:99–106

Heyrman J, Logan NA, Busse HJ, Balcaen A, Lebbe L, Rodriguez-Diaz M, Swings J, De Vos P (2003) *Virgibacillus carmonensis* sp. nov., *Virgibacillus necropolis* sp. nov. and *Virgibacillus picturae* sp. nov., three novel species isolated from deteriorated mural paintings, transfer of the species of the genus *Salibacillus* to *Virgibacillus*, as *Virgibacillus marismortui* comb. nov. and *Virgibacillus salexigens* comb.nov., and emended description of the genus *Virgibacillus*. Int J Syst Evol Microbiol 53:501–511

Hirrel MC, Gerdemann JW (1980) Improved growth of onion and bell pepper in saline soils by two vesicular-arbuscular mycorrhizal fungi. Soil Sci Soc Am J 44:654–665

Jamal Y, Shafi M, Bakht J, Arif M (2011) Seed priming improves salinity tolerance of wheat varieties. Pak J Bot 43(6):2683–2686

Jeon CO, Lim JM, Lee JC, Lee GS, Lee JM, Xu LH, Jiang CL, Kim CJ (2005a) *Lentibacillus salarius* sp. nov., isolated from saline sediment in China, and emended description of the genus *Lentibacillus*. Int J Syst Evol Microbiol 55:1339–1343

Jeon CO, Lim JM, Lee JM, Xu LH, Jiang CL, Kim CJ (2005b) Reclassification of *Bacillus haloalkaliphilus* Fritze 1996 as *Alkalibacillus haloalkaliphilus* gen. nov., comb. nov. and the description of *Alkalibacillus salilacus* sp. nov., a novel halophilic bacterium isolated from a salt lake in China. Int J Syst Evol Microbiol 55:1891–1896

Juniper S, Abbott L (1993) Vesicular and arbuscular mycorrhizae and soil salinity. Mycorrhiza 4:45–57

Khan AG (1974) The occurrence of mycorrhizae in halophytes, hydrophytes and xerophytes, and of endogone spores in adjacent soils. J Gen Microbiol 81:7–14

Khan AG, Belik M (1994) Occurrence and ecological significance of mycorrhizal symbiosis in aquatic plants. In: Varma A, Hock B (eds) Mycorrhiza: function, molecular biology and biotechnology. Springer, Heidelberg

Kim BY, Weon HY, Yoo SH, Kim JS, Kwon SW, Stackebrandt E, Go SJ (2006) *Marinobacter koreensis* sp. nov., isolated from sea sand in Korea. Int J Syst Evol Microbiol 56:2653–2656

Kushner DJ (1993) Growth and nutrition of halophilic bacteria. In: Vreeland RH, Hochstein LI (eds) The biology of halophilic bacteria. CRC Press, Boca Raton, pp 87–89

Landwehr M, Hilderbrandt U, Wilde P, Nawrath K, Toth T, Biro B, Bothe H (2002) The arbuscular mycorrhizal fungus *Glomus geosporum* in Europaen saline, sodic and gypsum soils. Mycorrhiza 12:199–211

Lee JS, Lim JM, Lee KC, Lee JC, Park YH, Kim CJ (2006) *Virgibacillus koreensis* sp. nov., a novel bacterium from salt field, and transfer of *Virgibacillus picturae* to the genus *Oceanobacillus* as *Oceanobacillus picturae* comb. nov. with emended descriptions. Int J Syst Evol Microbiol 56:251–257

Li MG, Li WJ, Xu P, Cui XL, Xu LH, Jiang CL (2003a) *Nocardiopsis xinjiangensis* sp. nov., a halophilic actinomycete isolated from a saline soil sample in China. Int J Syst Evol Microbiol 53:317–321

Li WJ, Tang SK, Stackebrandt E, Kroppenstedt RM, Schumann P, Xu LH, Jiang CL (2003b) *Saccharomonospora paurometabolica* sp. nov., a moderately halophilic actinomycete isolated from soil in China. Int J Syst Evol Microbiol 53:1591–1594

Li WJ, Park DJ, Tang SK, Wang D, Li JC, Lee JC, Xu LH, Kim CJ, Jiang CL (2004) *Nocardiopsis salina* sp. nov., a novel halophilic actinomycete isolated from saline soil in China. Int J Syst Evol Microbiol 54:1805–1809

Li WJ, Schumann P, Zhang YQ, Chen GZ, Tian XP, Xu LH, Stackebrandt E, Jiang CL (2005a) *Marinococcus halotolerans* sp. nov., isolated from Qinghai, north-west China. Int J Syst Evol Microbiol 55:1801–1804

Li WJ, Chen HH, Kim CJ, Park DJ, Tang SK, Lee JC, Xu LH, Jiang CL (2005b) *Microbacterium halotolerans* sp. nov., isolated from a saline soil in the west of China. Int J Syst Evol Microbiol 55:67–70

Li WJ, Kroppenstedt RM, Wang D, Tang SK, Lee JC, Park DJ, Kim CJ, Xu LH, Jiang CL (2006) Five novel species of the genus *Nocardiopsis* isolated from hypersaline soils and emended description of *Nocardiopsis salina* Li et al. 2004. Int J Syst Evol Microbiol 56:1089–1096

Marquez MC, Ventosa A, Ruiz-Berraquero F (1992) Phenoypic and chemotaxonomic characterization of *Marinococcus halophilus*. Syst Appl Microbiol 15:63–69

Martin S, Marquez MC, Sanchez-Porro C, Mellado E, Arahal DR, Ventosa A (2003) *Marinobacter lipolyticus* sp. nov., a novel moderate halophile with lipolytic activity. Int J Syst Evol Microbiol 53:1383–1387

Martinez-Canovas MJ, Bejar V, Martinez-Checa F, Quesada E (2004a) *Halomonas anticariensis* sp. nov., from Fuente de Piedra, a saline-wetland wildfowl reserve in Malaga, southern Spain. Int J Syst Evol Microbiol 54:1329–1332

Martinez-Canovas MJ, Quesada E, Martinez-Checa F, del Moral A, Bejar V (2004b) *Salipiger mucescens* gen. nov., sp. nov., a moderately halophilic, exopolysaccharide-producing bacterium isolated from hypersaline soil, belonging to the alpha-Proteobacteria. Int J Syst Evol Microbiol 54:1735–1740

Martinez-Checa F, Quesada E, Martinez-Canovas MJ, Llamas I, Bejar V (2005) *Palleronia marisminoris* gen. nov., sp. Nov., a moderately halophilic, exopolysaccharide producing bacterium belonging to the 'Alphaproteobacteria', isolated from the saline soil. Int J Syst Evol Microbiol 55:2525–2530

Mayak S, Tirosh T, Glick BR (2004) Plant growth promoting bacteria confer resistance in tomato plants to salt stress. Plant Physiol Biochem 42:565–572

Mevarech M, Frolow F, Gloss LM (2000) Halophilic enzymes: proteins with a grain of salt. Biophys Chem 86:155–164

Meyer J (1976) *Nocardiopsis dassonvillei*, a new genus of the order *Actinomycetales*. Int J Syst Bacteriol 26:487–493

Murugan M, Saju KA, Michael Babu M, Thiravia RS (2011) Survey on halophilic microbial diversity of Kovalam saltpans in Kanyakumari District and its industrial applications. J Appl Pharm Science 01(05):160–163

Nielsen P, Rainey FA, Outtrup H, Priest FG, Fritze D (1994) Comparative 16S rDNA sequence analysis of some alkaliphilic bacilli and the establishment of a sixth rRNA group within the genus *Bacillus*. FEMS Microbiol Lett 117:61–66

Nieto JJ, Fernandez-Castillo R, Marquez MC, Ventosa A, Quesada E, Ruiz-Berraquero F (1989) Survey of metal tolerance in moderately halophilic eubacteria. Appl Environ Microbiol 55:2385–2390

Nonomura H, Ohara Y (1971) Distribution of actinomycetes in soil. X New genus and species of monosporic actinomycetes. J Ferment Technol 49:895–903

Ojala JC, Jarrell WM, Menge JA, Johnson ELV (1983) Influence of mycorrhizal fungi on the mineral nutrition and yield of onion in saline soil. Agron J 75:255–259

Okamoto T, Taguchi H, Nakamura K, Ikenaga H, Kuraishi H, Yamasato K (1993) *Zymobacter palmae* gen. nov., sp. nov., a new ethanol-fermenting peritrichous bacterium isolated from palm sap. Arch Microbiol 160:333–337

Olivera N, Sineriz F, Breccia JD (2005) *Bacillus patagoniensis* sp. nov., a novel alkalitolerant bacterium from the rhizosphere of *Atriplex lampa* in Patagonia, Argentina. Int J Syst Evol Microbiol 55:443–447

Oren A (1999) Bioenergetic aspects of halophilism. Microbiol Mol Biol Rev 63:334–348

Oren A (2002) Diversity of halophilic microorganisms: environments, phylogeny, physiology, and applications. J Ind Microbiol Biotechnol 28:56–63

Oren A, Mana L (2002) Amino acid composition of bulk protein and salt relationships of selected enzymes of *Salinibacter ruber*, an extremely halophilic bacterium. Extremophiles 6:217–223

Pond EC, Menge JA, Jarrell WM (1984) Improved growth of tomato in salinized soil by vesicular arbuscular mycorrhizal fungi collected from saline sites. Mycologia 76:74–84

Porras-Soriano A, Soriano-Martin ML, Porras-Piedra A, Azco'n R (2009) Arbuscular mycorrhizal fungi increased growth, nutrient uptake and tolerance to salinity in olive trees under nursery conditions. J Plant Physiol 166:1350–1359. https://doi.org/10.1016/j.jplph.2009.02.010

Quesada E, Ventosa A, Rodriguez-Valera F, Megias L, Ramos-Cormenzana A (1983) Numerical taxonomy of moderate halophiles from hypersaline soils. J Gen Microbiol 129:2649–2657

Quillaguaman J, Hatti-Kaul R, Mattiasson B, Alvarez MT, Delgado O (2004) *Halomonas boliviensis* sp. nov., an alkalitolerant, moderate halophile isolated from soil around a Bolivian hypersaline lake. Int J Syst Evol Microbiol 54:721–725

Regvar M, Vogel K, Irgel N, Wraber T, Hildebrandt U, Wilde P, Bothe H (2003) Colonization of pennycress (*Thlaspi* spp.) of the Brassicaceae by arbuscular mycorrhizal fungi. J Plant Physiol 160:615–626

Ren PG, Zhou PJ (2005) *Tenuibacillus multivorans* gen. nov., sp. nov., a moderately halophilic bacterium isolated from saline soil in Xin-Jiang, China. Int J Syst Evol Microbiol 55:95–99

Rios M, Nieto JJ, Ventosa A (1998) Numerical taxonomy of heavy metal-tolerant nonhalophilic bacteria isolated from hypersaline environments. Int Microbiol 1:45–51

Rodriguez-Valera F (1988) Characteristics and microbial ecology of hypersaline environments. In: Rodriguez-Valera F (ed) Halophilic bacteria, vol 1. CRC Press, Boca Raton, pp 3–30

Rodriguez-Valera F (1993) Introduction to saline environments. In: Vreeland RH, Hochstein LI (eds) The biology of halophilic bacteria. CRC Press, Boca Raton, pp 1–12

Romanenko LA, Schumann P, Rohde M, Zhukova NV, Mikhailov VV, Stackebrandt E (2005) *Marinobacter bryozoorum* sp. nov. and *Marinobacter sediminum* sp. nov., novel bacteria from the marine environment. Int J Syst Evol Microbiol 55:143–148

Rosendahl CN, Rosendahl S (1991) Influence of vesicular-arbuscular mycorrhizal fungi (Glomus spp.) on the response of cucumber (*Cucumis sativus* L.) to salt stress. Environ Exp Bot 31:313–318

Ross IL, Alami Y, Harvey PR, Achouak W, Ryder MH (2000) Genetic diversity and biological control activity of novel species in South Australia. Appl Environ Microbiol 66:1609–1616

Ruan JS, Al-Tai AM, Zhou ZH, Qu LH (1994) *Actinopolyspora iraqiensis* sp. nov., a new halophilic actinomycete isolated from soil. Int J Syst Bacteriol 44:759–763

Saqib ZA, Akhtar J, Ul-Haq MA, Ahmad I (2012) Salt induced changes in leaf phenology of wheat plants are regulated by accumulation and distribution pattern of Na^{+}ion. Pak J Agri Sci 49:141–148

Saum SH, Muller V (2008) Regulation of osmoadaptation in the moderate halophile *Halobacillus halophilus*: chloride, glutamate and switching osmolyte strategies. Saline Syst 4:4

Schlesner H, Lawson PA, Collins MD, Weiss N, Wehmeyer U, Volker H, Thomm M (2001) *Filobacillus milensis* gen. nov., sp. nov., a new halophilic spore-forming bacterium with Orn-D-Glu-type peptidoglycan. Int J Syst Evol Microbiol 51:425–431

Schobert B, Lanyi JK (1982) Halorhodopsin is a light-driven chloride pump. J Biol Chem 257:10306–10313

Seigle-Murandi F, Guiraud P, Croize J, Falsen E, Eriksson KL (1996) Bacteria are omnipresent on *Phanerochaete chrysosporium* Burdsall. Appl. Environ Microbiol 62:2477–2481

Shivanand P, Mugeraya G (2011) Halophilic bacteria and their compatible solutes – osmoregulation and potential applications. Curr Sci 100(10):1516–1521

Smith R (1992) Legume inoculant formulation and application. Can J Microbiol 38:485–492. https://doi.org/10.1139/m92-080

Spring S, Ludwig W, Marquez MC, Ventosa A, Schleifer KH (1996) *Halobacillus* gen. nov., with descriptions of *Halobacillus litoralis* sp. nov., and *Halobacillus trueperi* sp. nov., and transfer of *Sporosarcina halophila* to *Halobacillus halophilus* comb. nov. Int J Syst Bacteriol 46:492–496

Stackebrandt E, Koch C, Gvozdiak O, Schumann P (1995) Taxonomic dissection of the genus *Micrococcus*: *Kocuria* gen. nov., *Nesterenkonia* gen. nov., *Kytococcus* gen. nov., *Dermacoccus* gen nov., and *Micrococcus* Cohn 1872 gen. emend. Int J Syst Bacteriol 45:682–692

Stahl PO, Williams SE (1986) Oil shale process water affects activity of vesicular-arbuscular fungi and *Rhizobium* four years after application to soil. Soil Biol Biochem 18:451–455

Szabolcs I (1989) Salt affected soils. CRC Press Inc, Boca Raton ISBN 0-8493-4818-8

Tang SK, Li WJ, Wang D, Zhang YG, Xu LH, Jiang CL (2003) Studies of the biological characteristic of some halophilic and halotolerant actinomycetes isolated from saline and alkaline soils. Actinomycetologica 17:6–10

Thijs S, Dillewijn PW, Sillen W, Truyens S, Holtappels M, Haen JD (2014) Exploring the rhizospheric and endophytic bacterial communities of Acer pseudoplatanus growing on a TNT-contaminated soil: Towards the development of a rhizocompetent TNTdetoxifying plant growth promoting consortium. Plant Soil 385:15–36

Tressner HD, Hayes JA (1971) Sodium chloride tolerance of terrestrial fungi. Appl Microbiol 22:210–213

Tripathi AK, Mishra BM, Tripathy P (1998) Salinity stress responses in the plant growth promoting rhizobacteria, *Azospirillum* spp. J Biosci 23:463–471

Trivedi R, Arora S (2013) Characterization of acid and salt tolerant *Rhizobium* sp. isolated from saline soils of Gujarat. Int Res J Chem 3(3):8–13

Ventosa A, Ramos-Cormenzana A, Kocur M (1983) Moderately halophilic gram-positive cocci from hypersaline environments. Syst Appl Microbiol 4:564–570

Ventosa A, Nieto JJ, Oren A (1998) Biology of moderately halophilic aerobic bacteria. Microbiol. Mol Biol Rev 62:504–544

Ventosa A, Sanchez-Porro C, Martin S, Mellado E (2005) Halophilic archaea and bacteria as a source of extracellular hydrolytic enzymes. In: Gunde-Cimerman A, Oren A, Plemenitas A (eds) Adaptation of life at high salt concentrations in archaea, bacteria and eukarya. Springer, Heidelberg, pp 337–354

Wagner G, Hartmann R, Oesterhelt D (1978) Potassium uniport and ATP synthesis in *Halobacterium halobium*. Eur J Biochem 89:169–179

Weyens N, van der Lelie D, Taghavi S, Newman L, Vangronsveld J (2009) Exploiting plant-microbe partnerships to improve biomass production and remediation. Trends Biotechnol 27:591–598

Wilde P, Manal A, Stodden M, Sieverding E, Hilderbrandt U, Bothe H (2009) Biodiversity of arbuscular mycorrhizal fungi in roots and soils of two salt marshes. Env Microbiol 11:1548–1561

Yadav RS, Mahatma MK, Thirumalaisamy PP, Meena HN, Bhaduri D, Arora S, Panwar J (2017) Arbuscular Mycorrhizal Fungi (AMF) for sustainable soil and plant health in salt-affected soils. In: Sanjay A, Singh AK, Singh YP (eds) Bioremediation of salt affected soils: an Indian perspective. Springer, Switzerland, pp 133–156

Yancey PH, Clark ME, Hand SC, Bowlus RD, Somero GN (1982) Living with water stress: evolution of osmolyte systems. Science 217:1214–1216

Yang LF, Jiang JQ, Zhao BS, Zhang B, Feng DQ, Lu WD, Wang L, Yang SS (2006) A Na^+/H^+ antiporter gene of the moderately halophilic bacterium *Halobacillus dabanensis* D-8T: cloning and molecular characterization. FEMS Microbiol Lett 255:89–95

Yildirim E, Taylor AG (2005) Effect of biological treatments on growth of bean Plants under Salt Stress. Annu Rep Bean Improv Coop 48:176–177

Yoon JH, Weiss N, Lee KC, Lee IS, Kang KH, Park YH (2001) *Jeotgalibacillus alimentarius* gen. nov., sp. nov., a novel bacterium isolated from jeotgal with L-lysine in the cell wall, and reclassification of *Bacillus marinus* Rueger 1983 as *Marinibacillus marinus* gen. nov., comb. nov. Int J Syst Evol Microbiol 51:2087–2093

Yoon JH, Kang KH, Park YH (2002) *Lentibacillus salicampi* gen. nov., sp. nov., a moderately halophilic bacterium isolated from a salt field in Korea. Int J Syst Evol Microbiol 52:2043–2048

Yoon JH, Kim IG, Kang KH, Oh TK, Park YH (2003) *Bacillus marisflavi* sp. nov. and *Bacillus aquimaris* sp. nov., isolated from sea water of a tidal flat of the Yellow Sea in Korea. Int J Syst Evol Microbiol 53:1297–1303

Yoon JH, Oh TK, Park YH (2004) Transfer of *Bacillus halodenitrificans* Denariaz et al. 1989 to the genus *Virgibacillus* as *Virgibacillus halodenitrificans* comb. nov. Int J Syst Evol Microbiol 54:2163–2167

Yoshida M, Matsubara K, Kudo T, Horikoshi K (1991) *Actinopolyspora mortivallis* sp. nov., a moderately halophilic actinomycete. Int J Syst Bacteriol 41:15–20

Yumoto I, Yamaga S, Sogabe Y, Nodasaka Y, Matsuyama H, Nakajima K, Suemori A (2003) *Bacillus krulwichiae* sp. nov., a halotolerant obligate alkaliphile that utilizes benzoate and mhydroxybenzoate. Int J Syst Evol Microbiol 53:1531–1536

Yumoto I, Hirota K, Goto T, Nodasaka Y, Nakajima K (2005) *Bacillus oshimensis* sp. nov., a moderately halophilic, non-motile alkaliphile. Int J Syst Evol Microbiol 55:907–911

Zahran HH, Räsänen LA, Karsisto M, Lindström K (1994) Alteration of lipopolysaccharide and protein profiles in SDSPAGE of rhizobia by osmotic and heat stress. World J Microbiol Biotechnol 10:100–105

Zhang Z, Wang Y, Ruan J (1998) Reclassification of *Thermomonospora* and *Microtetraspora*. Int J Syst Bacteriol 48:411–422

Chapter 8
Bioindicators of Degraded Soils

Debarati Bhaduri, Dibyendu Chatterjee, Koushik Chakraborty, Sumanta Chatterjee, and Ajoy Saha

Abstract Bioindicators are used to identify and monitor soil quality. Degraded soils refers to various altered properties such as physically-poor soils, C- and nutrient-deficient soils, waterlogged soils, salt-affected soils, and soils polluted by heavy metals and pesticides. Here we review degraded soils and bioindicators using plants, microbes and other living cells.

Keywords Soil biological indicators · Soil ecology · Environmental remediation · Soil pollution · Restoring problem soils · Environmental engineering

8.1 Introduction

Soil is a living-dynamic, and non-renewable resource for human societies and ecosystems. Due to the increasing pressure from agricultural production and urban and industrial development the quality of soil is depredating day by day (Dick 1997; Doran and Zeiss 2000). As soil is a non-renewable resource, it is our duty to ensure its protection. To instigate, pursue, and guarantee suitable protecting actions and management of soils, it is worthwhile to identify and define tools for efficient decision making. The currently used tool based on physical and chemical properties of soils are not able to appropriately address the whole effect of an environmental stress and must be complemented by biological data as soil health integrates both physico-chemical and biological properties. Monitoring biological indicators which give information on living organisms in soil (Hodkinson and Jackson 2005) and biomarker which gives information on measurable change in a biological organism in soil at a molecular, biochemical, cellular, physiological or behavioral level

D. Bhaduri (✉) · D. Chatterjee · K. Chakraborty · S. Chatterjee
ICAR-National Rice Research Institute, Cuttack, Odisha, India

A. Saha
ICAR-Central Inland Fisheries Research Institute, Bangalore Research Centre, Bangalore, Karnataka, India

E. Lichtfouse (ed.), *Sustainable Agriculture Reviews 33*, Sustainable Agriculture Reviews 33, https://doi.org/10.1007/978-3-319-99076-7_8

(Kammenga et al. 2000) related to changes in management or environmental variations in soil will definitely complement the physico-chemical tools available.

The use of bioindicators in soil is multi-faced. Thus it is one of the most happening and emerging areas of soil and environmental researches. Moreover, here basic soil researches meet with ecological perspectives creating interests among many. This is one of the probable reasons to develop the topic as fascinating as it could be. Scientists around the globe are in search of new bioindicators of soil and their rapid yet precise estimation procedures. However, the purposes and technological advancement may vary. Moreover, there are lots of hidden potentiality among wild and naturally cultivated plants which can serve as suitable soil bioindicators.

Restoration of degraded soils remained a burning research topic as well as concern for the landholders. The degradation in soil intrudes in both ways, natural and anthropogenic. Natural degradation occurs either by the major events like, climatic factors, desertification, landslides, salt deposits, acidification, submergence that creates havoc losses of nutrients and soil physical structure, whereas artificial or man-made degradation existed at contaminated soil sites mostly by heavy metals and radioactive elements accompanied by mining activities, deforestation, waste disposal, and industrial and commercial activities. Another principal source of problem soil is the intense agricultural practices like frequent tillage, mono-crop rotation, improper use of agricultural chemicals and many more. Bioindicators are generally considered as a good tool for monitoring both state of the soil environment, either toxic or healthy. The application of bioindicators was initiated in 1960s accounting the fact that bio-based indicators can solve the extent of difficulty to some extent; hence their role in management of degraded soils can be vital. Basically, any living organism can be an efficient bioindicator if that would enough sensitive to provide information of soil (or any ecosystem) health. Additionally, indicator species must respond well under presence or absence of other species as well as the presence of heavy metals or any pollutants (Stankovic and Stankovic 2013). However, optimum functioning of any bioindicator is regulated by number of environmental factors.

In this chapter, we will confine our discussion only for developmental research in terms of identifying soil bioindicators/biomarkers pertaining to restore degraded soils, based on the researches highlighted on this aspect so far. Undoubtedly there are lots of technological advancement over the years which not only generated high throughput analytical results, but also able to clear many ambiguities.

8.2 Definition of Degraded Soils

There is no definition of degraded soil available in the scientific literature. The degraded soil may be understood as a soil which is not suitable for normal cultivation of the dominant crops of that region in its natural states, but can be made suitable for

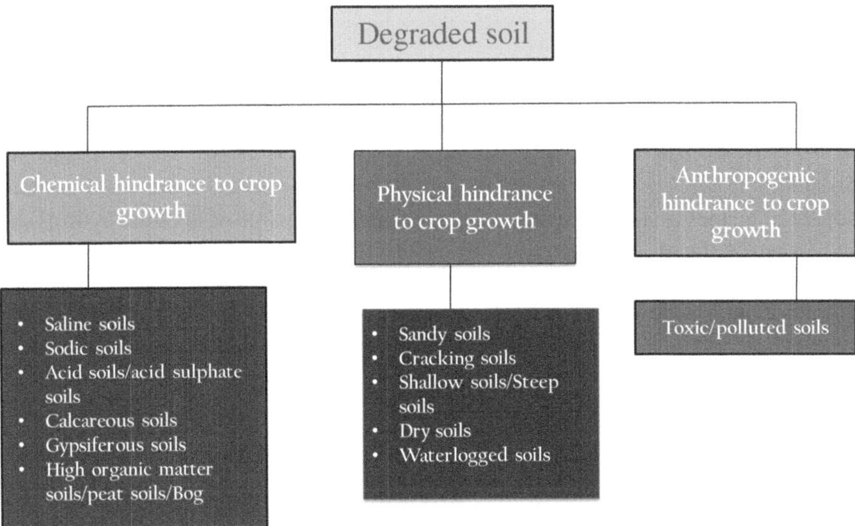

Fig. 8.1 Types of degraded soils. Note that degraded soils are broadly classified into three groups. Although polluted soils exert chemical hindrance to crop growth, it is originated due to anthropogenic activity, hence comes under the broad group of anthropogenic hindrance to crop growth

cultivation after proper application of amendments and conditioners through a suitable reclamation pathway. Typically all the soils in nature may have some problem. For instance, a soil has good drainage capability may be unsuitable for growing paddy. An acidic soil suitable for growing tea may be unsuitable for many other crops. Our concept is the degraded soil should be considered as those soil in which the majority of the vegetation of that region fails to grow. There are many types of degraded soils available in nature (Fig. 8.1).

8.2.1 Saline Soils

Saline soils have an excess amount of neutral soluble salts (chlorides and sulphates of sodium, calcium, and magnesium) which can adversely affect the root growth of most of the crop plants due to exo-osmosis. In the viewpoint of soil scientists, saline soils are defined as the soils which have an electrical conductivity of the saturation extract (ECs) of more than 4 dS m^{-1} at 25 °C (Richards 1954). However, the pH value of the saturated soil paste is less than 8.2 and exchangeable sodium percentage (ESP) is less than 15 (Abrol et al. 1980). Sodium is the dominant cation present in such type of soil.

8.2.2 Sodic Soils

Sodic soils contain a large amount of exchangeable sodium which can adversely affect the plant growth. These soils have an exchangeable sodium percentage (ESP) of more than 15. The ECs is mostly less than 4.0 dS m^{-1} at 25 °C and pH of saturated soil pastes is 8.2 or more. Unlike saline soils, these soils contain appreciable quantities of sodium carbonate which is capable of alkaline hydrolysis. Because of high Na concentration, these soils are highly dispersed and have poor soil structure. Due to high sodicity, the organic matter present in the soil get dissolved and deposited on the soil surface, resulting in a black colour on the soil surface.

8.2.3 Acid Soils and Acid Sulphate Soils

Acid soils have the pH less than 7 and hamper crop growth due to the high concentration of H^+ and Al^{3+}. A total of 3950 million ha of arable land is affected by soil acidity and it accounts 38% of Southeast Asia, 31% of Latin America, 20% of East Asia, 56% of Sub-Saharan Africa and parts of North America (Hoekenga et al. 2006). Soil pH adversely affects the availability of plant nutrients, increase the solubility of toxic heavy metals, interrupt soil microbial activity, and cation exchange capacity in soils, which lead to yield reduction (Merino-Gergichevich et al. 2010; Samac and Tesfaye 2003; Rousk et al. 2009; Bian et al. 2013). Acid sulphate soils contain iron sulphide, which upon oxidation produce sulphuric acid and the pH dropped to below 4. Actually under such situation cultivation is practically impossible except some acid tolerant cultivar of rice.

8.2.4 Calcareous Soils

These soils contain more than 15% free calcium carbonate in the surface layer. The calcium carbonate may be present in several forms like nodules, crust, powdery etc. These type of soils are mostly present in the arid and semiarid region of the earth. The major problems in this soil are the absence of sufficient quantity of organic matter, more loss of N in the form of ammonia volatilization due to its alkaline pH, deficiency of Zn and Fe etc.

8.2.5 *Gypsiferous Soils*

These soils contain a large quantity of calcium sulphate (anhydrite and gypsum) that negatively interfere crop growth. Moreover, the fragile nature of calcium sulphate in presence of water is problematic for engineering structures.

8.2.6 *Rich Organic Matter Soils/Peat Soils/Bog Soils*

These soils are excessively enriched with organic matter which is more than 20% by weight, roughly equivalent to 30–35% by volume (Driessen et al. 2001; Kroetsch et al. 2011). Commonly they are known as peat, muck or bog. In taxonomic classification, they are coming under Histosol. The reason of such high accumulation of organic materials is related to higher rate of input of plant remains than the rate of decomposition (Richardson and Vepraskas 2001).

8.2.7 *Physically Poor Soils*

8.2.7.1 Sandy Soils

Due to the high porosity of these soil, they are poor in nutrient and water holding capacity. These soils contain more than 65% sand and less than 18% clay.

8.2.7.2 Cracking Soils

These soils have high clay content, plasticity, and form crack of about 1 cm wide and 50 cm depth during the dry season. Any agricultural operation is very difficult in such soils. Pedoturbation and formation of gilgai microrelief are common in these soils.

8.2.7.3 Shallow Soils/Steep Soils

Due to steep slope (>30%) the soil depth is limited in the hilly region (around 50 cm), which cause improper plant stand. Generally, below the shallow surface these soils have hard pan or rocks.

8.2.7.4 Dry Soils

Mostly desert soil, remain dry for most of the time in a year.

8.2.8 Waterlogged Soils

Due to lower topographic position these soils remain waterlogged in some part of the year. This situation arises when the infiltration of water from rainfall or flooding surpasses the rate of subsurface drainage and evapotranspiration (Bramley et al. 2011). Oxygen diffusion under flooding or waterlogged soil is approximately 10,000 times slower in water than in air (Elzenga and Veen 2010). Due to this oxygen demand from root and microbial respiration cannot be fulfilled in such soils and deficiency of oxygen in rhizosphere affect plant growth by limiting root respiration (Vartapetian and Jackson 1997).

8.2.9 Toxic/Polluted Soils

These soils are polluted with heavy metals, organic pollutants, radioactive material etc. A large area of the world is affected by heavy metals. The sources of heavy metal pollutants are waste disposal, metallurgical industries, metal mining, fossil fuel combustion, mines, phosphatic and micronutrient fertilizers, pesticides, battery, metal coating, sewage, sludge distillery, pharmaceutical, thermal power etc. Agriculture in proximity of such sources faces major problem due to transfer of heavy metals from such sources to crops and subsequently into the food chain as a sink (Yu et al. 2012).

The main concern about organic pollutants is persistent, toxic, highly stable, biomagnified in food web and transported over long distances. Soil organic pollutants include pesticides, polycyclic aromatic hydrocarbons (PAHs), dioxins and and other minor organic pollutants like chlorophenols, BTEX (Benzene, toluene, xylene,), phthalic acid esters etc. Many of these compounds are carcinogenic.

Radionuclides release α, β, γ rays which possess a potential risk for the living cell. Gamma radiation emitted from naturally occurring radioisotopes, terrestrial background radiation, represents the main external source of irradiation of the human body (Abdi et al. 2008).

Fertilizers supply macronutrients (N, P, K, Ca, Mg, and S) and micronutrients (Fe, Mn, Zn, B, Mo, Cu, Cl, Ni) to soil. These micronutrients help plants to complete their life cycle at lower concentration, but the same may act as a heavy metal pollutant at higher concentration when applied in excess. Application of excess N-fertilizers result in nitrate pollution in groundwater, eutrophication in surface water bodies and harmful greenhouse gas emission like nitrous oxide in atmosphere. Application of some phosphatic fertilizer adds Cd and other potentially toxic elements to the soil, including F, Hg, and Pb (Raven et al. 1998). Organic manures, compost, vermicompost also contain heavy metal which could be released after its decomposition in the soil. Recently increased application of industrial wastewater, sewage-sludge and different effluents result in enrichment of carcinogenic materials and toxic elements in soil.

8.3 Bioindicators

'Bioindicator' has been termed and defined in many ways. One group mentioned, a bioindicator is a species or a group of species that reflects biotic and/or abiotic levels of contamination of an environment (Hodkinson and Jackson 2005). Stankovic and Stankovic (2013) described: a bioindicator is an organism or a part of an organism or a community of organisms, which contains information on the quantitative aspects of the quality of the environment; exposure of organisms can be measured by either levels or effects. In real sense, bioindicators can be any viable biological substance, either animal, plant, or microorganisms, by using that a valid conclusion can be made about the present environment they are exposed of. The broad classification of bioindicators with suitable explanation is given in Fig. 8.2.

A number of bioindicators were investigated and used indexing of soil quality for long-term agro-ecosystem for judging the sustainability (Masto et al. 2007; Bhaduri and Purakayastha 2014; Bhaduri et al. 2017a). There are several other instances where soil quality has been well interpreted using soil bioindicators (biological and biochemical) under altered management practices in agricultural soils (Koper and Piotrowska 2003; Kang et al. 2005; Bhattacharjya et al. 2017), among many others which has included bioindicators for indexing of soil quality. An explicit discussion was made for all possible relations between soil quality and plant-microbe interactions in the rhizosphere with a special mention of soil bioindicators (Bhaduri et al. 2015; Bhaduri et al. 2017b). However, a less attention was paid to bioindicators for stressed or problem soils. Though that needs equal focus to indulge since soil remediation or restoration has been a globally important issue.

These days, an extensive number of techniques are applied to estimate common soil microbial parameters (bioindicators). These indicators can be either of these types:

- *Physiological*, e.g. chloroform fumigation extraction for measuring microbial biomass carbon (MBC), phosphorous (MBP) and nitrogen (MBN), and CO_2 evolution by substrate induced respiration (Gonzalez-Quiñones et al. 2011)
- *Metabolic*, e.g. enzymatic activities (Izquierdo et al. 2005; Dimitriu et al. 2010), more commonly dehydrogenase, phosphatases, urease, aryl sulfatase, fluorescein diacetate hydrolyses

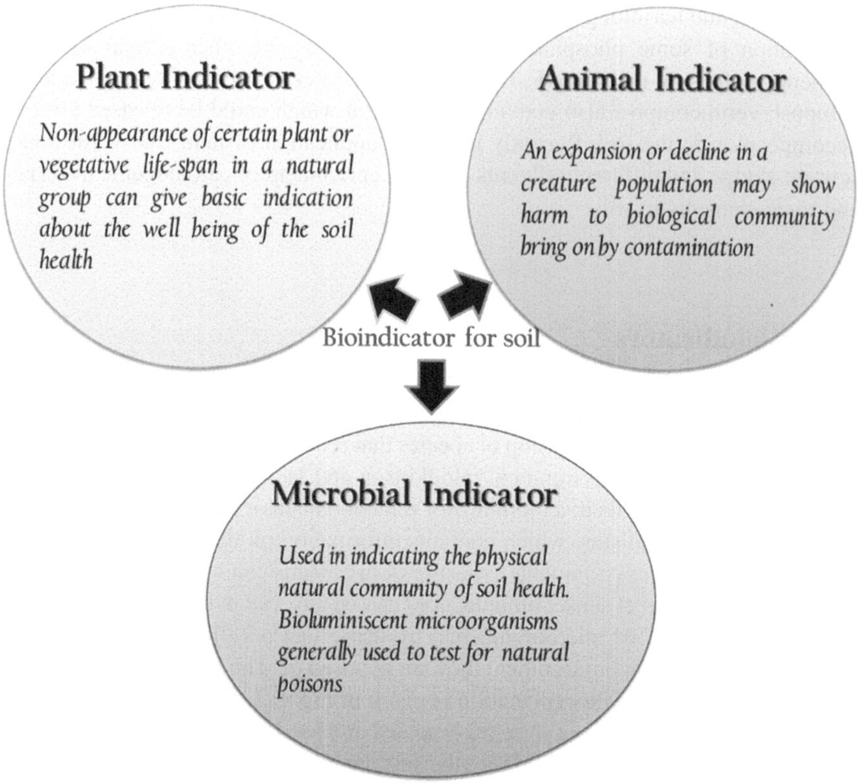

Fig. 8.2 Broad classes of bioindicators for assessing soil quality

- *Functional,* i.e. phospholipid fatty acid (PLFA), as a biogeochemical approach that offers information on both microbial taxonomic and functional diversity (Dimitriu et al. 2010)
- *Molecular* analysis of soil extracted nucleic acid sequences (DNA, RNA) (Banning et al. 2011)

Many of the above mentioned sensitive bioindicators have frequently been used to assess the soil resilience and restoration capability especially under problem (naturally degraded or contaminated) soil sites. To assess the soil degradation due to intense agricultural practices, including crop density and the application of organic fertilisers in different locations of Italy much emphasis has been given to different soil bioindicators to establish three 'alteration indices' (enzyme activities) (Puglisi et al. 2006). These three indices included altogether seven soil enzyme activities i.e. arylshulphtase, β-glucosidase, phosphatase, urease, invertase, dehydrogenase and phenoloxidase.

Both the metabolic (qCO_2, determined as microbal respiration or CO_2-C evolution per unit of MBC) and the microbial (Cmic: Corg, MBC to organic carbon ratio) quotients suitably represent the energy optimization and more often used as the bioindicators for ecosystems recovery (Anderson and Domsch 1990, 1993). They have been efficiently used in a recently conducted study under restored soils in degraded semiarid ecosystems of Western Australia. They also proposed a rapid and low-cost technique, the 1-day CO_2 test as another important bioindicator for measuring soil microbial activity in turn assessing the soil quality and functionality in restoration programmes (Muñoz Rojas et al. 2016). Bastida et al. (2006) established a 'Microbiological Degradation Index' based on varied soil bioindicators for natural soils in a semiarid climate in south-eastern Spain with a prominent influence of climate that may turn to desertification and ill-effect on soil quality.

For a reclaimed mine soils, MBC and dehydrogenase activity were considered as bioindicators along with total carbon (TC), labile C, and rhizosphere N as other important indicators for carbon sequestration (Mukhopadhyay et al. 2016). Dawson et al. (2007) proposed a 'Biological Soil Quality Index' using a number of soil bioindicators to understand the the impact of hydrocarbon-polluted soils.

In some recent studies, the impact of residual pesticides' contaminations on soil ecology was evaluated through some well-defined soil bioindicators. Tebuconazole, a fungicide, resulted in a short-lived and transitory toxic effect on soil microbial properties and enzymatic activities at lower dose but persisted longer at increased dose. Soil ergosterol content, dehydrogenase and nitrate reductase activity decreased sharply under tebuconazole application and revealed as important soil health indicators (Saha et al. 2016a). Two post-emergence herbicides (imazethapyr and quizalofop-*p*-ethyl) exhibited transient harmful effects on most of the soil bioindicators including microbial biomass C, fluorescein diacetate hydrolyzing activity, dehydrogenase activity, acid and alkaline phosphatase activity (Saha et al. 2016b).

By far it is established that soil microbial parameters which provide information on the biomass, activity or functionality and diversity of microbial communities have been widely proposed as bioindicators of soil health (Gómez-Sagasti et al. 2012), play a substantial role together with common soil chemical indicators in reclamation of soil health.

Biomarker, comparatively a newer concept of bioindicator, serves the same function. Two key properties are underlying to understand the biomarkers: it should be responsive only to the biologically active fraction of accumulated body burden among one or more toxicants; hence biomarkers typically characterize the bioavailable fraction of any environmental chemical of interest; secondly, biomarkers integrate the interactive effects of complex mixtures of chemicals faced by organisms impacted by modern industrial and agricultural chemicals in ecosystems (Ricketts et al. 2003). Thus biomarkers deal with single-chemical, single-species laboratory toxicity bioassays for ecological risk assessment.

There are quite a few numbers of novel tools evolved for analysis of soil biomarkers. PLFA profiles, a commonly used biomarker for identifying environmental monitoring and assessment more specifically to soil stress, and gives a broad picture of diversity of soil microbial community structure. Being a key component of cell membrane of microbes, this bioassay provides phenotypic expression at both intracellular and extracellular levels (Kaur et al. 2005). PLFAs are methylated to produce fatty acid methyl esters (FAMEs) that can be easily measured by gas chromatography with a flame ionization detector (GC-FID) is another viable and frequently-used option of soil biomarker (Dowling et al. 1986).

Molecular genetic biomarkers are also potentially used for monitoring of soil pollution in terms of heavy metal toxicity. Real-time quantitative PCR was used to measure gene transcription in earthworms (*Lumbricus rubellus*) for Cd and Cu enriched soil and revealed both qualitative and quantitative differences in the expression of two target genes in the responses to the two metal ions (Galay-Burgos et al. 2003). The use of biomarkers in combination with stable isotope analysis can also be another effective option to study organic matter sources utilised by microorganisms in complex ecosystems and identify few specific groups of microbes like methanotrophic bacteria (Boschker and Middelburg 2002). Combining PLFA with stable isotope analysis is another advanced biomarker soil microbial functioning thus characterized by both GC and isotope-ratio mass spectrometry (IRMS) (Watzinger 2015). Another recent study at a mixed aged and polycyclic aromatic hydrocarbons contaminated soil sites identified biomarkers and confirmed the level of pollutant by using gas chromatography/mass spectrometry (GC/MS) screening (Kao et al. 2015).

8.4 Plant Biomarkers

Plants can be conceptualized as important bioindicator for any ecosystem; similarly, for the problem soils; they also readily change their metabolic and growth behaviour under most of the abiotic/biotic stresses. Plant-tissue based biosensor/biomarker was first developed as early as 1980s by immobilizing slices of yellow squash tissue as a CO_2 gas sensor (Kuriyama and Rechnitz 1981). Since then a variety of plant tissue or field grown plants were started to be used as biomarkers for number of purposes. Despite the structural and metabolic differences of plants, they can substantially play the role as a bioindicator upon exposure to cytotoxic, genotoxic and mutagenic hazardous compounds in soil. Moreover an easy maintenance and low cost of cultivation made them demanding (Rodrigues et al. 1997). Exposure to an unfavourable environment often leads to metabolic changes in plant tissues leading to synthesis of key organic metabolites which can further be used as biomarkers. Plant wax and lipids, lignins and phenols are the most commonly formed molecular tracer for terrestrial plants (Drenzek et al. 2007; Ohkouchi and Eglinton 2008).

Although, the physiological processes, biochemical response and mechanisms of adaptation or mortality could well be used to evaluate the quality of plant's growing condition, but despite of its crucial role in both aquatic and/or terrestrial ecosystems, it has been underemployed as bioindicator for the diagnosis or prediction of the negative consequences of external factors (Vangronsveld et al. 1998). Plants are generally sensitive to environmental variations and react more rapidly to the presence of pollutants than other environmental hazards (Lovett Doust et al. 1994).

Technical advantages	*Cytological advantages*	*Economic advantages*
• Can face direct exposure to contaminant • Simple and agile analysis • Tested under varied environmental conditions, pH and temperature • In-situ monitoring • Easy regeneration • Higher reliability and replicability of results	• Higher plants are eukaryotes • Cellular structure and organization are close to humans and thus comparable to animals • Able to assess the genotoxic potential of simple substances or even complex mixtures • Can work together with promutagens and show higher sensitivity towards carcinogenic agents	• Low cost • Easy maintainence

Fig. 8.3 Three dimensions of advantages for using plant biomarkers

Recently (about two decades back), researchers have cited several advantageous features that plant biomarkers can cater (Grant 1994; Steinkellner et al. 1999). These potential benefits can be suitably highlighted to understand the usefulness and applicability of plant biomarkers (Fig. 8.3).

However, few limitations still exist while using plants as bioindicators, such as, plants failed to show sensitivity towards certain classes of pro-mutagens such as nitrosamines, heterocyclic amines, some classes of PAH (Majer et al. 2005) and BTEX (Mazzeo et al. 2010), nitro aminobenzene particularly in onion (*Allium cepa*) (Ventura 2009). Plants utilized as biomarkers and their specific features are shown in (Table 8.1).

8.4.1 Identification and Quantification of Plant Biomarkers

Identification of biomarkers and its further quantification and analysis for plants has been standardized pretty recently (Schudoma et al. 2012) in recognition for its applications to studies on growth (Bates et al. 2009), nutrient effects (Yang et al. 2011), responses and tolerances to biotic and abiotic stressors (Garg et al. 2002; Wang et al. 2011) and for marker-assisted breeding (Deyanira et al. 2012). Still there are many challenges and obstacles particularly with the integrated systems biology approaches for the 'Omics', *i.e.* genomics, transcriptomics, proteomics, ionomics, metabolomics and phenomics (Degenkolbe et al. 2013).

Initially, to establish utility of biomarkers as predictors, in future instances on classifying individual samples, the application of detection theory is critically required for biomarker quantification for numerous individual samples as universal standard (Schudoma et al. 2012). Also there is severe limitation for computational

Table 8.1 Plants utilized as biomarkers and their specific features

Plant	Advantageous features as biomarker	Reference
Onion (*A. cepa*)	Rapid root growth	Grant (1994)
	Faster cell division	
	Easy detection of mitotic activity and abnormality in meristematic root cells	
	Fast and sensitive towards genoxotic and mutagenic substances	Leme and Marin-Morales (2009)
Tradescantia	Detection of mutations induced by contamination agents	Shima et al. (1997)
	Analysis of micronuclei in the mother cell of the pollen grain	
	Easy handling and relatively low maintenance cost	
Vicia faba	Fit for radiobiological tests to detect chromosomal aberrations by ionizing radiation	Read (1959), Kihlman (1975), Kihlman and Kronborg (1976), Kihlman and Andersson (1984)
	Meristematic cell bioassay	
	Genotoxicity studies	
	Cytological and physiological studies	Kanaya et al. (1994)
Phytochelatins in maize (*Zea mays*) and wheat (*Triticum aestivum*)	Assessed the heavy metal stress under combined effect of Cu and Cd by using phytochelatins as biomarkers	Keltjens and Van Beusichem (1998)

modelling of metabolic pathways simulation of individual organism specific models for biomarker studies (Chen et al. 2013).

8.4.2 Measurable Responses of Plant Biomarkers

The measurable response of biomarkers can be achieved through monitoring of changes in photosynthetic activities of plant, its chlorophyll fluorescence pattern, changes in enzyme dependent nutritional processes in plants, production of reactive oxygen species and oxidative stress status of the plants (Ferrat et al. 2003). Chlorophyll fluorescence measurement is a way to evaluate the biochemical and physiological state of the plants. In recent times, chlorophyll content and chlorophyll fluorescence are used to highlight stress due to a single environmental factor or due to a combined stress effect of different environmental factors. But on the other hand, they could also be used as potential biomarkers of anthropogenic and/or environmental stresses (Vangronsveld et al. 1998).

Another measurable response is changes in photosynthetic pigment concentration (Ralph 2000). In many instances, under continuous exposure trace metals can substitute for the magnesium ion in the chlorophyll molecule, leading to its inability to catch photons and thus to a decrease in the photosynthetic activity. Under different environmental stresses plants generally increase their carotenoid concentration in order to impart better photo-protection against the formation of free radicals. Thus, a net decrease in total chlorophyll concentration and in the of ratio chlorophyll/carotenoids is often observed as an indicator for unfavourable environment (Ferrat et al. 2003).

Nutrient use efficiency or the plant nutrient metabolism can also be influenced by various biotic or abiotic stressors. So the activity of enzymes involved in the assimilation of key plant nutrients get affected, which can be used as bioindicator for certain kind of situations. For example, changes in the activities of glutamine synthetase (GS) and nitrate reductase (NR), key enzymes involved in N-metabolism can indicate N-status of the soil where the plant is growing (Schwalbe et al. 1999). Similarly, phosphate metabolizing enzymes (alkaline phosphatase, an enzyme hydrolysing organic phosphate monoesters to inorganic phosphate) can be an interesting biomarker for their capacity to highlight a nutrient deficiency for plants under stressing conditions (Invers et al. 1995).

The phenomenon of production of reactive oxygen species as a result of oxidative burst had been studied widely in the case of the invasion of terrestrial plants by pathogens and also few studies have been carried out on the mechanisms of defence against pathogens in the marine environment (Potin et al. 1999). Lipid peroxidation leading to formation of lipid degradation products especially from membranes serve as important bioindicator for different stressors. Thiobarbituric acid reactive substances (TBARs) are used a bioindicator for metal induced oxidative stress (Vavilin et al. 1998).

8.5 Biosensors: Concept and Applications

In recent times, this biosensor technology has been emerged as a powerful alternative over conventional analytical techniques, harnessing the specificity and sensitivity of biological systems in small size and cheaper devices. Biosensor can be defined as a compact analytical device, incorporating a biological or biomimetic sensing element, either closely connected to, or integrated within, a transducer system (Velasco-Garcia and Mottram 2003).

Biosensing devices have more advanced technological base and thus more advantageous over conventional analytical techniques. These kind of highly specific devices are meant for real-time analysis in complex mixtures, without the need for large volume of sample or sample pre-treatment. Overall the biosensors also offer highly sensitive, rapid, reproducible and easy-to-operate analytical tools.

The underlying principle of this biosensor for detection is based on the specific binding of the analyte to the complementary bio-recognition element immobilised on a support medium. Nevertheless, this specific binding or interaction brings few changes in either one or more physico-chemical properties; pH change, electron transfer, mass change, heat transfer, uptake or release of gases or specific ions, are most common. These physio-chemical alterations are detected and mostly signalled by the transducer. The biosensor technology uses the biological materials, like the couples of (enzyme+substrate), (antibody+antigen) and (nucleic acids+complementary sequences). Moreover, whole plant cells or tissue slices, microorganisms, animal are also embedded in the biosensing device.

Several published reports claimed that the biosensors has efficiently been employed for detecting pesticide residues in soil, and also from the crops grown over it, and thus established its role in environmental management. For identifying the extent of hazards in problem soils, either toxic by heavy metals or pesticides, the main principle acts here is the correlation between toxicity of a pesticide and a decrease in the activity of a biomarker such as an enzyme. While, this activity can be registered by using different transducers (amperometry, potentiometry, spectrometry, fluorimetry or thermometry) for real-time detection of different substrates or products of enzymatic reaction (Velasco-Garcia and Mottram 2003).

There are few reports found where biosensors have successfully used as indicators for heavy metal pollution in soils. Mostly the bioluminescence-based biosensors are used for this purpose which provides an inexpensive and rapid technique to evaluate the bioavailability of metals in soil. The response of this kind of biosensors can be related to measurements of soil solution speciation, and thus gives an expression of toxic thresholds in soils. A lux-based biosensor was applied to detect availability of Zn; and it was found that bioluminescence response declined as the free Zn^{2+} in soil solution increased (McGrath et al. 1999). While Rhizotox-C, another biosensor was tested for both Zn and Cu (as total, soil solution and soil solution free forms) in soils of long-term sewage sludge treated field experiment (Chaudri et al. 2000). Both the studies established the potentiality of luminescence-based biosensors for identifying and monitoring of the contaminated soil.

Much advanced analytical tools using biosensors for detection of pesticides using molecular imprinted polymers based biomimetic sensors have also been figured out. Using the similar techniques, herbicides such as atrazine (Sergeyera et al. 1999) and 2,4-dichlorophenoxyacetic acid (Kroger et al. 1999) was detected. This has been found as highly sensitive (detection limit 5 nM), rapid (response in few minutes) and much more stable (>6 months). While some biosensors worked on the reduction of intensity of certain natural processes such as photosynthesis and bioluminescence property under influence of toxic compounds. This particular biosensor detected traces of herbicide residues in soil measuring the inhibited oxygen levels based on a chlorophyll–protein reaction center complex. The test has been ultra-sensitive, with detection limits comparable to highly sensitive enzyme-linked immunosorbent assay (ELISA) (Koblizek et al. 1998). Another biosensor was developed for monitoring the level toxicity of polycylic aromatic hydrocarbons contaminated soil which used an immobilized recombinant bioluminescent bacterium, GC2. This system using a biosurfactant is effective as an in-situ biosensor for detecting the hydrophobic contaminants in soils and determined the PAH degradation in soils (Gu and Chang 2001). Lui et al. (1997) reported the biosensor (based on acetylcholinesterase immobilised onto magnetic particles in a photometric flow injection system) and detected methamidophos in lettuce and cabbage at 12 and 3.0 mg kg^{-1}, respectively. Among other types of biosensors, Chemiluminiscence-based technique was used for detecting organophorus and carbamate pesticides (Roda et al. 1994), while the cellular and immunological biosensors effectively quantified the toxicity and concentrations of four popular herbicides (atrazine, diuron, mecoprop, paraquat) in soil extracts (Strachan et al. 2002).

8.6 Role of Bioindicators in Stress Management

8.6.1 Managing Pollution Stress

Aquatic pollution can be measured using aquatic fauna like freshwater planarians which is only present in unpolluted streams or lakes (Knakievicz 2014). Some more commonly used bioindicators are mollusks bivalves, mollusks gastropods and fishes. Some zooplanktons accumulate and metabolize pollutants and they may be used as bioindicators of water quality (Zhoua et al. 2008).

Organisms belonging to Isopoda, Collembola, Oligochaeta and Diplopoda are proposed as bioindicator organisms due to their close contact with soil (Fontanetti et al. 2011). Moreover, higher plants such as *Arabidopsis thaliana, A. cepa, Hordeum vulgare*, *Tradescantia* sp., *Vicia faba* and *Zea mays* are also used in the assessment of soil toxicity (White and Claxton 2004).

8.6.2 Managing Stress of Soil Acidity and Alkalinity

Acidophilus species of collembolan (*Tomocerus flavescens*), mites (*Hypochthonius rufulus* and *Adoristes ovatus*) and isopod (*Oniscus asellus*) groups are used as an indicator for acid soils. While the alkalophilous species belonging to cllembola (*Isotoma notabilis*, *Entomobrya corticalis*), mites (*Pelops occultus*, *Platynothrus peltifer*) and isopod (*Armadillidium vulgare*) are used (van Straalen and Verhoef 1997).

8.6.3 Managing Stress of Soil Salinity

Ruprechtia triflora is a tree species in dry forests of Paraguay. It shows extreme osmotic adaptability. In three greenhouse experiments with NaCl application,

Ruprechtia triflora and *Eucalyptus dunnii* seedlings showed highly significant responses to their soil salinities (Mitloehner and Koepp 2007).

8.6.4 Managing Stress of Water

In a long-term study (8–10 years), it was observed that the abundance of enchytraeids, mesostigmatid mites and macroarthropod predators were lowest in the drought plots. Drought decreased the abundance and diversity of Oribatida and Collembola. Thus the soil microarthropods can be used as environmental indicators for drought (Lindberg et al. 2002).

8.6.5 Managing Heavy Metals Stress

An organism can serve as a metal pollution bioindicators only if it meets certain criteria: (i) the body must constantly accumulate and tolerate large amounts of toxic metals, (ii) it must be tied to a single place to make it a true 'representative' for the soil, air, and water environmental area, (iii) it must be available for collection, identification, and handling, (iv) it must have sufficient tissue for chemical analysis and a long life span to ensure sampling over a longer period of time (Stankovic et al. 2014).

Among many, the main anthropogenic sources of toxic metals are fertilizers, pesticides, contaminated irrigation water, combustion of coal and oil, vehicular emissions, incineration of urban and industrial wastes and, mining and smelting (Tavares and Carvalho 1992).

More or less the common symptoms observed in the susceptible plants grown in soils contaminated with heavy metals are- reduced root growth, reduced seed sprouting, necrosis, and chlorosis (Park et al. 2011). Earthworms (*Eisenia foetida*) has the capacity of toxic metals' accumulation (Hg, Cd, Cu, Pb, and Zn), and hence most exploited soil invertebrate bioindicators. Few researchers showed a significant positive correlation between metal concentrations in the earthworm and in the soil (Hirano and Tamae 2010; Olayinka et al. 2011). Even in aquatic ecosystems, earthworms can be served as equally suitable bioindicator. Aquatic and marine

Table 8.2 Normal natural concentration intervals for toxic metals in terrestrial plants (Yildiz et al. 2010)

Metals	Concentration intervals (μg/g)
Cadmium (Cd)	0.2–2.4
Zinc (Zn)	20–400
Iron (Fe)	70–700
Nickel (Ni)	1–5
Lead (Pb)	1–13
Manganese (Mn)	20–700

gastropods and terrestrial gastropods (snails) are also recognized as adequate bioindicators for toxic metals like Pb, Zn, Cu, and Cd (Madoz-Escande and Simon 2006) (Table 8.2).

8.6.6 Managing Pesticide Load

Due to intense agricultural production, soil has been heavily loaded with pesticides. Even after its action, pesticide residues remain in the soil in significantly fatal quantity. To detect the potential hamper out of it, bioindicators may prove helpful.

Among plant indicators, onion (*A. cepa*) has been successfully used for determining the mutagenic and genotoxic potential of trifluralin, a known herbicide (Fernandes et al. 2007, 2009). However in presence of certain pesticides, the terrestrial invertebrates showed some common symptoms of alterations in the biomass, reproduction, behaviour, survival and tissular/cellular lesions (Fontanetti et al. 2011).

8.7 Bioindicators for Climate Change

Soil biota considered as 'the biological engine of the earth' drives many fundamental nutrient cycling processes, soil structural dynamics, degradation of toxic pollutants, etc. and maintains the ecosystem sustainability and microorganisms are main players in these services. Hence, it is logical that biological health of soil ecosystem has great potential as indicator of ecosystem health, which can be of use in environmental analysis. The indicators used for monitoring the state of the environment should be able to reflect the structure and function of ecosystem processes sensitive to variations in management and climate, reproducible, easily measurable and widely applicable from local to national scale (Neher 2001). Bioindicators may be used as an indirect measure of soil function, helping to measure soil quality or health and its course of change with time, by involving functional relationships among measurable traits and monitoring for sustainable land management which also include ecological and climatic effects (Dalal et al. 2003a, b; Doran 2002; Doran and Zeiss 2000).

Soil microbial properties react much quicker to environmental and climatic disturbances and perturbations, and these changes need to be reckoned for maintenance of an ecosystem. Climate change impacts such as higher CO_2 concentration, elevated temperature, atmospheric N deposition and changes in seasonal and total distribution of rainfall and extreme weather events such as droughts and floods will effect soil microbial community (IPCC 2007), C and N cycling, and therefore, on soil structure and erosion occasions, availability of nutrients and plant diseases, and hence on and agricultural productivity and ecosystem functionality.

Due to global warming and climate change the average temperature of the globe has been increasing. Soil microbes of single species could adapt this change by physiological means (Malcolm et al. 2008) or by shifting of species within microbial community. With temperature the respiration rate of soil microbe increases so does the organic matter decomposition rate too.

There is proof that due to warming the decomposer physiology changes potentially and therefore, the CO_2 efflux from soil (Allison et al. 2010; Bárcenas-Moreno et al. 2009; Balser and Wixon 2009; Bradford et al. 2008). So soil microbial respiration study could be a great tool to quantify the impacts of global warming and climate change and specific bioindicators could be of immense use in this regard.

In an attempt to simulate anticipated global warming (1–5 °C) a large number of field and laboratory experiments have been carried out worldwide and some general patterns were recorded. Almost in all the studies it was observed that soil microbial biomass did not increase by warming (Feng and Simpson 2009; Biasi et al. 2008; Rinnan et al. 2007, 2008, 2009; Vanhala et al. 2011; Zhang et al. 2005). Results showed that the microbial biomass either remained at steady levels or reduced depending on the length of the warming treatment, however, the picture was more complex regarding the composition of microbial community. Some changes in composition of microbial community were reported by scientists globally such as increased as well as decreased in abundance of fungal community and some Gram-positive bacteria, decreased abundance of Gram-negative bacteria, or no change in abundance of these soil microbial communities at all (Castro et al. 2010; Karhu et al. 2010; Feng and Simpson 2009; Frey et al. 2008; Rinnan et al. 2007, 2008, 2009; Vanhala et al. 2011; Biasi et al. 2005).

There are some other studies where biomarker levels for microbial stress were observed significantly higher in plots where warming treatments were done. Positive correlation between microbial stress biomarker during soil warming and microbial metabolic activity was also observed (Schindlbacher et al. 2011). Environmental stress such as warming stress enhances microbial maintenance demand (respiration per unit biomass), therefore, microbial stress biomarker levels were observed higher in warmed soil (Anderson and Domsch 2010).

PLFA biomarkers specific to certain soil microorganisms could be helpful to assess soil microbial population under climate changed conditions. The sum of PLFA biomarker concentrations (ΣPLFA) in soil is sometimes used to estimate differences in the size of the soil microbial community (mainly fungi and bacteria) under different climatic treatment such as elevated CO_2 and increased temperature. It was observed that warming treatments reduces Σ PLFA biomarker in general and the effect found to be alike for fungi and different bacterial groups (Andresen et al. 2014).

Under climate changed condition, biological indicators play an important role in soil health assessment, since they include intricate adaptive systems (i.e. the biota) by integrating important soil processes (Ritz et al. 2009). Latest studies suggest that climatic changes, such as global warming, will have an intense impact on the rhizosphere, heterotrophic community structure of soil and other soil processes such as soil respiration, mineralisation of N and ecosystem functioning of carbon (Bardgett et al. 2008; Briones et al. 2009). Altering climatic factors is also a matter of concern due to possible evolutionary changes, which countenance the spread of many virulence factors and genes that aid in environmental existence (French et al. 2009). Understanding the impacts of climate change on soil health especially for problem soils is promising through the use of bioindicators which relate soil physico-chemical and biological properties to ecological functions and which can be examined in the context of changing climatic scenario.

8.8 Conclusion

Soil contamination is unavoidable. It affects soil health and quality and disbalance the whole ecosystem. More emphasis has been given to restoration of problem soils or uncultivated soil or less productive soils (in agriculture point of view), considering the fact that our land resources are limited and sufficiency of food production is yet to be achieved. Hunger is still inevitable and population pressure is alarming at many parts of the world.

As an important strategy to characterize the soil contamination, the role of bioindicators and biomarkers are vital. Several biological materials like different plant, animals, microbes in parts or whole and their cellular, physiological and molecular functions forms the basis of applying bioindicators and biomarkers. In various parts of the world, researchers are engaged in identifying suitable bioindicators/biomarkers and also devising the biosensor with major focus to problem/contaminated soils.

Though studying the soil contamination is vast, researchers should confine their goals and concentrate on the locally or regionally relevant soil problems. Moreover, many bio-materials are plenty in diversified nature, the focus should not be diverted to select the ideal bioindicator(s) in line with the research or applied purposes. Secondly, standardization of analytical techniques for evaluating a specific soil bioindicator should be another major research theme, since it will provide more accuracy in estimation. Thirdly, a balanced comparison should be made for assessing the performance of bioindicators for problem soils would be another researchable issue. More often, the problem soils are co-contaminated with two or more toxic metals, so there should be constant search for such bioindicators for detecting multiple ion toxicity. Once the approach of bioindicators is efficiently executed for identifying a specific problem then it would promisingly facilitate to follow the remediation or revegetation strategies for any problem soils.

References

Abdi MR, Kamali M, Vaezifar S (2008) Distribution of radioactive pollution of ^{238}U, ^{232}Th, ^{40K} and ^{137}Cs in northwestern coasts of Persian Gulf, Iran. Mar Pollut Bull 56:751–757

Abrol IP, Chhabra R, Gupta RK (1980) A fresh look at the diagnostic criteria for sodic soils. In: International symposium on salt affected soils. Central Soil Salinity Research Institute, Karnal. pp 142–147, February 18–21, 1980

Allison SD, Wallenstein MD, Bradford MA (2010) Soil-carbon response to warming dependent on microbial physiology. Nat Geosci 3:336–340. https://doi.org/10.1038/ngeo846

Anderson TH, Domsch KH (1990) Application of eco-physiological quotients (qCO_2 and qD) on microbial biomasses from soils of different cropping histories. Soil Biol Biochem 22:251–255. https://doi.org/10.1016/0038-0717(90)90094-G

Anderson TH, Domsch KH (1993) The metabolic quotient for CO_2 (qCO_2) as a specific activity parameter to assess the effects of environmental conditions, such as pH, on the microbial biomass of forest soils. Soil Biol Biochem 25:393–395. https://doi.org/10.1016/0038-0717(93)90140-7

Anderson TH, Domsch KH (2010) Soil microbial biomass: the eco-physiological approach. Soil Biol Biochem 42:2039–2043. https://doi.org/10.1016/j.soilbio.2010.06.026

Andresen LC, Dungait JAJ, Bol R, Selsted MB, Ambus P, Michelsen A (2014) Bacteria and Fungi respond differently to multifactorial climate change in a temperate heathland, traced with ^{13}C Glycine and FACE CO_2. PLoS One 9(1):e85070. https://doi.org/10.1371/journal.pone.0085070

Balser TC, Wixon DL (2009) Investigating biological control over soil carbon temperature sensitivity. Glob Chang Biol 15:2935–2949. https://doi.org/10.1111/j.1365-2486.2009.01946.x

Banning N, Gleeson DB, Grigg AH, Grant CD, Andersen GL, Brodie EL, Murphy D (2011) Microbial community successional patterns during forest ecosystem restoration. Appl Environ Microbiol 77:6158–6164. https://doi.org/10.1128/AEM.00764-11

Bárcenas-Moreno G, Gómez-Brandón M, Rousk J, Bååth E (2009) Adaptation of soil microbial communities to temperature: comparison of fungi and bacteria in a laboratory experiment. Glob Chang Biol 15:2950–2957. https://doi.org/10.1111/j.1365-2486.2009.01882.x

Bardgett RD, Freeman C, Ostle NJ (2008) Microbial contributions to climate change through carbon cycle feedbacks. ISME J 2:805–814

Bastida F, Moreno JL, Hernández T, García C (2006) Microbiological degradation index of soils in a semiarid climate. Soil Biol Biochem 38:3463–3473. https://doi.org/10.1016/j.soilbio.2006.06.001

Bates PD, Durrett TP, Ohlrogge JB, Pollard M (2009) Analysis of acyl fluxes through multiple pathways of triacylglycerol synthesis in developing soybean embryos. Plant Physiol 150:55–72. https://doi.org/10.1104/pp.109.137737

Bhaduri D, Purakayastha TJ (2014) Long-term tillage, water and nutrient management in rice–wheat cropping system: assessment and response of soil quality. Soil Tillage Res 144:83–95. https://doi.org/10.1016/j.still.2014.07.007

Bhaduri D, Pal S, Purakayastha TJ, Chakraborty K, Yadav RS, Akhtar MS (2015) Soil quality and plant-microbe interactions in the rhizosphere. In: Sustainable agriculture reviews. Springer International Publishing, Cham, pp 307–335. https://doi.org/10.1007/978-3-319-16742-8_9

Bhaduri D, Purakayastha TJ, Patra AK, Singh M, Wilson BR (2017a) Biological indicators of soil quality in a long-term rice–wheat system on the Indo-Gangetic plain: combined effect of tillage–water–nutrient management. Environ Earth Sci *76*(5):202. https://doi.org/10.1007/s12665-017-6513-0

Bhaduri D, Pramanik P, Ghosh S, Chakraborty K, Pal S (2017b) Agroforestry for improving soil biological health. In: Gupta SK, Panwar P, Kaushal R (eds) Agroforestry for increased production and livelihood security. New India Publishing Agency, New Delhi, pp 465–491 ISBN: 9789385516764

Bhattacharjya S, Bhaduri D, Chauhan S, Chandra R, Raverkar KP, Pareek N (2017) Comparative evaluation of three contrasting land use systems for soil carbon, microbial and biochemical indicators in North-Western Himalaya. Ecol Engg 103:21–30. https://doi.org/10.1016/j.ecoleng.2017.03.001

Bian M, Zhou M, Sun D, Li C (2013) Molecular approaches unravel the mechanism of acid soil tolerance in plants. Crop J 1:91–104. https://doi.org/10.1016/j.cj.2013.08.002

Biasi C, Rusalimova O, Meyer H, Kaiser C, Wanek W, Barsukov P, Junger H, Richter A (2005) Temperature-dependent shift from labile to recalcitrant carbon sources of arctic heterotrophs. Rapid Commun Mass Spectrom 19:1401–1408. https://doi.org/10.1002/rcm.1911

Biasi C, Meyer H, Rusolimova O, Hämmerle R, Kaiser C, Baranyi C, Daims H, Lashchinsky N, Barsukov P, Richter A (2008) Initial effects of experimental warming on carbon exchange rates, plant growth and microbial dynamics of lichen-rich dwarf shrub tundra in Siberia. Plant Soil 307:191–205. https://doi.org/10.1007/s11104-008-9596-2

Boschker HTS, Middelburg JJ (2002) Stable isotopes and biomarkers in microbial ecology. FEMS Microbiol Ecol 40(2):85–95. https://doi.org/10.1111/j.1574-6941.2002.tb00940.x

Bradford MA, Davies CA, Frey SD, Maddox TR, Melillo JM, Mohan JE, Reynolds JF, Treseder KK, Wallenstein MD (2008) Thermal adaptation of soil microbial respiration to elevated temperature. Ecol Lett 11:1316–1327. https://doi.org/10.1111/j.1461-0248.2008.01251.x

Bramley H, Tyerman SD, Turner DW, Turner N (2011) Root growth of lupins is more sensitive to waterlogging than wheat. Funct Plant Biol 38:910–918. https://doi.org/10.1071/FP11148

Briones MJI, Ostle NJ, McNamara NP, Poskitt J (2009) Functional shifts of grassland soil communities in response to soil warming. Soil Biol Biochem 41:315–322. https://doi.org/10.1016/j.soilbio.2008.11.003

Castro HF, Classen AT, Austin EE, Norby RJ, Schadt CW (2010) Soil microbial community response to multiple experimental climate change drivers. Appl Environ Microbiol 76:999–1007. https://doi.org/10.1128/AEM.02874-09

Chaudri AM, Lawlor K, Preston S, Paton GI, Killham K, McGrath SP (2000) Response of a Rhizobium-based luminescence biosensor to Zn and Cu in soil solutions from sewage sludge treated soils. Soil Biol Biochem 32(3):383–388. https://doi.org/10.1016/S0038-0717(99)00166-2

Chen PW, Fonseca LL, Hannun YA, Voit EO (2013) Coordination of rapid sphingolipid responses to heat stress in yeast. PLoS Comp Biol 9:e1003078. https://doi.org/10.1371/journal.pcbi.1003078

Dalal RC, Eberhard R, Grantham T, Mayer DG (2003a) Application of sustainability indicators, soil organic matter and electrical conductivity, to resource management in the northern grains region. Aust J Exp Agric 43:253–259. https://doi.org/10.1071/EA00186

Dalal RC, Wang WJ, Robertson GP, Parton WJ (2003b) Nitrous oxide emission from Australian agricultural lands and mitigation options: a review. Aust J Soil Res 41:165–195. https://doi.org/10.1071/SR02064

Dawson JJC, Godsiffe EJ, Thompson IP, Ralebitso-Senior TK, Killham KS, Paton GI (2007) Application of biological indicators to assess recovery of hydrocarbon impacted soils. Soil Biol Biochem 39:164–177. https://doi.org/10.1016/j.soilbio.2006.06.020

Degenkolbe T, Do PT, Kopka J, Zuther E, Hincha DK, Kohl KI (2013) Identification of drought tolerant markers in a diverse population of rice cultivars by expression and metabolite profiling. PLoS ONE 8:e63637. https://doi.org/10.1371/journal.pone.0063637

Deyanira QM, Estrada-Luna AA, Altamirano-Hernandez J, Pena-Cabriales JJ, de Oca-Luna RM, Cabrera-Ponce JL (2012) Use of trehalose metabolism as a biochemical marker in rice breeding. Mol Breed 30:469–477. https://doi.org/10.1007/s11032-011-9636-0

Dick RP (1997) Soil enzyme activities as integrative indicators of soil health. In: Pankhurst CE, Doube BM, Gupta VVSR (eds) Biological indicators of soil health. CABI Publishing, Wallingford

Dimitriu PA, Prescott C, Quideau SA, Grayston SJ (2010) Impact of reclamation of surface-mined boreal forest soils on microbial community composition and function. Soil Biol Biochem 42:2289–2297. https://doi.org/10.1016/j.soilbio.2010.09.001

Doran JW (2002) Soil health and global sustainability: translating science into practice. Agric Ecosyst Environ 88:119–127. https://doi.org/10.1016/S0167-8809(01)00246-8

Doran JW, Zeiss MR (2000) Soil health and sustainability: managing the biotic component of soil quality. Appl Soil Ecol 15:3–11. https://doi.org/10.1016/S0929-1393(00)00067-6

Dowling NJE, Widdel F, White DC (1986) Phospholipid ester-linked fatty acid biomarkers of acetate-oxidizing sulfate reducers and sulfide forming bacteria. J Gen Microbiol 132:1815–1825. https://doi.org/10.1099/00221287-132-7-1815

Drenzek NJ, Montluçon DB, Yunker MB, Macdonald RW, Eglinton TI (2007) Constraints on the origin of sedimentary organic carbon in the Beaufort Sea from coupled molecular ^{13}C and ^{14}C measurements. Marine Chem 103(1):146–162. https://doi.org/10.1016/j.marchem.2006.06.017

Driessen P, Deckers J, Spaargaren, O (eds) (2001) Lecture notes on the major soils of the world, world soil resources reports 94. Food and Agriculture Organization of the United Nations, Rome, 334 p

Elzenga JTM, Veen HV (2010) Waterlogging and plant nutrient uptake. In: Mancuso S, Shabala S (eds) Waterlogging tolerance and signaling in plants. Springer, Heidelberg, pp 23–35. https://doi.org/10.1007/978-3-642-10305-6_2

Feng X, Simpson MJ (2009) Temperature and substrate controls on microbial phospholipid fatty acid composition during incubation of grassland soils contrasting in organic matter quality. Soil Biol Biochem 41:804–812. https://doi.org/10.1016/j.soilbio.2009.01.020

Fernandes TCC, Mazzeo DEC, Marin-Morales MA (2007) Mechanism of micronuclei formation in polyploidizated cells of *Allium cepa* exposed to trifluralin herbicide. Pestic Biochem Physiol 88:252–259. https://doi.org/10.1016/j.pestbp.2006.12.003. ISSN 00483575

Fernandes TCC, Mazzeo DEC, Marin-Morales MA (2009) Origin of nuclear and chromosomal alterations derived from the action of an aneugenic agent Trifluralin herbicide. Ecotoxicol Environ Saf 72:1680–1686. https://doi.org/10.1016/j.ecoenv.2009.03.014. ISSN 01476513

Ferrat L, Pergent-Martini C, Roméo M (2003) Assessment of the use of biomarkers in aquatic plants for the evaluation of environmental quality: application to seagrasses. Aquat Toxicol 65 (2):187–204. https://doi.org/10.1016/S0166-445X(03)00133-4

Fontanetti CS, Nogarol LR, de Souza RB, Perez DG, Maziviero GT (2011) Bioindicators and biomarkers in the assessment of soil toxicity. In: Pascucci S (ed) Soil contamination. Intechopen, Rijeka, pp 143–168. https://doi.org/10.5772/25042

French S, Levy-Booth D, Samarajeewa A, Shannon KE, Smith J, Trevors JT, (2009) Elevated temperatures and carbon dioxide concentrations: effects on selected microbial activities in temperate agricultural soils. World J Microbiol Biotechnol 25(11):1887–1900

Frey SD, Drijber R, Smith H, Melillo JM (2008) Microbial biomass, functional capacity, and community structure after 12 years of soil warming. Soil Biol Biochem 40:2904–2907. https://doi.org/10.1016/j.soilbio.2008.07.020

Galay-Burgos M, Spurgeon DJ, Weeks JM, Stürzenbaum SR, Morgan AJ, Kille P (2003) Developing a new method for soil pollution monitoring using molecular genetic biomarkers. Biomarkers 8(3–4):229–239. https://doi.org/10.1080/354750031000138685

Garg AK, Kim JK, Owens TG, Ranwala AP, Choi YD, Kochian LV, Wu RJ (2002) Trehalose accumulation in rice plants confers high tolerance levels to different abiotic stresses. Proc Natl Acad Sci USA 99:15898–15903. https://doi.org/10.1073/pnas.252637799

Gómez-Sagasti MT, Alkorta I, Becerril JM, Epelde L, Anza M, Garbisu C (2012) Microbial monitoring of the recovery of soil quality during heavy metal phytoremediation. Water Air Soil Pollut 223(6):3249–3262. https://doi.org/10.1007/s11270-012-1106-8

Gonzalez-Quiñones V, Stockdale EA, Banning NC, Hoyle FC, Sawada Y, Wherrett AD, Jones DL, Murphy DV (2011) Soil microbial biomass—interpretation and consideration for soil monitoring. Soil Res 49(4):287–304. https://doi.org/10.1071/SR10203

Grant WF (1994) The present status of higher plant bioassays for the detection of environmental mutagens. Mutation Res 310(2):175–185. https://doi.org/10.1016/0027-5107(94)90112-0

Gu MB, Chang ST (2001) Soil biosensor for the detection of PAH toxicity using an immobilized recombinant bacterium and a biosurfactant. Biosen Bioelectronics 16(9):667–674. https://doi.org/10.1016/S0956-5663(01)00230-5

Hirano T, Tamae K (2010) Heavy metal-induced oxidative DNA damage in earthworms: a review. Appl Environ Soil Sci 2010:1–7. https://doi.org/10.1155/2010/726946

Hodkinson ID, Jackson JK (2005) Terrestrial and aquatic invertebrates as bioindicators for environmental monitoring, with particular reference to mountain ecosystems. Environ Manag 35 (5):649–666. https://doi.org/10.1007/s00267-004-0211-x

Hoekenga OA, Maron LG, Pineros MA, Cancado GMA, Shaff J, Kobayashi Y, Ryan PR, Dong B, Delhaize E, Sasaki T, Matsumoto H, Yamamoto Y, Koyama H, Kochian LV (2006) AtALMT1, which encodes a malate transporter, is identified as one of several genes critical for aluminum tolerance in Arabidopsis. Proc Natl Acad Sci U S A 103:9738–9743. https://doi.org/10.1073/pnas.0602868103

Intergovernmental Panel on Climate Change (IPCC) (2007) Climate change 2007: mitigation. In: Metz B, Davidson OR, Bosch PR, Dave R, Meyers LA (eds) Contribution of working group III to the fourth assessment report of the intergovernmental panel on climate change. Cambridge University Press, Cambridge/New York

Invers O, Perez M, Romero J (1995) Alkaline phosphatase activity as a tool for assessing nutritional conditions in the seagrass *Posidonia oceanica* (L) Delile. Sci Mar 59(1):41–47. http://hdl.handle.net/2445/32430

Izquierdo I, Caravaca F, Alguacil MM, Hernandez G, Roldan A (2005) Use of microbiological indicators for evaluating success in soil restoration after revegetation of a mining area under subtropical conditions. Appl Soil Ecol 30:3–10. https://doi.org/10.1016/j.apsoil.2005.02.004

Kammenga JE, Dallinger R, Donker MH, Köhler HR, Simonsen V, Triebskorn R, Weeks JM (2000) Biomarkers in terrestrial in terrestrial invertebrates for ecotoxicological soil risks assessment. Rev Environ Contam Toxicol 164:93–147

Kanaya N, Gill BS, Grover IS, Murin A, Osiecka R, Sandhu SS, Andersson HC (1994) *Vicia faba* chromosomal aberration assay. Mutat Res Fundam Mol Mech Mutagen 310(2):231–247. https://doi.org/10.1016/0027-5107(94)90116-3

Kang GS, Beri V, Sidhu BS, Rupela OP (2005) A new index to assess soil quality and sustainability of wheat-based cropping systems. Biol Fertil Soils 41:389–398. https://doi.org/10.1007/s00374-005-0857-4

Kao NH, Su MC, Fan JR, Chung YY (2015) Identification and quantification of biomarkers and polycyclic aromatic hydrocarbons (PAHs) in an aged mixed contaminated site: from source to soil. Environ Sci Pollut Res 22(10):7529–7546. https://doi.org/10.1007/s11356-015-4237-9

Karhu K, Fritze H, Tuomi M, Vanhala P, Spetz P, Kitunen V, Liski J (2010) Temperature sensitivity of organic matter decomposition in two boreal forest soil profiles. Soil Biol Biochem 42:72–82. https://doi.org/10.1016/j.soilbio.2009.10.002

Kaur A, Chaudhary A, Kaur A, Choudhary R, Kaushik R (2005) Phospholipid fatty acid-A bioindicator of environment monitoring and assessment in soil ecosystem. Curr Sci 89 (7):1103–1112

Keltjens WG, Van Beusichem ML (1998) Phytochelatins as biomarkers for heavy metal stress in maize (*Zea mays* L.) and wheat (*Triticum aestivum* L.): combined effects of copper and cadmium. Plant Soil 203(1):119–126. https://doi.org/10.1023/A:1004373700581

Kihlman BA (1975) Root tips of *Vicia faba* for the study of the induction of chromosomal aberrations. Mutat Res Environ Mutagen Relat Subj 31(6):401–412. https://doi.org/10.1016/0165-1161(75)90050-3

Kihlman BA, Andersson HC (1984) Root tips of *Vicia faba* for the study of the induction of chromosomal aberrations and sister chromatid exchanges. In: Kilbey BJ (ed) Handbook of mutagenicity test procedures. Elsevier, Amsterdam, New York, Oxford

Kihlman BA, Kronborg D (1976) Sister chromatid exchanges in Vicia faba. I. Demonstration by a modified fluorescent plus Giemsa (FPG) technique. Chromosoma 51:1–10. https://doi.org/10.1007/BF00285801

Knakievicz T (2014) Planarians as invertebrate bioindicators in freshwater environmental quality: the biomarkers approach. Ecotoxicol Environ Contam 9:01–12. https://doi.org/10.5132/eec.2014.01.001

Koblizek M, Masojidek J, Komenda J, Kucera T, Pilloton R, Mattoo AK, Giardi MT (1998) A sensitive photosystem II based biosensor for detection of a class of herbicides. Biotechnol Bioeng 60(6):664–669

Koper J, Piotrowska A (2003) Application of biochemical index to define soil fertility depending on varied organic and mineral fertilization. Electron J Pol Agric Univ 6, #06. Available Online: http://www.ejpau.media.pl/volume6/issue1/agronomy/art-06.html

Kroetsch DJ, Geng X, Chang SX, Saurette DD (2011) Organic soils of Canada: part 1. Wetland organic soils. Can J Soil Sci 91:807–822. https://doi.org/10.1139/CJSS10043

Kroger S, Turner APF, Mosbach K, Haupt K (1999) Imprinted polymer based sensor system for herbicides using differential pulse voltammetry on screen-printed electrodes. Anal Chem 71(17): 3698–3702

Kuriyama S, Rechnitz GA (1981) Plant tissue-based bioselective membrane electrode for glutamate. Anal Chim Acta 131:91–96

Leme DM, Marin-Morales MA (2009) *Allium cepa* test in environmental monitoring: a review on its application. Mutat Res 682:71–81. https://doi.org/10.1016/j.mrrev.2009.06.002

Lindberg N, Engtsson JB, Persson T (2002) Effects of experimental irrigation and drought on the composition and diversity of soil fauna in a coniferous stand. J Appl Ecol 39:924–936. https://doi.org/10.1046/j.1365-2664.2002.00769.x

Lovett Doust J, Schmidt M, Lovett Doust L (1994) Biological assessment of aquatic pollution: a review, with emphasis on plants as biomonitors. Biol Rev 69:147–186. https://doi.org/10.1111/j.1469-185X.1994.tb01504.x

Lui J, Gunther A, Bilitewski U (1997) Detection of methamidophos in vegetables using a photometric flow injection system. Environ Monit Assess 44(1–3):375–382. https://doi.org/10.1023/A:1005704017083

Madoz-Escande C, Simon O (2006) Contamination of terrestrial gastropods, *Helix aspersa* Maxima, with ^{137}Cs, ^{85}Sr, ^{133}Ba, ^{123}mTe by direct, trophic and combined pathways. J Environ Radioact 89:30–47. https://doi.org/10.1016/j.jenvrad.2006.03.004

Majer B, Grummt T, Uhl M, Knasmüller S (2005) Use of plant bioassays for the detection of genotoxins in the aquatic environment. Acta Hydrochim Hydrobiol 33(1):45–55. https://doi.org/10.1002/aheh.200300557

Malcolm GM, López-Gutiérrez JC, Koide RT, Eissenstat DM (2008) Acclimation to temperature and temperature sensitivity of metabolism by ectomycorrhizal fungi. Glob Chang Biol 14:1–12. https://doi.org/10.1111/j.1365-2486.2008.01555.x

Masto RE, Chhonkar PK, Singh D, Patra AK (2007) Soil quality response to long-term nutrient and crop management on a semi-arid Inceptisol. Agric Ecosyst Environ 118(1):130–142. https://doi.org/10.1016/j.agee.2006.05.008

Mazzeo DEC, Levy CE, Angelis DF, Marin-Morales MA (2010) BTEX biodegradation by bactéria from effluents of petroleum refinery. Sci Total Environ 408(20):4334–4340. https://doi.org/10.1016/j.scitotenv.2010.07.004

McGrath SP, Knight B, Killham K, Preston S, Paton GI (1999) Assessment of the toxicity of metals in soils amended with sewage sludge using a chemical speciation technique and a lux based biosensor. Environ Toxicol Chem 18(4):659–663. https://doi.org/10.1002/etc.5620180411

Merino-Gergichevich C, Alberdi M, Ivanov AG, Reyes-Diaz M (2010) Al^{3+}–Ca^{2+} interaction in plants growing in acid soils: Al-phytotoxicity response to calcareous amendments. J Soil Sci Plant Nutr 10:217–243

Mitloehner R, Koepp R (2007) Bioindicator capacity of trees towards dryland salinity. Trees 21:411–419. https://doi.org/10.1007/s00468-007-0133-3

Mukhopadhyay S, Masto RE, Cerdà A, Ram LC (2016) Rhizosphere soil indicators for carbon sequestration in a reclaimed coal mine spoil. Catena 14:100–108. https://doi.org/10.1016/j.catena.2016.02.023

Muñoz Rojas M, Erickson TE, Dixon KW, Merritt DJ (2016) Soil quality indicators to assess functionality of restored soils in degraded semiarid ecosystems. Restor Ecol 24:S43–S52. https://doi.org/10.1111/rec.12368

Neher DA (2001) Role of nematodes in soil health and their use as indicator. J Nematol 33:161–168

Ohkouchi N, Eglinton TI (2008) Compound-specific radiocarbon dating of Ross Sea sediments: a prospect for constructing chronologies in high-latitude oceanic sediments. Quaternary Geochronology 3(3):235–243. https://doi.org/10.1016/j.quageo.2007.11.001

Olayinka OT, Idowu AB, Dedeke GA, Akinloye OA, Ademolu KO, Bamgbola AA (2011) Earthworm as bio-indicator of heavy metal pollution around Lafarge, Wapco Cement Factory, Ewekoro, Nigeria. In: Proceedings of the environmental man conference, Federal University of Agriculture, Abeokuta, Nigeria

Park JH, Bolan NS, Chung JW, Naidu R, Megharaj M (2011) Environmental monitoring of the role of phosphate compounds in enhancing immobilization and reducing bioavailability of lead in contaminated soils. J Environ Monit 13:2234–2242

Potin P, Bouarab K, Kupper F, Kloareg B (1999) Oligosaccharide recognition signals and defence reactions in marine plant-microbe interactions. Curr Opin Microbiol 2:276–283. https://doi.org/10.1016/S1369-5274(99)80048-4

Puglisi E, Del Re AAM, Rao MA, Gianfreda L (2006) Development and validation of numerical indices integrating enzyme activities of soils. Soil Biol Biochem 38:1673–1681. https://doi.org/10.1016/j.soilbio.2005.11.021

Ralph PJ (2000) Herbicide toxicity of *Halophila ovalis* assessed by chlorophyll a fluorescence. Aquat Bot 66:141–152. https://doi.org/10.1016/S0304-3770(99)00024-8

Raven PH, Berg LR, Johnson GB (1998) Environment, 2nd edn. Saunders College Publishing, New York

Read J (1959) Radiation biology of *Vicia faba* in relation to the general problem. Blackwell, Oxford

Richards LA (ed) (1954) Diagnosis and improvements of saline and alkali soils. USDA Agriculture Handbook 60. USDA, Washington, DC. 160 p

Richardson JL, Vepraskas MJ (2001) Wetland soils; genesis, hydrology, landscapes and classification. CRC Press, Boca Raton, 417 p

Ricketts HJ, Morgan AJ, Spurgeon DJ, Kille P (2003) Measurement of annetocin gene expression: a new reproductive biomarker in earthworm toxicology. Ecotox Environ Saf 57:4–10. https://doi.org/10.1016/j.ecoenv.2003.08.008

Rinnan R, Michelsen A, Bååth E, Jonasson S (2007) Fifteen years of climate change manipulations alter soil microbial communities in a subarctic heath ecosystem. Glob Chang Biol 13:28–39. https://doi.org/10.1111/j.1365-2486.2006.01263.x

Rinnan R, Michelsen A, Jonasson S (2008) Effects of litter addition and warming on soil carbon, nutrient pools and microbial communities in a subarctic heath ecosystem. Appl Soil Ecol 39:271–281. https://doi.org/10.1016/j.apsoil.2007.12.014

Rinnan R, Stark S, Tolvanen A (2009) Responses of vegetation and soil microbial communities to warming and simulated herbivory in a subarctic heath. J Ecol 97:788–800. https://doi.org/10.1111/j.1365-2745.2009.01506.x

Ritz K, Black HIJ, Campbell CD, Harris JA, Wood C (2009) Selecting biological indicators for monitoring soils: a framework for balancing scientific and technical opinion to assist policy development. Ecol Indic 9:1212–1221. https://doi.org/10.1016/j.ecolind.2009.02.009

Roda A, Rauch P, Ferri E, Girotti S, Ghini S, Carrea G, Bovara R (1994) Chemiluminiscent flow sensor for the determination of paraoxon and aldicarb pesticides. Anal Chim Acta 294(1):35–42. https://doi.org/10.1016/0003-2670(94)85043-7

Rodrigues GS, Ma TH, Pimentel D, Weinstein LH (1997) Tradescantia bioassays as monitoring systems for environmental mutagenesis: a review. Crit Rev Sci 16:325–359. https://doi.org/10.1080/07352689709701953

Rousk J, Brookes PC, Baath E (2009) Contrasting soil pH effects on fungal and bacterial growth suggest functional redundancy in carbon mineralization. Appl Environ Microbiol 75:1589–1596. https://doi.org/10.1128/AEM.02775-08

Saha A, Pipariya A, Bhaduri D (2016a) Enzymatic activities and microbial biomass in peanut field soil as affected by the foliar application of tebuconazole. Environ Earth Sci 75(7):1–13. https://doi.org/10.1007/s12665-015-5116-x

Saha A, Bhaduri D, Pipariya A, Jain NK (2016b) Influence of imazethapyr and quizalofop-p-ethyl application on microbial biomass and enzymatic activity in peanut grown soil. Environ Sci Poll Res 23(23):23758–23771. https://doi.org/10.1007/s11356-016-7553-9

Samac DA, Tesfaye M (2003) Plant improvement for tolerance to aluminum in acid soils– a review. Plant Cell Tissue Organ Cult 75:189–207. https://doi.org/10.1023/A:1025843829545

Schindlbacher A, Rodler A, Kuffner M, Kitzler B, Sessitsch A, Zechmeister-Boltenstern S (2011) Experimental warming effects on the microbial community of a temperate mountain forest soil. Soil Biol Biochem 43:1417–1425. https://doi.org/10.1016/j.soilbio.2011.03.005

Schudoma C, Steinfath M, Sprenger H, van Dongen JT, Hincha D, Zuther E, Geigenberger P, Kopka J, Köhl K, Walther D (2012) Conducting molecular biomarker discovery studies in plants. In: High-throughput phenotyping in plants: methods and protocols. Humana Press, New York, pp 127–150. https://doi.org/10.1007/978-1-61779-995-2_10

Schwalbe M, Teller S, Oelmuller R, Appenroth KJ (1999) Influence of UVB irradiation on nitrate and ammonium assimilating enzymes in *Spirodela polyrhiza*. Aquat Bot 64:19–34. https://doi.org/10.1016/S0304-3770(99)00005-4

Sergeyera TA, Piletsky SA, Brovko AA, Slinchenko EA, Sergeeva LM, El'skaya AV (1999) Selective recognition of atrazine by molecularly imprinted polymer membranes. Development of conductimetric sensor for herbicides detection. Anal ChimActa 392(2–3):105–111

Shima N, Xiao LZ, Sakuramoto F, Ichikawa S (1997) Young inflorescence-bearing shoots with roots of *Tradescantia* clone BNL 4430 cultivated on nutrient solution circulating systems: an alternative to potted plants and cuttings for mutagenicity tests. Mutat Res 395:199–208. https://doi.org/10.1016/S1383-5718(97)00169-1

Stankovic S, Stankovic AR (2013) Bioindicators of toxic metals. In: Green materials for energy, products and depollution. Springer, Dordrecht, pp 151–228. https://doi.org/10.1007/978-94-007-6836-9_5

Stankovic S, Kalaba P, Stankovic AR (2014) Biota as toxic metal indicators. Environ Chem Lett 12 (1):63–84. https://doi.org/10.1007/s10311-013-0430-6

Steinkellner H, Kassie F, Knasmüller S (1999) *Tradescantia*-micronucleus assay for the assessment of the clastogenicity of Austrian water. Mutat Res 426:113–116. https://doi.org/10.1016/S0027-5107(99)00051-2

Strachan G, Capel S, Maciel H, Porter AJR, Paton GI (2002) Application of cellular and immunological biosensor techniques to assess herbicide toxicity in soils. Eur J Soil Sci 53(1):37–44. https://doi.org/10.1046/j.1365-2389.2002.00421.x

Tavares TM, Carvalho FM (1992) Avaliação de exposição de populações humanas a metais pesados no ambiente: exemplos do recôncavo baiano. Química Nova 15(2):147–154

van Straalen NM, Verhoef HA (1997) The development of a bioindicator system for soil acidity based on arthropod pH preferences. J Appl Ecol 34:217–232. https://doi.org/10.2307/2404860

Vangronsveld J, Mench M, Mocquot B, Clijsters H (1998) Biomarqueurs d'exposition des ve'ge'taux terrestres aux polluants. Application a` la pollution par les me'taux. In: Lagadic L, Caquet T, Amiard JC, Ramade F (eds) Utilisation de biomarqueurs pour la surveillance de la qualite´ de l'environnement. Lavoisier publ, Tec & Doc 320 p

Vanhala P, Karhu K, Tuomi M, Björklöf K, Fritze H, Hyvärinen H, Liski J (2011) Transplantation of organic surface horizons of boreal soils into warmer regions alters microbiology but not the temperature sensitivity of decomposition. Glob Chang Biol 17:538–550. https://doi.org/10.1111/j.1365-2486.2009.02154.x

Vartapetian B, Jackson MB (1997) Plant adaptations to anaerobic stress. Ann Bot 79(Suppl. A):3–20. https://doi.org/10.1093/oxfordjournals.aob.a010303

Vavilin DV, Ducruet JM, Matorin DN, Venediktov PS, Rubin AB (1998) Membrane lipid peroxidation, cell viability and photosystem II activity in the green alga Chlorella pyrenoidosa subjected to various stress conditions. J Photochem Photobiol B Biol 42:233–239. https://doi.org/10.1016/S1011-1344(98)00076-1

Velasco-Garcia MN, Mottram T (2003) Biosensor technology addressing agricultural problems. Biosyst Eng 84(1):1–12. https://doi.org/10.1016/S1537-5110(02)00236-2

Ventura BC (2009) Investigação da mutagenicidade do azocorante comercial BDCP (Black Dye Commercial Product), antes e após tratamento microbiano, utilizando o sistema – teste de *Allium cepa*. Tese de Doutorado. Instituto de Biociências. Universidade Estadual Paulista. Rio Claro-SP, 205 p

Wang D, Pan Y, Zhao X, Zhu L, Fu B, Li Z (2011) Genome-wide temporal-spatial gene expression profiling of drought response in rice. BMC Genomics 12:149. https://doi.org/10.1186/1471-2164-12-149

Watzinger A (2015) Microbial phospholipid biomarkers and stable isotope methods help reveal soil functions. Soil Biol Biochem 86:98–107. https://doi.org/10.1016/j.soilbio.2015.03.019

White PA, Claxton LD (2004) Mutagens in contaminated soil: a review. Mutat Res 567:227–345. https://doi.org/10.1016/j.mrrev.2004.09.003

Yang XS, Wu J, Ziegler TE, Yang X, Zayed A, Rajani MS, Zhou D, Basra AS, Schachtman DP, Peng M, Armstrong CL, Caldo RA, Morrell JA, Lacy M, Staub JM (2011) Gene expression biomarkers provide sensitive indicators of in planta nitrogen status in Maize. Plant Physiol 157:1841–1852. https://doi.org/10.1104/pp.111.187898

Yildiz D, Kula I, Ay G, Baslar S, Dogan Y (2010) Determination of trace elements in the plants of Mt. Bozdag, Izmir, Turkey. Arch Biol Sci Belgrade 62(3):731–738. https://doi.org/10.2298/ABS1003731Y

Yu T, Zhang Y, Hu XN, Meng W (2012) Distribution and bioaccumulation of heavy metals in aquatic organisms of different trophic levels and potential health risk assessment from Taihu lake, China. Ecotoxicol Environ Saf 81:55–64. https://doi.org/10.1016/j.ecoenv.2012.04.014

Zhang W, Parker KM, Luo Y, Wan S, Wallace LL, Hu S (2005) Soil microbial responses to experimental warming and clipping in a tall grass prairie. Glob Chang Biol 11:266–277. https://doi.org/10.1111/j.1365-2486.2005.00902.x

Zhoua Q, Zhanga J, Fua J, Shi J, Jiang G (2008) Biomonitoring: an appealing tool for assessment of metal pollution in the aquatic ecosystem. Anal Chim Acta 606:135–150. https://doi.org/10.1016/j.aca.2007.11.018

Chapter 9
Groundwater Irrigated Agriculture Evolution in Central Punjab, Pakistan

Muhammad Usman, Rudolf Liedl, Fan Zhang, and Muhammad Zaman

Abstract Irrigation water for agriculture in Pakistan is an issue due to a significant difference between rainfall and crop water needs. Irrigation water is either coming from snowmelt and rainfall in the northern mountains, or being pumped from groundwater. Canal water is limited, and water distribution using the warabandi system, a fixed canal water rotation system among water users on a particular irrigation channel, is not adequate and flexible. The result is overdependence on groundwater, which has impaired crop growth, notably in regions of bad groundwater quality.

The history of groundwater use is not very old in Punjab, Pakistan. By the end of 1990s, canal irrigation was dominant, which was then surpassed by groundwater at the start of 1991s. Since then the groundwater development has expanded exponentially, and recently the groundwater share in irrigated agriculture of the country is about 50%. By the end of 2013, more than one million tubewells are operational in the country and most of them are located in the Punjab province. The consequence is a drop of groundwater level in majority canal commands including the lower Chenab canal irrigation system. Evapotranspiration is the major outflow from the water balance in the region. Cultivation of high delta crops during kharif seasons including rice, cotton and sugarcane are responsible, which is triggered by elevated temperatures. During rabi seasons, wheat is the single major crop all over the lower Chenab canal with its coverage on more than 50% area. The overall recharge results showed that rainfall is the major inflow during kharif seasons, while during rabi canal

M. Usman (✉)
Department of Irrigation & Drainage, University of Agriculture, Faisalabad, Pakistan

Institute for Groundwater Management, Technical University Dresden, Dresden, Germany
e-mail: musman@uaf.edu.pk

R. Liedl
Institute for Groundwater Management, Technical University Dresden, Dresden, Germany

F. Zhang
Institute for Tibetan Research Plateau, Chinese Academy of Sciences, Beijing, China

M. Zaman
Department of Irrigation & Drainage, University of Agriculture, Faisalabad, Pakistan

E. Lichtfouse (ed.), *Sustainable Agriculture Reviews 33*, Sustainable Agriculture Reviews 33, https://doi.org/10.1007/978-3-319-99076-7_9

seepage dominates all other recharge sources. During kharif, the other major sources of recharge are field percolations, canal seepage, watercourse losses and distributary losses. Rainfall recharge, field percolation, watercourse losses and distributary losses are considered major recharge sources during rabi seasons.

Keywords Groundwater · Recharge · Remote sensing · Modelling · Kharif · Rabi · Climate change · Punjab · Pakistan

9.1 Introduction

9.1.1 Groundwater for Agriculture Worldwide

Irrigated agriculture is the largest consumer of groundwater resource accounting for about 70% of the global fresh water abstraction and 90% of consumptive water use (FAO 2010; Döll 2009). According to Llamas and Martinez-Santos (2005), during the last 20–30 years, there is a boom in the utilization of groundwater resources for irrigation in areas subject to extended dry seasons and/or regular droughts. Globally, an area of about 300 million ha (Mha) is under irrigation and 38% of this land are equipped for irrigation with groundwater amounting to 545 km^3/year (Siebert et al. 2010). Extended groundwater use is not only restricted to semi-arid regions, but also occurs in many humid areas (Fig. 9.1). It is envisaged that groundwater use for irrigated agriculture will continue to expand due to many possible reasons including:

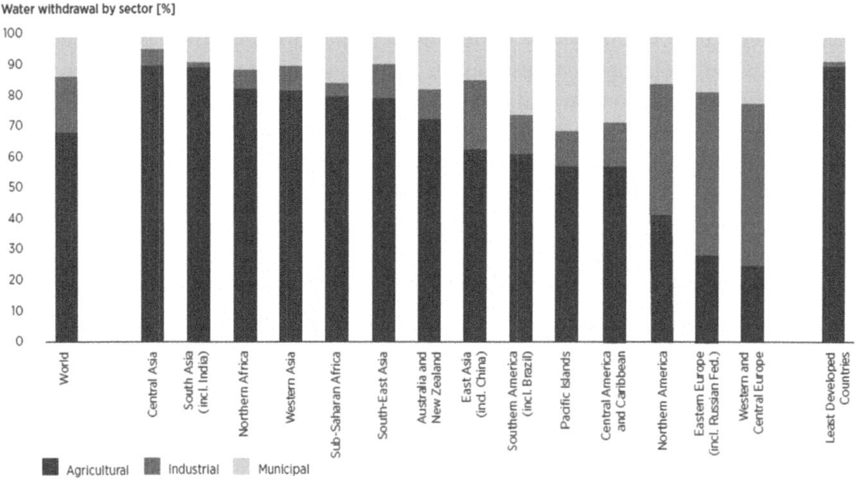

Fig. 9.1 Water withdrawal by sector in different world regions (2005). Agricultural water use is highest in South Asia, nearly 90%, followed by least developed world countries

(a) it is usually found close to point of use, (b) it can be developed quickly by individual private investment at low capital cost, (c) it is available directly for crop needs, (d) it is suited to pressurized irrigation and, (e) it has permitted irrigated agriculture outside of canal command regions (Shah et al. 2007).

Groundwater use has been the crux of the green revolution in agriculture across many Asian nations. Currently, the nations with highest groundwater use are India (39 Mha), China (19 Mha) and USA (17 Mha) (Siebert et al. 2010; Madramootoo 2012). Groundwater use in developing countries is likely to continue and the pressure on groundwater resources over next 25 years in Asia will come from demographic increase, agriculture and increasing water demand per capita, industrial activity and energy demand. It is predicted that the world population will increase from 6.9 billion in 2010 to 8.3 billion in 2030 and to 9.1 billion in 2050, most of which will occur in Asia (Christmann et al. 2009; UNDESA 2009). This increase in population will expand food demand by 50% in 2030 and by 70% in 2050. Nevertheless, Ayars et al. (2006) reported that future scenarios predict a worldwide fresh irrigation water scarcity which is even higher in arid and semi-arid regions. This fact emphasizes that the role of water should be properly regarded as socio-economic and life sustaining commodity demanding management procedures and be implemented through water conservation and resource assessment and reuse (UNCED 2002). Otherwise, poor management of groundwater resources will nullify the social gains made so far (Mukherji and Shah 2005).

9.1.2 Development of Groundwater Use in Pakistan

The Indus Basin Irrigation System (IBIS) of Pakistan was designed about a century ago and is one of the largest contiguous irrigation systems in the world. Its design objectives were to prevent crop failure, avoid famine and expand settlement opportunities (Jurriens and Mollinga 1996) by constructing reservoirs, barrages and main canals which are now serving an area of 16 Mha with some 172 billion m^3 of river water flow per year (Aslam and Prathapar 2006). The IBIS is supported by the basin of the Indus river and its tributaries including the Kabul, Jehlum, Chenab, Ravi and Satluj rivers. The irrigation system is comprised of three major storage reservoirs, 19 barrages, 12 link canals, 45 major irrigation canal commands and over 120,000 field water channels. The total canal length is about 60,000 km, with additional 1.8 million km comprising of watercourses, farm channels and field ditches (COMSATS 2003). The rivers of IBIS have glaciated headwaters and snowfields that provide about 50–80% of surface water flow out of the total volume of 137×10^9 m^3. The remaining volume is due to monsoon runoff. It is estimated that effective rainfall contributes about 200–300 mm in total crop water availability in the north of the country and some 50 mm in the south (Qureshi et al. 2010).

The IBIS was designed for an annual cropping intensity (ratio of effective crop area harvested to the physical area) of about 75% with the intention of spreading the irrigation water over large areas to expand settlement opportunities (Qureshi et al.

2010), and has grown up to 200% (Kazmi et al. 2012) because more than one crop cycle per year has become possible. Also many canals have lost their design capacity over time due to siltation and erosion of their banks (Badruddin 1996). The result is further limitation of canal water availability per unit of irrigated land (Sarwar 2000).

Huge crop yield losses, land degradation and social instability were observed during the 1970s due to inadequacy, inequity and unreliability of surface water supplies, which resulted in large scale migration of populations from rural areas to cities (Postel 2003). Nevertheless, large farming communities also came forward to rescue themselves against this situation and huge investment is made to extract groundwater by installing agricultural wells for crops. The government also helped farmers by subsidizing the power supply after realizing the benefits of groundwater irrigation for expansion of irrigated areas and to maintain higher crop production levels. In the early days, open wells, Persian wheels, karezes, hand pumps and reciprocating pumps were used for groundwater abstraction. Introduction of indigenous small diesel engines and subsidized energy supply caused a dramatic increase in the number of private tubewells (i.e. individual farmer owned) in the country. By the end of the 1990s, canal irrigation dominated the irrigated agriculture in the country, but in the early 1990s, groundwater irrigation had surpassed canal irrigation (Van der Velde and Kijne 1992). According to Chaudhary et al. (2002), more than 50% of irrigated lands in the country are irrigated by groundwater wells. More than 70% of the farmers in the Punjab province depend directly or indirectly on groundwater for agriculture (Qureshi et al. 2003). About 80% of total tubewells in the country are private owned. According to some estimates, the investment in the private tubewells is of the order of Rs. 25 billion (US$ 400 million) whereas, the annual benefits are of the order of Rs. 150 billion (US$ 2.3 billion) in the form of agricultural production (Shah 2003; Qureshi et al. 2010). According to Government of Pakistan, on average, every fourth farming family owns a tubewell and a large proportion of farmers without tubewell ownership purchase water through local groundwater markets (Government of Pakistan (GOP) 2000). Figure 9.2 depicts the development and distribution of tubewells in each province of Pakistan.

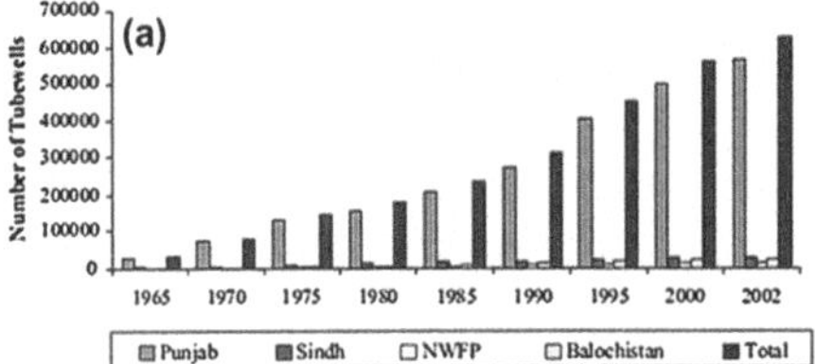

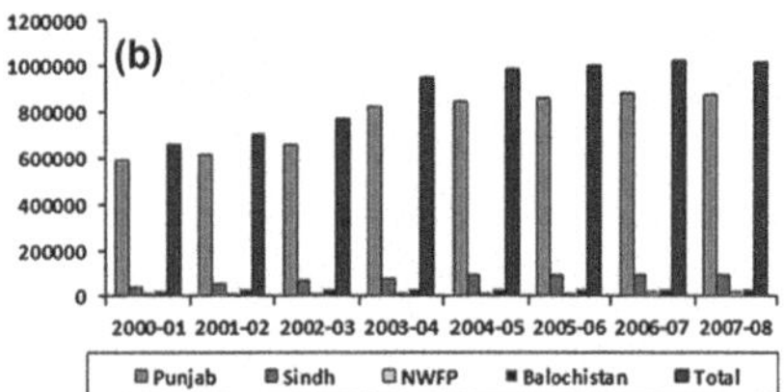

Fig. 9.2 Number of tubewells in Pakistan (**a**) Qureshi et al. (2010), (**b**) Agricultural Statistics of Pakistan (2008–2009). The data show that over the years, the growth of tubewells have been exponential in Pakistan particularly in the province of Punjab

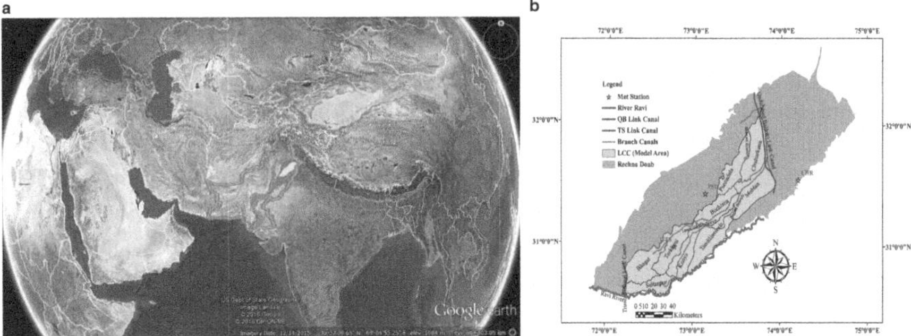

Fig. 9.3 Location of lower Chenab canal irrigation system in Rechna Doab, Punjab, Pakistan. Lower Chenab canal irrigation system is subdivided into ten irrigation subdivisions, a smallest administrative unit for better irrigation management

9.1.3 Study Site

Lower Chenab canal irrigation system, Punjab, Pakistan has been chosen as the study region (Fig. 9.3). The lower Chenab canal irrigation system originates at the Khanki headworks which distribute water to its eastern and western parts through seven branch canals. This irrigation system was designed in 1892–1898 and its command area lies in Rechna Doab which comprises of the land mass between rivers Ravi and Chenab.

The location of the area is between latitude 30° 36′ and 32° 09′ N and longitude 72° 14′ and 77° 44′ E. The present study mainly focuses on the eastern part of lower Chenab canal irrigation system. Two link canals namely Qadirabad-Balloki and Trimu-Sidhnai flow from north to south and fall into river Ravi. Major part of lower Chenab canal irrigation system (east) lies in the districts of Faisalabad and Toba Tek Singh. Administratively, the entire study area is split into ten irrigation subdivisions: Sagar, Chuharkana, Paccadala, Mohlan, Buchiana, Tandlianwala, Kanya, Tarkhani, Bhagat and Sultanpur. Irrigation subdivision is considered as the smallest management unit of the irrigation system. The structuring of these irrigation subdivisions ensures the equitable distribution of canal water among different consumers.

9.1.3.1 History of Groundwater Use in Lower Chenab Canal Irrigation System

Punjab province of Pakistan is called 'the land of five rivers' and covers an area of about 127,000 km^2. These rivers include Indus, Jhelum, Chenab, Ravi and Sutlej from west to east. The land between any two rivers is known as 'doab'. These doabs include Thal (between rivers Indus and Jhelum), Chaj (between rivers Chenab and Jhelum), Rechna (between rivers Chenab and Ravi) and Bari (between rivers Ravi and Sutlej) and the plains of these doabs have been formed by alluvial deposits and are very fertile. During the 1900s, during the rule of British over the subcontinent, an extensive network of irrigation canal was constructed in order to develop the barren land and to utilize the water of these five rivers. These practices paid off the investment of millions of rupees for construction of canals and headworks within a few years as the area converted into lush green fields (Hassan and Bhutta 1996). The period of prosperity proved very short as intensive irrigation application coupled with poor subsurface drainage resulted in a gradual increase of groundwater. By the late 1930s and early 1940s, several million acres of land had been affected by waterlogging and salinization, both of which were spreading alarmingly every year (Malmberg 1975). In some areas the groundwater rise was about 24.4 m with an average rate of rise of 0.46 m/year (Hassan and Bhutta 1996). According to Soomro (1975), waterlogging was first identified in the upper parts of Rechna doab within few years of opening of the Lower Chenab Canal.

A comprehensive study of the geology and hydrology of the Indus Plain was carried out in 1954 by the Government of Pakistan in cooperation with the U.S. International Cooperation Administration to assess the groundwater potential of the Northern Plain in order to formulate reclamation measures that would solve the problems of waterlogging and salinity and restore the productive capacity of the land (Malmberg 1975). The results of these studies provided the basis for the reclamation projects utilizing deep tubewells to lower the groundwater level and supplement the canal water supply. The launch of the first project phase took place in 1960 with the first salinity control and reclamation project. The interfluvial area between the Ravi and Chenab rivers was selected for construction of first salinity control and reclamation project. This project was the first of 18 planned reclamation projects that ultimately included about 21 million acres and more than 28,000 production and drainage wells (Malmberg 1975).

Large scale groundwater extraction for irrigated agriculture in Rechna started by the launching of salinity control and reclamation project. Thousands of large capacity tubewells were installed under this program. In the initial phase more than 10,000 public tubewells (supplying an area of 2.6 Mha) with an average discharge capacity of 80 l/s were installed (Bhutta and Smedema 2007; Kazmi et al. 2012). The project resulted not only in the lowering of the water table but also in supplemented irrigation. This also encouraged farmers to own their individual tubewells and led to a proliferation with a typical tubewell discharge capacity of 28 l/s (i.e. 0.03 m^3s^{-1}) or less. The results of first salinity control and reclamation project indicated that it managed to lower the groundwater level below 1.5 m over an area of 2 Mha and below 3 m over 4 Mha, thereby overcoming the problem of waterlogging significantly. It also reclaimed salt affected area from 4.5 to 7.0 Mha (Qureshi et al. 2010). Moreover, the additional groundwater abstraction by salinity control and reclamation project tubewells increased cropping intensities from 80 to 120% in most of the salinity control and reclamation project areas (IWASRI 1998).

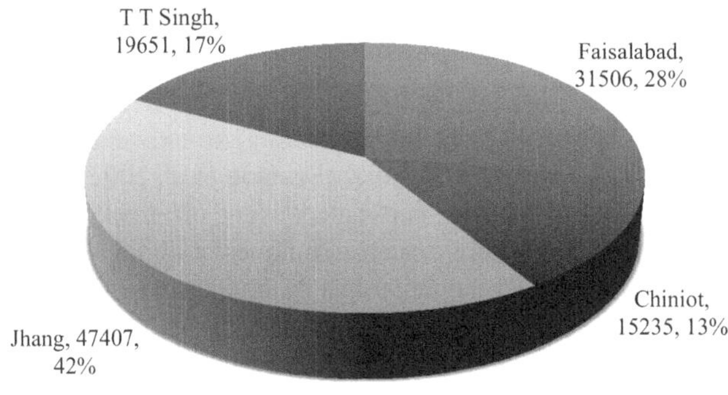

Fig. 9.4 Number of tubewells installed in different districts of Faisalabad division including Faisalabad, Toba Tek Singh, Chiniot and Jhang. The area mainly consist of agricultural lands with mixed cropping system. Major crops grown in the division include rice, cotton, sugarcane, fodder and wheat with higher dependence on groundwater use. (Source: Figure based on own research)

9.1.3.2 Current Status of Tubewells and Groundwater Use

Currently, groundwater is exploited at a huge quantity in the whole country. The total irrigated agricultural area is about 16 Mha and about 9 Mha is located in Punjab, and total share of groundwater pumping in agricultural water use is about 50% (Usman et al. 2015a). According to Punjab Development Statistics Report (2015), there were 1,049,000 tubewells installed in the whole country and alone Punjab is having 867,369 tubewells functional by the end of 2013. As LCC-east is mainly located in the districts of Faisalabad and Toba Tek Singh and 51,157 tubewells were installed in these two districts by the end of 2013. This figure reflects the higher dependence of farmers on groundwater use in the region. The distribution of tubewells in different districts of Faisalabad division can be seen from Fig. 9.4.

The result is drop of groundwater levels in majority parts of the lower Chenab canal irrigation system as can be read from Hassan and Bhuttah (1996) and Usman et al. (2015a), which threatens the sustainability of this resource. To apprehend this phenomenon, there is a dire need to assess the water resources in detail where comprehensive estimation and evaluation of irrigation system components, that contribute to recharge and discharge from the aquifer need to be performed.

The following two case studies were performed in the lower Chenab canal irrigation system. In the first study, a comprehensive assessment of recharge was made, while in the second one, effects of climate change were performed.

9.2 Case Study – I: Estimating Net Groundwater Recharge in Lower Chenab Canal Irrigation System

Sustainability of groundwater is highly dependent on the amount of water that replenishes aquifers, known as recharge. Estimating groundwater recharge is crucial in assessment of water resource availability (Scanlon et al. 2002).

Estimation of recharge may be performed using different methods including direct measurement by lysimeters, tracer techniques, and stream gauging (Lerner et al. 1990). Groundwater modeling approach is another way of estimating recharge if various input components are known with fair accuracy. However, as hydraulic parameters are rarely known and therefore recharge is inaccurately estimated as a lumped parameter. In the current study, groundwater recharge estimation is done by employing water balance approach and different inflow and outflow parameters of water balance equation are worked out separately. The brief description of recharge estimation methodology is presented as follows; however, detailed description of different water balance approaches and its comparison with water table fluctuation method may be seen from Usman et al. (2015a).

9.2.1 Water Balance Approach

Water balance approaches have been widely used for estimating groundwater recharge (Sarwar and Eggers 2006; Maréchal et al. 2006). The water budget for irrigated agricultural regions can be written by the equation (9.1) devised by Singh (2011) and Schicht and Walton (1961).

$$I + R + G_{in} - G_{out} - ET - RO - G_p - G_s = dS \tag{9.1}$$

where I and R are total water from canal supply and rainfall, respectively, G_{in} and G_{out} are lateral groundwater inflow and outflow along a boundary; ET is loss due to evapotranspiration; RO is runoff at surface; G_p is groundwater abstraction by pumping; G_s is groundwater discharge to stream; and dS is the change in groundwater storage. The units of all components are mm per time period.

The simplified version of the above equation may be written if we ignore the surface runoff and groundwater outflow to stream considering the local hydrological conditions in the lower Chenab Canal irrigation system. The above general water budget equation now reads as below:

$$I + R + G_{in} - G_{out} - ET - G_p = dS \tag{9.2}$$

The above equation describes the water budget by incorporating all surface and subsurface parameters for estimation of recharge ignoring detail of each parameter. According to Scanlon et al. (2002), water balance can also be written for saturated zone (below groundwater). The description of saturated groundwater storage by this principle can be written as below.

$$L_s + G_{in} - G_{out} + IR + RR - G_p = dS \quad (9.3)$$

where L_s is seepage water loss from irrigation canal network, IR is return flow of irrigation from agricultural field and RR is rainfall recharge. The units of all components are mm per time period.

Both equations (9.2) and (9.3) are identical except they offer utilizing of different input data types. For the present study, equation (9.3) was used and remote sensing data and different spatial techniques was applied for input data.

9.2.2 *Estimation of Different Recharge Input/Output Parameters*

For irrigated agricultural regions, various data types are required including crop, soil, weather, water flow, geology, shape files and piezometric observations. Following is the brief description of recharge estimation from various recharge sources.

9.2.2.1 Canal Geometry and Flow Data

The canal flow and geometry data of different irrigation streams are collected from Irrigation Department, Punjab, Pakistan. Daily data are converted to monthly and then seasonal after integrated for each canal tributary. The irrigation system of lower Chenab canal is designed in a way that separate canals provide irrigation to each irrigation subdivision and hence apartheid of data is easy. Other data include shape geometries of different canals and information about lining/un-lining canal sections.

9.2.2.2 Crop Inventory

Land us land cover mapping at spatial resolution of 250 m for both rabi and kharif cropping seasons were performed using MODIS normalized difference vegetation index spatial data. A number of normalized difference vegetation index composite images were retrieved from both terra and aqua sensors, and after preprocessing of these images, unsupervised classification using *ISODATA* algorithm was performed (Usman et al. 2015b). The temporal profiles of normalized difference vegetation index trends were utilized to identify different major crops in the region. The results of land use land cover classification were validated by constructing error matrixes.

Table 9.1 Summary of MODIS remote sensing data used for soil energy balance algorithm

Data layer	Spatial resolution	Temporal resolution	Purpose
Surface reflectance, band (1–7)	500 m	8 days	Surface albedo, Vegetation index, Land surface water index
Land surface temperature & Emissivity, band (31–32)	1000 m	daily	Land surface temperature, Emissivity

9.2.2.3 Actual Evapotranspiration

Actual evapotranspiration was calculated using soil energy balance algorithm devised by Bastiaanssen et al. (1998) which used different satellite data products for estimation of different variables. The detail of these products can be seen in Table 9.1.

The detailed methodology of soil energy balance algorithm adopted for the current study can be found in Usman et al. (2014, 2015c). The soil energy balance algorithm give daily values of actual evapotranspiration, and from these daily information about actual evapotranspiration, monthly and seasonal maps of consumptive water use were prepared for both rabi and kharif cropping seasons.

9.2.2.4 Net Recharge Components for Water Balance Approach

Different water inflow and outflow components for WBA were worked out at seasonal temporal scales for estimation of net recharge. The detailed description of which is presented as under:

(a) **Rainfall**

Raster based monthly rainfall data were retrieved from Tropical Rainfall Measurement Mission (TRMM). Spatial data with a resolution of 25 km were downloaded, which were downscaled to 1 km for further use. Local calibration of this data was performed with data from three weather stations located at different places of lower Chenab canal irrigation system. This monthly rainfall data summed up to get seasonal rainfall for all kharif and rabi seasons. Mean, maximum, and minimum rainfalls were calculated in different irrigation subdivision of lower Chenab canal irrigation system along with the standard deviations and coefficients of variation for each season. Lastly, the effective rainfall was estimated by following the United States Department of Agriculture, Soil Conservation Services method. Recharge from rainfall estimated by following guidelines proposed by many different researchers. For instance, Maasland (1968) as reported in Ahmed and Chaudhry (1988) has considered about 20% of total rainfall as recharge. Ahmad and Chaudhry (1988) have reported about 17% to 22% of total annual rainfall as recharge in lower Chenab canal irrigation system. Ashraf and Ahmad (2008) have used 17.9% of total rainfall as groundwater recharge in nearby Chaj Doab, Punjab, Pakistan. All the proposed guidelines can only yield rough estimates, therefore, in the current study; rainfall recharge is estimated by subtracting monthly effective rainfall plus 5% of total rainfall as unseen losses from total monthly rainfall (Usman et al. 2015a).

(b) **Main canal seepage**

Estimation of seepage from main irrigation canals was done using two different empirical models developed by the Punjab Private Sector Groundwater Development Project Consultants (1998) and Irrigation Department (2008). These models utilize data of discharge, canal length, wetted perimeter, number of canal operational days in a season and seepage factor. They can be written as:

$$S = 0.052\,(Q)^{0.658} \tag{9.4}$$

and

$$RC = 86400.LC.WP.N.SF \tag{9.5}$$

where S is seepage loss or recharge ($ft^3/s/mile$), Q is canal discharge (ft^3/s), RC is recharge due to canal seepage (m^3), LC is length of canal (m), WP is canal wetted perimeter during its run (m), N is canal running time in a season (d), and SF is seepage factor with recommended values of 0.62–0.75 and 2.5–3.0 $m^3/s/10^6\ m^2$ of wetted area for lined and unlined canal sections, respectively.

(c) **Recharge from distributaries and watercourses**

Equation (9.4) can also be used for recharge estimation from distributaries. However, for the current study, a fixed percent of head water diversion was considered as recharge as reported in different studies conducted in the study region (Kennedy 1890; Benton 1904; Blench 1941–1942 and Khangar 1946 (cited in Ahmed and Chaudhry 1988)). Maasland (1968) as cited in Ahmed and Chaudhry (1988) estimated watercourse losses and recommended that about 10–20% of total delivery head is a seepage loss. A value of 10% is considered for the current study.

(d) **Recharge from field percolations**

Considerable part of irrigation water returns to groundwater through infiltration. A number of researchers has devised different fixed percent values of applied water as recharge. For instance, about 15% of irrigation water is recharge regardless of crop type, according to Maasland (1968). Similarly, Ashraf and Ahmad (2008) assumed 25% of irrigation water as recharge. According to other researchers including Jalota and Arora (2002), Tyagi et al. (2000a, b) and Maréchal et al. (2006) recharge is variable according to crop type and hence they have proposed different coefficients of irrigation return flow for different crops. For the current study, later approach was employed and satellite images were processed to differentiate different crop areas and assignment of different coefficients of irrigation return flows were done. Following relationship was used for estimation of recharge:

$$IR = TI.F_c \tag{9.6}$$

where IR is recharge from field percolation (mm), TI is the total irrigation water applied at farm gate both from canal and groundwater sources (mm), and F_c is the fraction of water contributing to recharge.

The information about F_c (i.e field application efficiency) for different crops can be taken from Jalota and Arora (2002), and Tyagi et al. (2000a, b). Field percolation loss is about 50% for rice, while it is about 5.6%, 31.2%, 15%, 20% for wheat, kharif fodder, cotton and, rabi fodder and sugarcane crops, respectively.

(e) **Estimation of total irrigation water at farm**

Total irrigation is equal to estimated actual evapotranspiration under the ideal conditions, where all water supplied at farms is available for the crop utilization; nevertheless, in reality applied water at farm is always greater than actual evapotranspiration due to limited application efficiencies and conveyance efficiencies that are never 100%. Thus, total irrigation may be calculated from the following equation:

$$TI = \left[\frac{(ET - ER)x\ 100}{AE}\right] + RO + US \tag{9.7}$$

where AE, ER, RO and UWS are irrigation application efficiency, effective rainfall, runoff and unsaturated water storage, respectively. Effective rainfall is excluded from actual evapotranspiration because total irrigation is total water available at farms from only canal and groundwater sources. Runoff and unsaturated water storage are ignored in equation (6) as very little runoff occur due to bunds on irrigated fields and, for long time periods steady state conditions prevails and the soil water content is constant and, hence, changes can be ignored (Usman et al. 2015a; Yin et al. 2011).

(f) **Groundwater pumping**

Groundwater pumping was estimated using the following relationship:

$$G_p = TI - CI \tag{9.8}$$

where CI is the net water availability from canal irrigation after incorporating conveyance losses.

(g) **Lateral inflow and outflow of groundwater**

Inverse modeling using FEFLOW 6.1 groundwater software and PEST was used for estimation of later groundwater inflow and outflow. The modeling results were verified with Darcy's law (Darcy 1856). For Darcy's law inflow and outflow cross sections were identified by plotting piezometric data and borelog information in Voxler geo-statistical software from Golden software's. The cross sectional flow areas for each irrigation subdivision were estimated as suggested by Baalousha (2005). Groundwater contour maps were prepared using Surfer 8.0. The flow area of each sub section contributing to inflow and outflow was estimated to be the area below the intersection of the regional potentiometric surface.

9.2.3 Results of Case Study – I

9.2.3.1 Land Use Land Cover Mapping

Error matrixes were constructed for different land use land cover types by incorporating 250 control points and polygons. From Table 9.2, it can be observed that overall accuracy for rabi seasons fluctuated between 79.5% and 87.4%. During kharif seasons, the efficiency varied from 76.2% to 80.08%. The overall average accuracies for rabi and kharif were found to be 82.8% and 78.2%, respectively. The results are in concurrence with some other studies of this kind (Thi et al. 2012; Wardlow et al. 2007). Bastiaanssen et al. (1998) reported that overall accuracy generally ranges between 49% and 96% depending upon the spatial data coverage and the field size under study. According to Giri and Clinton (2005), the overall efficiency gives a crude measure of accuracy as its measurement by error matrix is dependent on sampling size. Small sampling size may lead to assign the correct class by chance (Foody 2002), and therefore the kappa coefficient is necessary to measure (Congalton 1996). The value of kappa coefficient results the agreement after removing the errors come by chance by incorporating the off diagonal elements of the error matrices (Yuan et al. 2005). The values of kappa coefficient for each individual season were calculated and shown in Table 9.2. The average kappa coefficient value for rabi seasons was 0.73 with minimum and maximum values of 0.66 and 0.77. For

Table 9.2 Average classification accuracies and kappa coefficient for different land use classes during rabi (i.e. November to April) and kharif (May to October) cropping seasons

		Rabi		Kharif	
No	Year	Overall accuracy	K	Overally accuracy	K
1	2005–2006	87.4	0.77		
2	2006–2007	79.5	0.66	76.2	0.69
3	2007–2008	83.6	0.74	79.3	0.73
4	2008–2009	81.9	0.71	78.1	0.71
5	2009–2010	81.2	0.71	77.0	0.70
6	2010–2011	83.8	0.75	78.6	0.72
7	2011–2012	82.5	0.74	80.1	0.74
Average		82.8	0.73	78.2	0.71

Table 9.3 Coverage of major land use land cover types during rabi (i.e. November to April) and kharif (May to October) cropping seasons

	Rabi season			Kharif season	
Year	Class	Area (hectares)	Year	Class	Area (hectares)
2005–2006	Fodder	227,020	2006	Rice	336,947
	Sugarcane	128,257		Fodder	319,260
	Wheat	548,635		Sugarcane	99,519
	Fallow/Barren	22,160		Cotton	136,191
				Fallow/Barren	34,156
2006–2007	Fodder	230,029	2007	Rice	253,017
	Sugarcane	124,575		Fodder	375,766
	Wheat	551,216		Sugarcane	198,080
	Fallow/Barren	20,251		Cotton	76,758
				Fallow/Barren	22,451
2007–2008	Fodder	281,160	2008	Rice	251,654
	Sugarcane	126,627		Fodder	414,699
	Wheat	497,546		Sugarcane	106,228
	Fallow/Barren	20,738		Cotton	131,034
				Fallow/Barren	22,457
2008–2009	Fodder	289,135	2009	Rice	362,218
	Sugarcane	105,722		Fodder	323,286
	Wheat	512,320		Sugarcane	87,172
	Fallow/Barren	18,895		Cotton	116,974
				Fallow/Barren	36,422
2009–2010	Fodder	213,847	2010	Rice	349,121
	Sugarcane	97,907		Fodder	231,065
	Wheat	598,403		Sugarcane	137,540
	Fallow/Barren	15,915		Cotton	179,155
				Fallow/Barren	29,190
2010–2011	Fodder	299,478	2011	Rice	342,038
	Sugarcane	78,435		Fodder	169,562
	Wheat	534,474		Sugarcane	137,791
	Fallow/Barren	13,685		Cotton	259,529
				Fallow/Barren	17,152
2011–2012	Fodder	248,550			
	Sugarcane	113,555			
	Wheat	545,132			
	Fallow/Barren	18,836			

kharif, the average value was 0.71 with a minimum value of 0.69 and a maximum value of 0.74. The findings are in agreement with the results of Altman (1991).

Table 9.3 depicts the distribution of different land use land cover classes from 2005 to 2012 for rabi and kharif seasons. Wheat was found to be major crop during rabi seasons with overall average area of more than 50%. The other dominant crops were rabi fodder and sugarcane with percent areas of 32.3% in 2010–2011, and

13.8% in 2005–2006, respectively. During kharif seasons, rice was found to be highest grown crops with minimum volatility in its command area for different seasons. Sugarcane had a maximum percent area of about 21.4% for 2007 with minimum value of 9.4% for 2009. The minimum and maximum percent values for kharif fodder were 18.3% for 2011 and 44.8% for 2008. In case of cotton, the minimum and maximum values were 8.3% in 2007 and 28% in 2011, respectively.

9.2.3.2 Validation of Soil Energy Balance Algorithm Based Actual Evapotranspiration

ET results using SEBAL are mandatory to validate prior to its use for hydrological studies. The results were validated with ET from advection-aridity approach (Brutsaert and Stricker 1979), as results from lysimeters, Bowen ration energy balance (Bowen 1926), and the eddy covariance (Wilson et al. 2002) were unavailable. The advection-aridity equation used was formulated as:

$$ET = (2\propto_e - 1)\frac{\Delta}{\Delta + \gamma}Q_{ne} - \frac{\gamma}{\Delta + \gamma}E_a \tag{9.9}$$

where ET is the actual evapotranspiration (mm d^{-1}), Δ is the slope of temperature versus vapour pressure (kPa $^{o}C^{-1}$), Q_{ne} is the ratio between R_n and λ, R_n is net radiation, γ is the psychometric constant (kPa $^{o}C^{-1}$) and Ea is the drying power of the air (Brutsaert and Stricker 1979; Brutsaert 2005). The effects of advection were scaled by the aerodynamic vapor transfer term Ea:

$$E_a = f\left(\bar{U_r}\right)(e_s - e_a) \tag{9.10}$$

where $f\left(\bar{U_r}\right)$ is the wind function, e_s and e_a are the saturation and actual vapor pressures in mmHg (Brutsaert and Stricker 1979). The wind function was a stelling-type standard equation written as devised by Brutsaert (2005):

$$f\left(\bar{U_r}\right) = 0.26\left(1 + 0.54f_{\bar{U_2}}\right) \tag{9.11}$$

Where $f_{\bar{U_2}}$ is the mean wind speed at 2 m height (ms^{-1}).

The validity results can be visualized in Fig. 9.5, which indicates satisfactory results for both rabi and kharif cropping seasons. Nevertheless, it is noted that higher Nash Sutcliffe efficiency and relatively low bias values were found during winter seasons (rabi) as compared to summer seasons (kharif). Advection-aridity method performs better under low temperatures as compared to hotter and arid environmental conditions (Hobbins and Ramirez 2001; Liu et al. 2010). Similar types of results were reported for the current study where quite high Nash Sutcliffe efficiency (i.e. 0.92), and relatively low bias (−13.32) were found for rabi seasons. However,

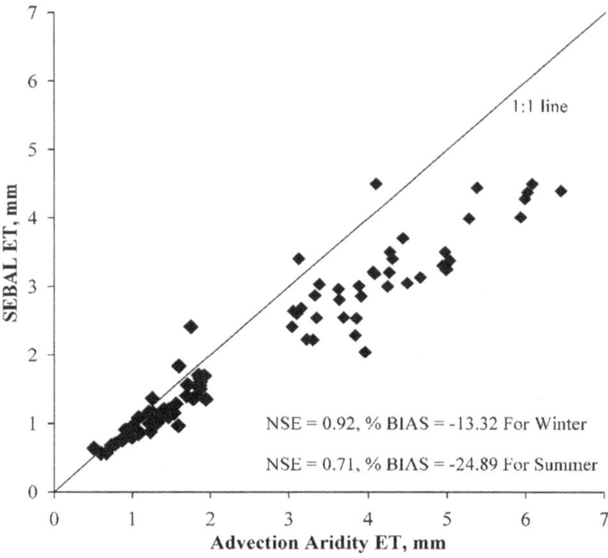

Fig. 9.5 Validation of soil energy balance algorithm based actual evapotranspiration with advection-aridity approach. During rabi seasons (November to April) results are better correlated as compared to kharif seasons (May to October). NSE Nash Sutcliffe efficiency, %BIAS percentage bias error, SEBAL soil energy balance algorithm, ET actual evapotranspiration. (Source: figure based on own research)

for kharif, the values of Nash Sutcliffe efficiency and bias were found to be 0.71 and −24.89, respectively (Usman et al. 2014).

The actual evapotranspiration during kharif seasons (546.2 mm ± 47.50 mm) was observed to be about two times higher than in rabi seasons (274 mm ± 20 mm). The main reason for this was high crop water requirements of rice, cotton, and sugarcane which were triggered by high temperatures during these seasons. Wheat was cultivated at larger areas along with fodder during rabi season, as opposed to this. There was an abrupt change in the trends of actual evapotranspiration for kharif from mid July to September because of higher availability of irrigation water due to monsoon rainfalls, which resulted in better vegetative growth for crops. A relatively steep rise in trend was observed from end of February to end of March, when temperature raised and sufficient soil moisture became available to crops after winter rainfall during the rabi seasons.

9.2.3.3 Rainfall Recharge

Table 9.4 shows the distribution of average rainfall in different irrigation subdivisions of lower Chenab canal irrigation system for both rabi and kharif seasons. It is visible that rainfall decreases from upper parts of lower Chenab canal irrigation

Table 9.4 Details of average rainfall, average effective rainfall and average annual recharge (mm) in different irrigation subdivisions of lower Chenab canal irrigation system during rabi (November to April) and kharif (May to October) cropping seasons. (Source: figure based on own research)

	Average rainfall recharge	
Irrigation sub-division	Kharif	Rabi
Sagar	127.5 ± 26.9	29.9 ± 10.7
Chuharkana	120.3 ± 26.1	24.3 ± 8.6
Paccadala	106.6 ± 22.9	23.2 ± 8.2
Mohlan	100.8 ± 27.3	20.1 ± 7.7
Buchiana	103.7 ± 23.5	19.0 ± 8.0
Tandilianwala	100.0 ± 24.4	18.7 ± 7.3
Tarkhani	77.5 ± 24.8	17.6 ± 6.5
Kanya	86.8 ± 23.5	16.1 ± 5.5
Bhagat	95.5 ± 28.6	20.7 ± 9.3
Sultanpur	84.8 ± 13.2	17.6 ± 6.9

Table 9.4 (continued)

Irrigation sub-division	Average rainfall recharge	
	Kharif	Rabi

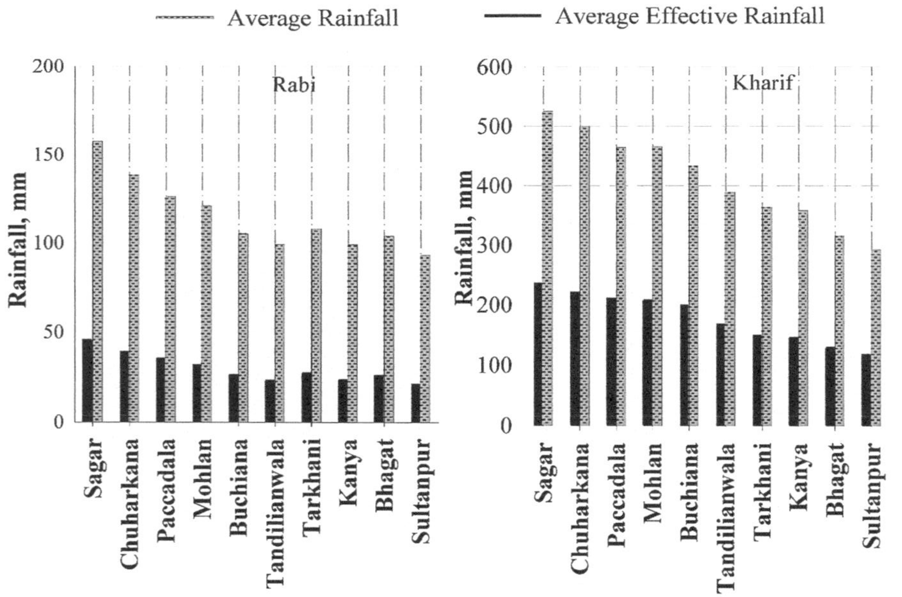

Average Rainfall — Average Effective Rainfall

Rabi

Rainfall, mm

200
150
100
50
0

Sagar
Chuharkana
Paccadala
Mohlan
Buchiana
Tandilianwala
Tarkhani
Kanya
Bhagat
Sultanpur

Kharif

Rainfall, mm

600
500
400
300
200
100
0

Sagar
Chuharkana
Paccadala
Mohlan
Buchiana
Tandilianwala
Tarkhani
Kanya
Bhagat
Sultanpur

system to lower parts. The rainfall was higher during the kharif seasons mainly due to monsoon rainfalls from mid-July to mid-September, which accounted for about 65% of total annual average rainfall. Similarly, the average effective rainfall can be seen for each individual irrigation subdivision, however, the individual average effective rainfall values for each cropping year vary due to variability in cropping, rainfall intensity and consumptive water requirements (Patwardhan et al. 1990).

The recharge results for different seasons and irrigation subdivisions indicate that generally higher recharge was found in the upper lower Chenab irrigation system regions as compared to lower ones. The highest recharge was estimated for Sagar with an average value of 127.5 mm ± 26.9 mm followed by Chuharkana, Paccadala, Buchiana, Mohlan, Tandlianwala, Bhagat, Kanya, Sultanpur and Tarkhani with average values of 120.3 mm ± 26.1 mm, 106.6 mm ± 22.9 mm, 103.7 mm ± 23.5 mm, 100.8 mm ± 27.3 mm, 100 mm ± 24.4 mm, 95.5 mm ± 28.6 mm, 86.8 mm ± 23.5 mm, 84.8 mm ± 13.2 mm, and 77.5 mm ± 24.8 mm, respectively, during the kharif seasons. Bhagat subdivisions showed relatively higher recharge in the lower reaches with more variability, which was mainly associated to higher rice cultivation in the area. For rabi seasons, the variability in recharge for different subdivisions was found low as compared to kharif seasons.

9.2.3.4 Recharge from Other Sources

Table 9.5 shows the detailed results of recharge from different components of irrigation system for each irrigation subdivision during both rabi and kharif seasons. During kharif seasons, after rainfall (i.e. 1003.5 mm), field percolation (809.2 mm) was considered to be major recharge sources in lower Chenab canal irrigation system during the study duration, however, its value was variable for individual irrigation subdivision. The other major sources of recharge were canal seepage (352.7 mm), watercourse losses (258 mm) and distributary losses (226 mm), respectively. The share of field percolation in recharge in comparison to other recharge sources, except rainfall, was almost equal. The lowest contribution was coming from lateral groundwater inflow and outflow (91 mm). For rabi, rainfall (207.1 mm) and field percolation (277.1 mm) were no more major recharge sources, but canal seepage was the dominant source (318.2 mm) followed by watercourse losses (174.2 mm), distributary losses (160.5 mm) and lateral groundwater inflow and outflow (91 mm). It was also to be noted that during the rabi season, rainfall recharge was less due to absent of monsoon rainfalls. The detailed information of different recharge components for each irrigation subdivisions can be found from the following Table 9.5, both for kharif and rabi seasons.

Table 9.5 Detailed groundwater recharge (mm) in different irrigation subdivisions of lower Chenab canal irrigation system by different sources during rabi (November to April) and kharif (May to October) cropping seasons

	Canal seepage		Distributary losses		Watercourse losses		Field percolation		Groundwater inflow and outflow	
Irrigation sub-division	Kharif	Rabi	Kharif	Rabi	Kharif	Rabi	Kharif	Rabi	Kharif	Rabi
Sagar	7.9 ± 2.2	5.1 ± 0.5	15.8 ± 2.1	1.6 ± 0.2	22.8 ± 3.1	2.9 ± 0.4	99.6 ± 13.5	15.2 ± 3.3	15.4 ± 2.2	16.0 ± 1.7
Chuharkana	59.7 ± 3.4	60.2 ± 12.3	14.7 ± 1.6	8.4 ± 1.4	18.2 ± 1.8	10.4 ± 1.9	106.2 ± 14.0	16.7 ± 2.5	3.0 ± 0.8	2.9 ± 0.5
Paccadala	10.9 ± 0.6	8.9 ± 0.8	28.9 ± 1.9	22.8 ± 2.1	27.5 ± 2.6	21.3 ± 3.4	91.0 ± 9.2	31.5 ± 7.6	3.1 ± 0.8	3.4 ± 0.7
Mohlan	44.9 ± 2.5	41.3 ± 4.1	30.4 ± 2.2	20.8 ± 2.1	33.1 ± 2.9	22.0 ± 2.8	118.1 ± 9.4	26.2 ± 4.8	7.0 ± 0.6	6.8 ± 1.0
Buchiana	103.6 ± 1.7	92.9 ± 8.4	23.9 ± 1.3	18.6 ± 1.7	27.1 ± 2.0	18.1 ± 2.1	68.8 ± 8.3	49.0 ± 4.5	−3.0 ± 0.9	−2.9 ± 0.5
Tandilianwala	20.9 ± 1.1	18.4 ± 1.5	21.4 ± 2.7	16.9 ± 2.4	25.8 ± 3.4	20.1 ± 3.9	66.5 ± 5.3	26.3 ± 3.4	0.0 ± 0.4	0.2 ± 0.1
Tarkhani	30.9 ± 1.3	26.6 ± 1.9	27.1 ± 1.8	21.8 ± 2.8	37.7 ± 2.3	29.7 ± 3.5	64.6 ± 10.1	29.0 ± 2.8	9.0 ± 0.7	8.6 ± 0.8
Kanya	41.3 ± 2.1	36.4 ± 3.0	26.0 ± 4.6	20.6 ± 2.3	28.0 ± 4.9	21.3 ± 2.7	60.2 ± 6.1	24.3 ± 3.5	6.5 ± 1.0	6.7 ± 0.7
Bhagat	14.5 ± 0.6	12.3 ± 0.8	18.5 ± 0.9	14.5 ± 1.3	14.9 ± 0.9	11.4 ± 1.4	73.0 ± 10.2	23.5 ± 5.0	22.3 ± 1.0	22.6 ± 1.0
Sultanpur	18.3 ± 0.9	16.1 ± 1.3	19.7 ± 4.1	14.8 ± 4.1	23.1 ± 4.9	17.1 ± 5.0	61.2 ± 3.5	35.3 ± 4.6	27.4 ± 1.8	26.8 ± 2.7

Table 9.6 Groundwater pumping and net recharge (mm) in different irrigation subdivisions of lower Chenab canal irrigation system during rabi (November to April) and kharif (May to October) cropping seasons

	Groundwater Pumping		Net recharge	
Irrigation sub-division	Kharif	Rabi	Kharif	Rabi
Sagar	193.8 ± 31.5	192.4 ± 32.6	95.2 ± 37.7	−121.9 ± 40.8
Chuharkana	225.3 ± 47.6	142.5 ± 24.0	96.9 ± 54.1	−19.6 ± 37.3
Paccadala	145.0 ± 48.6	83.2 ± 33.3	123.0 ± 59.5	27.9 ± 37.5
Mohlan	157.7 ± 41.4	83.5 ± 32.4	176.7 ± 50.1	53.7 ± 40.7
Buchiana	177.9 ± 47.8	116.7 ± 25.0	146.2 ± 50.6	78.0 ± 33.7
Tandilianwala	173.3 ± 29.8	90.5 ± 31.2	61.2 ± 46.1	10.0 ± 39.2
Tarkhani	108.1 ± 25.1	41.9 ± 19.0	138.6 ± 30.5	91.3 ± 31.7
Kanya	169.1 ± 24.0	85.3 ± 29.3	79.7 ± 38.8	40.2 ± 32.3
Bhagat	264.8 ± 33.1	158.7 ± 28.7	−26.1 ± 31.9	−53.8 ± 36.0
Sultanpur	228.7 ± 16.6	134.1 ± 44.1	5.8 ± 30.9	−6.4 ± 51.9

9.2.3.5 Groundwater Pumping and Net Recharge

Groundwater pumping was higher during kharif seasons as compared to rabi seasons at lower Chenab canal irrigation system scale. The detailed analysis of groundwater pumping data showed that highest groundwater pumping was found at two lower and one upper irrigation subdivisions including Bhagat, Sultanpur and Chuharkana with average values of 264.7 mm and 229 mm and 226 mm, respectively (Table 9.6). While, least pumping was found in Tarkhani with an average value of 108 mm, during different kharif seasons. The results of net recharge showed that it was generally positive for all the irrigation subdivisions during kharif seasons excluding two lower irrigation subdivisions including Bhagat and Sultanpur. For rabi seasons, groundwater pumping was observed to be highest in most of the upper irrigation subdivisions of lower Chenab canal irrigation system, which was mainly due to decreased canal flow in spite of less water requirements by the crops, particularly in the Sagar irrigation subdivision, where the net recharge is negative for the whole study period. It was also negative for Bhagat and Sultanpur irrigation subdivisions at the lower locations of lower Chenab canal irrigation system. For rest of the locations, the net recharge was variable from season to season.

The results of net recharge from the current study are comparable to previous studies including Hassan and Bhutta (1996), according to which annual recharge was found to be 60 mm for the Rechna Doab. They also concluded positive recharge for kharif seasons, whereas it was generally negative for rabi seasons. Similarly, Boonstra and Bhutta (1996) conducted a recharge study in the whole Rechna Doab and concluded that positive recharge (i.e. 73 mm) is observed annually. The other studies include Habib (2004) and Bhuttah and Alam (2005), according to which annual groundwater recharge is positive for majority parts of Indus plain.

9.3 Case Study – II: Recharge Under Changing Climate in Lower Chenab Canal Irrigation System

Climate change is important to study as it can induce changes in natural ecosystems (IPCC 2007). Particularly, groundwater has attained least attention in the past under climate change from scientific and other concerned communities (Taylor et al. 2012). According to IPCC (2008), there has been very limited research on the effects of climate change on groundwater although its contribution in world's water use is considerable. The effects of climate change on natural systems are quite complex, for instance, one may question that annual recharge is considerably affected by annual rainfall only (Crosbie et al. 2009), but recharge can also be affected by rainfall seasonality, intensity, humidity, air temperature, and actual evapotranspiration. In arid and semi-arid regions, increased variability in rainfall may increase groundwater recharge due to more frequent rainfalls. But at the same time, actual evapotranspiration can be higher due to elevated temperatures and hence, there is less net recharge (Hetze et al. 2008) and vice versa.

For irrigated regions such as lower Chenab canal irrigation system, the potential parameters of interest may be rainfall and evapotranspiration as both of these can behave differently for recharge due to climate change. Apart from many different water inflow and outflow fluxes, crop consumptive waiter use and rainfall are very important variables which control recharge rates particularly in the semi-arid and arid regions (Scanlon et al. 2002). Therefore, in the current study groundwater recharge was estimated under the influence of changing patterns of rainfall and actual evapotranspiration.

9.3.1 Downscaling Climatic Data

The outputs of global circulation models cannot be used directly to investigate the impact of climate change in environmental studies on local/regional scale, as their outputs are based on a large grid scale (250–600 km) (Wilby et al. 2000). The most decent strategy is to build a bridge between global circulation model scales (a coarse scale) and local scale (0–50 km) by downscaling (Wetterhall et al. 2006; Xu 1999).

Many downscaling methods, including dynamic and statistical downscaling, have been developed and implemented by utilizing the outputs of global circulation models to downscale climate variables at local/regional scales (Huth 2002; Hay and Clark 2003; Diaz-Nieto and Wilby 2005; Salzmann et al. 2007; Akhtar et al. 2008; Elshamy et al. 2009; Yang et al. 2010; Sunyer et al. 2011; Huang et al. 2011). Statistical downscaling approaches establish statistical links among the local scale and large scale variables and are computationally inexpensive and fast (Wilby et al. 2000). Such approaches are applied for a wide range of climate applications apart from its utility for numerical weather predictions. According to Giorgi et al. (2001), statistical downscaling approach provides local scale information for climate change impact assessment studies, although its main disadvantage is requirement of long term historical meteorological data.

Many statistical downscaling software tools have been developed to date and statistical downscaling model is one of them. It is widely used throughout the world to downscale important climatic variables like precipitation, temperature and actual evapotranspiration etc. (Chu et al. 2010). This tool is a hybrid of multiple linear regression and stochastic weather generator (Wilby et al. 2000; Wilby and Harris 2006). Multiple linear regression establishes a statistical relationship between gridded predictors (such as mean sea level pressure) and single site predictands (such as rainfall), and produces some calibration parameters. These parameters are then used by the stochastic weather generator to simulate up to one hundred daily time series to create a better correlation with the observed data (Wilby et al. 2000).

9.3.2 Data Requirements of Statistical Downscaling Model

Two types of daily time series, namely daily historic weather station data and large scale variables (National Center for Environmental Prediction (NCEP) daily predictors), were used to develop statistical downscaling model. There are four weather stations located in or near to the study area including Faisalabad, Toba Tek Singh, Lahore and Pindi Bhattian. Daily based long period weather data are required for statistical downscaling model, which were only available for Lahore and Faisalabad stations. From these two stations, daily data regarding maximum temperature, minimum temperature, relative humidity, sunshine hours, wind speed and rainfall were collected from 1960 to 2014. The daily data for Toba Tek Singh were available only from 2009 to 2014, while data were available from 2005 to 2014 for Pindi Bhattian. All such data were collected from Pakistan metrological department.

Statistical downscaling model produces output daily time series by forcing the NCEP or HadCM3 predictors (Mahmood and Babel 2013; Huang et al. 2011), the data of which were obtained cost free from *http://www.cics.uvic.ca/scenarios/index.cgi?Scenarios*, for the period of 1961–2010 and 1996–2050, respectively. H3A2 is the IPCC emission scenarios A2 of HadCM3. HadCM3 was selected for statistical downscaling model because it showed better agreement during evaluation of various global circulation models (Mahmood and Babel 2013; Akhtar et al. 2008; Huang et al. 2011).

9.3.3 Calibration, Validation and Bias Correction

Statistical downscaling models were developed by utilizing NCEP predictors screened for different variables at different locations. Daily data of rainfall and ET were used for calibration of statistical downscaling model from 1961 to 1995. Annual sub-models were developed individually for each predictand. Unconditional sub-models with fourth root transformation and conditional sub-models without transformation were used for rainfall and actual evapotranspiration, respectively. The calibrated models were used for simulation of the predictands from 1996 to 2010 using NCEP and H3A2 predictors by generating 20 ensembles and the means of these ensembles are used. Different statistical indicators are used for comparison of downscaling results with the observed data including coefficient of determination (R^2), root mean square error (RMSE), mean (M), and relative error in mean (RE_M) for the periods of calibration and validation.

Bias correction was applied to compensate for any tendency to over- or under-estimate the mean of conditional processes by statistical downscaling model (Wilby and Dawson 2013). For this purpose, the mean monthly bias factors for different variables were obtained from the calibration period of 1961–1995. Then, these biases are adjusted to downscaled data for the validated period from 1996 to 2010. The statistical comparison is performed between un-biased statistical downscaling model downscaled data of rainfall and actual evapotranspiration, and observed data. Following successful validation, adjusted bias factors are utilized to rectify the current and future downscaled data obtained from HadCM3 predictors to achieve a more realistic picture of future climate (Mahmood and Babel 2013). It is to be noted that the application of these bias corrections for rainfall are only valid to its intensity and also to remove any systematic error occurred by statistical downscaling model downscaling. However, it is assumed that rainfall frequency is accurately simulated by statistical downscaling model (Mahmood and Babel 2013).

9.3.4 Baselines and Utilization of Scenarios for Groundwater Modeling

Since specific change is always relative to some baseline time or period, therefore, a baseline is selected for climate change scenarios, which is from 2002 to 2012 for future periods 2016–2025, 2026–2035 and 2036–2045 by using the following relationship:

$$\%\text{change} = \left(\frac{x - y}{y}\right).100 \tag{9.12}$$

where x is the mean for the future period e.g., 2016–2025, and y is the mean for the baseline period of 2002–2012.

For the current study, future climate change data was dealt by considering a constant change because the objective is not to track the changes in groundwater recharge yearly; rather the general situation at the end of a particular time period was to be investigated.

Table 9.7 Statistics of observed and downscaled mean monthly actual evapotranspiration and rainfall for calibration period at Lahore and Faisalabad meteorological stations

Predictand	Station	Model	R^2	RMSE (mm)	M (mm)	RE_M (%)
Evapotranspiration	Lahore	Observed			128.5	
		NCEP	0.94	14.2	128.1	−0.3
Rainfall		Observed			38.9	
		NCEP	0.88	24.1	21.5	−44.8
Evapotranspiration	Faisalabad	Observed			132.7	
		NCEP	0.92	15.3	135.2	1.9
Rainfall		Observed			26.9	
		NCEP	0.84	18.1	18.5	−31.3

R^2 coefficient of determination, RMSE root mean square error, M mean, RE_M relative error in mean

9.3.5 *Results of Case Study – II*

9.3.5.1 Screening of Predictors

Temperature at 2 m height (temp) was observed to be super-predictor for actual evapotranspiration at Lahore and Faisalabad stations. For both stations, the other predictors included mean sea level pressure (mslp) and super-specific humidity (shum). Super-specific humidity was also a super-predictor for both Lahore and Faisalabad stations in case of rainfall along with zonal velocity at 500 hPa and vorticity at 500 hPa, respectively. The results were consistent with Mahmood and Babel (2013), according to which shum was one of the major super-predictors for the majority of precipitation stations. Similarly, temp was found to be the main predictor for maximum and minimum temperatures. Along with wind velocity, temperature has a high effect on actual evapotranspiration and this behaviour was also witnessed for the current study. The predictors selected for the current study were almost similar to those selected for some other studies with similar predictands (Mahmood and Babel 2013; Hashmi et al. 2011; Huang et al. 2011).

9.3.5.2 Calibration of Statistical Downscaling Model

The calibration period was 35 years from 1961 to 1995. The daily rainfall and actual evapotranspiration data were simulated by statistical downscaling using NCEP variables. The model performed reasonably well in the case of actual evapotranspiration, which can be seen from the results in Table 9.7. The mean simulated values of actual evapotranspiration for both stations were comparable to observed data. However, the results for rainfall are relatively weak. For both stations, the difference between modelled and observed mean rainfall was large. Relative errors of mean were much greater for rainfall than evapotranspiration. Different researchers have evaluated statistical downscaling mode for different variables including rainfall

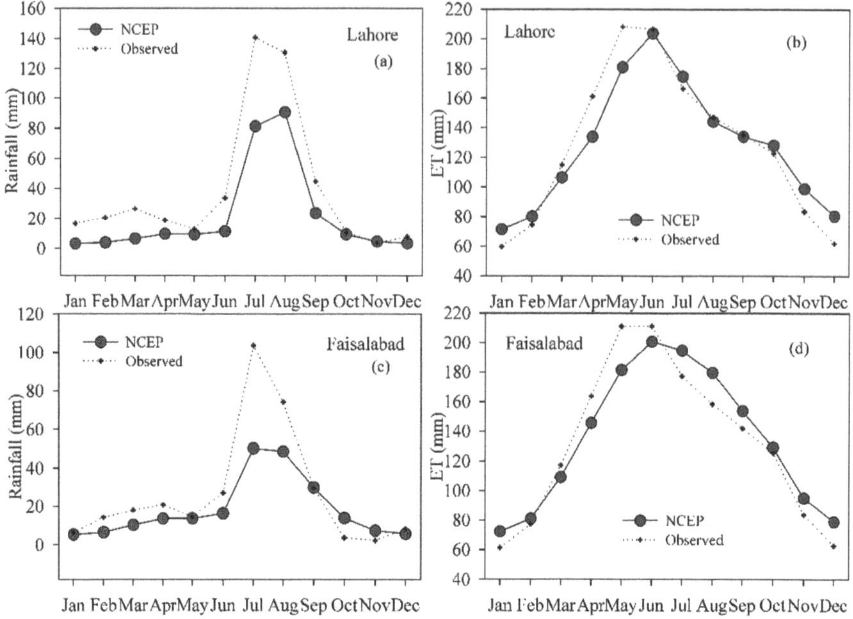

Fig. 9.6 Calibration results of statistical downscaling model. The model respond better for actual evapotranspiration as compared to rainfall due to its more variability in lower Chenab canal irrigation system. NCEP national centre for environmental protection. (Source: figure based on own research)

using different models and reported similar types of results. For example, Huang et al. (2011) developed different models for rainfall downscaling and their results vary from weak to medium values for different performance indicators (e.g. coefficient of determination ranges from 0.11 to 0.97). Whereas, Mahmood and Babel (2013) investigated statistical downscaling model for two models, monthly and annual, and according to them the monthly model performed better (coefficient of determination = 0.99) as compared to the annual model (coefficient of determination = 0.69). Overall, there is a consensus among different studies that temperature and actual evapotranspiration performed better than rainfall (Mahmood and Babel 2013; Dibike and Coulibaly 2005). The possible reason was the heterogeneous nature of the precipitation occurrence/amounts, which is therefore difficult to simulate accurately (Wilby et al. 2000). Moreover, the calibration process of rainfall could be biased by the large number of zero values entered in the multiple regressions (Huang et al. 2011).

Figure 9.6 indicates the graphical comparison between observed data and monthly mean output of statistical downscaling model. In case of actual evapotranspiration, statistical downscaling model underestimated the results from March to May for Lahore and from March to June for Faisalabad. However, it overestimated the results for November, December, January and February for both stations. It also

Table 9.8 Statistics of observed and downscaled mean monthly actual evapotranspiration and rainfall during validation (bias corrected) period for Lahore and Faisalabad meteorological stations

Predictand	Station	Model	R^2	RMSE (mm)	M (mm)	RE_M (%)
Evapotranspiration	Lahore	Observed			127.4	
		NCEP	0.98	7.3	124.2	−2.8
		H3A2	0.99	4.0	127.2	−0.2
Rainfall		Observed			29.2	
		NCEP	0.99	11.2	32.7	10.0
		H3A2	0.99	15.0	34.4	20.1
Evapotranspiration	Faisalabad	Observed			133.4	
		NCEP	0.99	8.4	128.1	−2.9
		H3A2	0.99	6.5	131.4	−1.5
Rainfall		Observed			25.1	
		NCEP	0.99	6.2	28.3	6.7
		H3A2	0.98	9.8	29.7	11.1

R^2 coefficient of determination, RMSE root mean square error, M mean, RE_M relative error in mean

overestimated results during July to October, which was more prominent for Faisalabad in comparison to Lahore. With regard to rainfall, observed data were underestimated by statistical downscaling model in the majority of months especially in rainy months (July to August). There was only a small overestimation by statistical downscaling model results during October and November in comparison to observed rainfall.

9.3.5.3 Validation of Statistical Downscaling Model with Bias Correction

The calibration of statistical downscaling model results indicated large biases, especially for rainfall, which should be removed for validation results. The current study has adopted the bias correction approach proposed by Mahmood and Babel (2013) and Salzmann et al. (2007). Different statistical downscaling model models for actual evapotranspiration and rainfall were corrected for biases and the detailed results are presented in Table 9.8.

Bias corrected downscaled mean results were also compared graphically with observed data as shown in Fig. 9.7. From all the statistical indicators, it is obvious that both evapotranspiration and rainfall results improved. Especially, the rainfall results improved significantly as coefficient of determination increased from 0.80–0.84% to 0.98–0.99%, root mean square values decreased from 26.52–28.08 mm to 4.04–7.30 mm and relative errors in mean 58.42–64.19% to 10.01–22.47%, for Lahore. Similarly, the results for rainfall at Faisalabad also indicated significant improvement as coefficient of determination increased from 0.76–0.80% to 0.98–0.99%, root mean square error decreased from 17.72–17.91 mm to 6.18–10.72 mm, and relative errors in mean were decreased

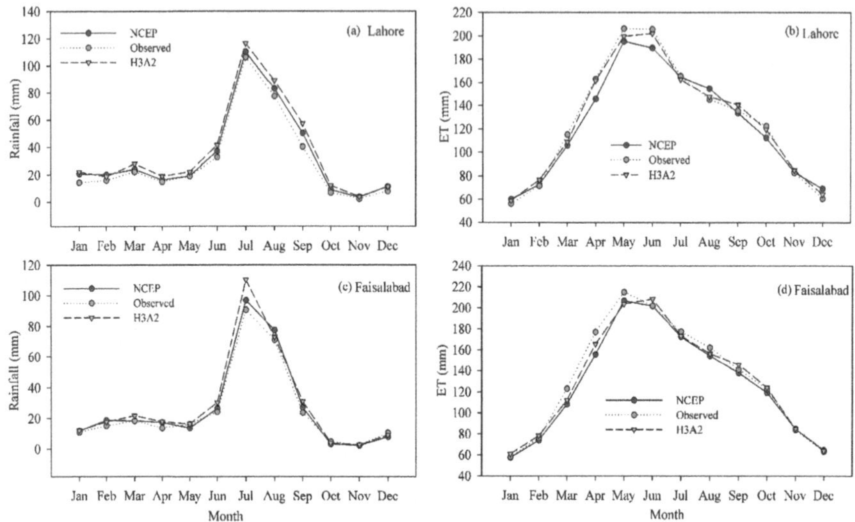

Fig. 9.7 Validation of bias corrected statistical downscaling model results. The results for both actual evapotranspiration and rainfall were improved after removing bias error. NCEP national centre for environmental protection. (Source: figure based on own research)

from 28.34–32.59% to 6.67–13.34%. The results from all variables including NCEP and H3A2 were satisfactory and indicate strong applicability of SDSM to downscale actual evapotranspiration and rainfall under H3A2 emission scenario.

9.3.5.4 Downscaling Results

Table 9.9 presents the projected results for different time durations (i.e. 2016–2025, 2026–2035 and 2036–2045) with reference to the base line period under emission scenario H3A2. According to this, the change in actual evapotranspiration is +2.23% in 2016–2025, +6.18% in 2026–2035 and +5.87% in 2036–2045 at Faisalabad for rabi seasons. The change during kharif seasons is +2.22% at Faisalabad whereas at Lahore the change is +1.36%. The detailed results for other periods can be seen from Table 9.9.

With regard to rainfall, the future positive change during rabi seasons is +1.68% in 2016–2025 at Faisalabad, +4.67% and+ 5.79% at Lahore in 2016–2025 and 2036–2045, respectively. Conversely, there is a decrease of −7.31% and −0.18% in 2026–2035 and 2036–2045 at Faisalabad. In case of kharif seasons, the change is positive for all future periods at Lahore.

The selection of Lahore and Faisalabad for downscaling of climatic data was done because long time series were available only for these two stations. The other two stations located inside the study region are Pindi Bhattian (upstream location) and Toba Tek Singh (downstream location). There was a need to investigate whether any

Table 9.9 Future changes (%) in actual evapotranspiration and rainfall at Lahore and Faisalabad meteorological stations till end of 2045 for rabi (November to April) and kharif (May to October) cropping seasons under A2 scenarios

			Faisalabad			Lahore		
Scenario	Predictor	Season	2016–2025	2026–2035	2036–2045	2016–2025	2026–2035	2036–2045
H3A2	Evapotranspiration	Rabi	2.23	6.19	5.87	2.20	9.46	7.48
		Kharif	−2.29	2.22	1.58	−10.08	1.36	−0.59
		Annual	−0.67	3.64	3.12	−5.62	4.30	2.34
H3A2	Rainfall	Rabi	1.68	−7.31	−0.18	4.67	−6.81	5.79
		Kharif	27.86	7.61	11.99	33.52	11.05	13.98
		Annual	25.56	4.12	9.14	31.81	7.26	14.60

significant relationship exists between different stations (i.e. between Lahore and Pindi Bhattian, and between Faisalabad and Toba Tek Singh). Owing to time series data, autocorrelations were worked out to see if there was any current time or lag time relationship between different stations. This analysis was based on the daily rainfall data from 2005 to 2012 for Pindi Bhattian and from 2009 to 2012 for Toba Tek Singh. The highest correlation (0.72) was found between Lahore and Pindi Bhattian, followed by 0.50 for Faisalabad and Toba Tek Singh. The correlations between other stations were not very strong, for example, correlations of 0.321, 0.30, 0.421, and 0.305 were found between Faisalabad and Lahore, Toba Tek Singh and Lahore, Faisalabad and Pindi Bhattian, and Pindi Bhattian and Toba Tek Singh, respectively. Generally, the lag time correlation relationship does not depict any strong relationship for any case.

The autocorrelation results show that better correlation is found between Lahore and Pindi Bhattian and between Faisalabad and Toba Tek Singh. Also, Pindi Bhattian and Toba Tek Singh are located at the central locations of upper and lower irrigation subdivisions, respectively. Therefore, it was decided to utilize climate change results of Lahore for upper irrigation subdivisions including Sagar, Chuharkana, Paccadala, Buchiana and Mohlan and results of Faisalabad for lower irrigation subdivisions including Tandlianwala, Tarkhani, Bhagat, Kanya and Sultanpur. Actual evapotranspiration is used directly for estimation of modified recharge while rainfall was first processed for effective rainfall before their utilization for estimation of recharge.

9.3.5.5 Projected Recharge Results

The projected results of recharge, in Fig. 9.8, show an expected increase during 2016–2025 for the kharif seasons. The maximum increase is up to an average value of 139.76 mm for Lahore and upper LCC regions. The increase of 43.51 mm is expected for Faisalabad and for lower LCC regions. Major groundwater recharge in lower Chenab canal irrigation system takes place during kharif seasons due to intensive monsoon rainfalls. The results of current study confirm the findings of Awan and Ismaeel (2015), according to which increase in recharge is expected due to more rainfall from 2012 to 2020 under changing climate conditions. For Lahore and upper regions, the recharge is expected to be stable or increasing for all time durations, whereas for Faisalabad and lower parts of lower Chenab canal irrigation system, the recharge is most likely to decrease during 2026–2045. During rabi seasons, the recharge is expected to be decreasing for all the time periods both for Lahore and Faisalabad regions, however, its quantum is variable and can be seen from the Fig. 9.8. The rate of decrease in recharge is higher during the later decades as compared to 2016–2025 period.

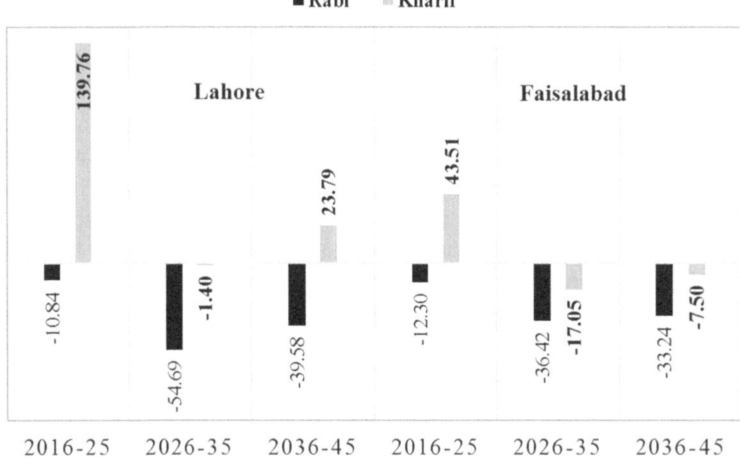

Fig. 9.8 Projected seasonal average recharge (mm) for Lahore and Faisalabad during 2016–2045. During kharif seasons (May to October), the recharge is expected to increase in the upper regions and decreasing in the lower ones contrast to rabi seasons (November to April) where recharge is expected to decrease throughout the lower Chenab canal irrigation system. (Source: figure based on own research)

9.4 Conclusion

Groundwater is an integral part of irrigation for successful agriculture in Pakistan. Present book chapter explores the historical and present trends of groundwater use in the country and particularly for the Punjab province. Moreover, results of two case studies are presented. Estimation of recharge in the lower Chenab canal irrigation system, Punjab, its important sources, and future trends are explored under changing climatic conditions. Following main conclusions are drawn and presented as below:

1. Waterlogging and salinity had been a major issue in Pakistan by late 1930s and early 1940s, due to construction of major canal network in the country and poor drainage network.
2. Launching of different salinity control and reclamation projects after 1960s helped to get rid of water logging and salinity issue in major parts of the country by installing many tubewells.
3. Canal water had been major contributor in irrigated agriculture by the end of 1990s, which was surpassed by groundwater afterwards due to promotion of private diesel operated small tubewells in the country overall, and particularly for Punjab.
4. By the end of 2013, the total number of functional tubewells were 1,049,000 in the country, 867,000 in Punjab and 51,157 in lower Chenab canal irrigation system, which shows the importance of groundwater use for agriculture.
5. Analysis of recharge in lower Chenab irrigation system shows that rainfall was the major source of recharge during kharif (May to October) seasons followed by

field percolation, canal seepage, watercourse losses and distributary losses. For rabi (November to April), canal seepage surpasses rainfall followed by field percolation, watercourse losses, distributary losses and later groundwater inflow and outflow.

6. Groundwater pumping was found higher during kharif seasons as compared to rabi seasons, However, net groundwater recharge found positive for kharif seasons whereas it was found negative during rabi seasons.
7. The daily rainfall and evapotranspiration data were simulated by statistical downscaling model using National Centers for Environmental Prediction (NCEP) variables. The calibration of statistical downscaling model showed reasonable results at both Lahore and Faisalabad weather stations in case of evapotranspiration, however for rainfall, the variation between observed and simulated values were higher which were removed by applying the bias corrections.
8. The projected recharge results showed that overall there will be increase in recharge during kharif seasons for lower Chenab canal irrigation system during 2016–2025 with the major contribution from monsoon rainfalls. For upper parts of lower Chenab canal irrigation system, recharge is expected to be increasing in future, whereas for lower regions of lower Chenab canal irrigation system, it is most likely decreasing during 2026–2045. During rabi seasons, the recharge is expected to be decreasing both for upper and lower regions of lower Chenab canal irrigation system.

References

Agricultural Statistics of Pakistan (2008–09) Government of Pakistan. Ministry of Food and Agriculture (Economic Wing), Islamabad, Pakistan

Ahmed N, Chaudhry GR (1988) Irrigated Agriculture of Pakistan. Shahzad Nazir Publisher, Lahore

Akhtar M, Ahmad N, Booij MJ (2008) The impact of climate change on the water resources of Hindukush–Karakorum–Himalaya region under different glacier coverage scenarios. J Hydrol 355(1–4):148–163. https://doi.org/10.1016/j.jhydrol.2008.03.015

Altman DG (1991) Practical statistics for medical research. Chapman and Hall, London. https://doi.org/10.1002/sim.4780101015

Ashraf A, Ahmad Z (2008) Regional Groundwater Flow Modelling of upper Chaj Doab of Indus Basin, Pakistan using Finite Element Model (Feflow) and Geoinformatics. Geophys J Int 173 (1):17–24

Aslam M, Prathapar SA (2006) Strategies to mitigate secondary salinization in the Indus Basin of Pakistan: a selective review. Research Report 97. Colombo, Sri Lanka: International Water Management Institute (IWMI). https://doi.org/10.3910/2009.097

Awan UK, Ismaeel A (2015) A new technique to map groundwater recharge in irrigated areas using a SWAT model under changing climate. J Hydrol 519:1368–1382. https://doi.org/10.1016/j.jhydrol.2014.08.049

Ayars JE, Christen EW, Soppe RW, Meyer WS (2006) The resource potential of in-situ shallow ground water use in irrigated agriculture: a review. Irrig Sci 24:147–160

Baalousha H (2005) Using CRD method for quantification of groundwater recharge in the Gaza Strip, Palestine. Environ Geol 48(7):889–900

Badruddin M (1996) Country profile. Pakistan Internal Report. International Water Management Institute (IWMI): Lahore, Pakistan

Bastiaanssen WGM, Menenti M, Feddes RA, Holtslag AAM (1998) A remote sensing surface energy balance algorithm for land (SEBAL) 1. Formulation. J Hydrol 212(213):198–212
Bhutta MN, Alam MM (2005) Perspective and limits of groundwater use in Pakistan. Groundwater Research and Management: Integrating Science into Management Decisions, Roorkee, India, pp 105–113
Bhutta MN, Smedema K (2007) One hundred years of waterlogging and salinity control in the Indus valley, Pakistan: a historical review. Irrig Drain 56:81–90
Bowen I (1926) The ratio of heat losses by conduction and by evaporation from any water surface. Phys Rev 27:779–787
Boonstra J, Bhutta MN (1996) Groundwater recharge in irrigated agriculture: the theory and practice of inverse modeling. J Hydrol 174(3–4):357–374
Brutsaert W (2005) Hydrology: an introduction. Cambridge University Press, New York, 605 pp, ISBN 0-521-82479-6
Brutsaert W, Stricker H (1979) An advection-aridity approach to estimate actual regional evapotranspiration. Water Resour Res 15(2):443–450
Chaudhary MR, Bhutta NM, Iqbal M, Subhani KM (2002) Groundwater resources: use and impact on soil and crops. Paper presented at the Second South Asia Water Forum, 14–16 December, Islamabad, Pakistan
Christmann S, Martius C, Bedoshvili D et al (2009) Food security and climate change in Central Asia and the Caucasus. Sustainable agriculture in Central Asia and the Caucasus Series No. 7. CGIAR-PFU, Tashkent, Uzbekistan
Chu JT, Xia J, Xu CY, Singh VP (2010) Statistical downscaling of daily mean temperature, pan evaporation and precipitation for climate change scenarios in Haihe River, China. Theor Appl Climatol 99:149–161
COMSATS (Commission on Science and Technology for Sustainable Development in the South) (2003) Water resources in the south: present scenario and future prospects. COMSATS, Islamabad
Congalton RG (1996) Accuracy assessment: a critical component of land cover mapping. In: Gap analysis: a landscape approach to biodiversity planning, a peer-reviewed proceedings of the ASPRS/GAP Symposium, 27 February – 2 March 1995, Charlotte, NC, pp 119–131
Crosbie RS, McCallum JL, Harrington GA (2009) Diffuse groundwater recharge modelling across northern Australia. A report to the Australian government from the CSIRO northern Australia sustainable yields project. CSIRO Water for a Healthy Country Flagship, Australia, p 56
Darcy H (1856) Les Fontaines Publiques de la Ville de Dijon. Dalmont, Paris
Diaz-Nieto J, Wilby RL (2005) A comparison of statistical downscaling and climate change factor methods: impacts on low flows in the River Thames, United Kingdom. Clim Change 69 (2):245–268. https://doi.org/10.1007/s10584-005-1157-6
Dibike YB, Coulibaly P (2005) Hydrologic impact of climate change in the Saguenay watershed: comparison of downscaling methods and hydrologic models. J Hydrol 307(1–4):145–163. https://doi.org/10.1016/j.hydrol.2004.10.012
Döll P (2009) Vulnerability to the impact of climate change on renewable groundwater resources: a global-scale assessment. Environ Res Lett 4:035006. https://doi.org/10.1088/1748-9326/4/3/035006
Elshamy ME, Seierstad IA, Sorteberg A (2009) Impacts of climate change on Blue Nile flows using bias-corrected GCM scenarios. Hydrol Earth Syst Sci 13:551–565. https://doi.org/10.5194/hess-13-551-2009
FAO (2010) AQUASTAT – FAO's global information system on water and agriculture, FAO. http://www.fao.org/nr/aquastat
Foody GM (2002) Status of land cover classification accuracy assessment. Remote Sens Environ 80:85–201
Giorgi F, Hewitson B, Christensen J, Hulme M, Von Storch H, Whetton P, Jones R, Mearns L, Fu C (2001) Regional climate information – evaluation and projections. In: Houghton JT, Ding Y, Griggs DJ, Noguer M, van der Linden PJ, Dai X, Maskell K, CA Johnson (eds) Climate change

2001: the scientific basis. contribution of Working Group I to the third assessment report of the intergovernmental panel on climate change. Cambridge University Press, Cambridge, New York, 881pp

Giri C, Clinton J (2005) Land cover mapping of greater mesoamerica using MODIS data. Remote Sens 31(4):274–282

GOP (2000) Agricultural statistics of Pakistan. Ministry of Food, Agriculture and Livestock, Economics Division, and Government of Pakistan, Islamabad

Habib Z (2004) Scope for reallocation of river waters for agriculture in the Indus Basin. Ph.D thesis l'Ecole Nationale du Génie Rural, des Eaux et Forêts Centre de Paris

Hashmi M, Shamseldin A, Melville B (2011) Comparison of SDSM and LARS-WG for simulation and downscaling of extreme precipitation events in a watershed. Stoch Env Res Risk A 25 (4):475–484. https://doi.org/10.1007/s00477-010-0416-x

Hassan ZH, Bhutta MN (1996) A water balance model to estimate groundwater recharge in Rechna Doab. Irrig Drain Syst 10:297–317

Hay LE, Clark MP (2003) Use of statistically and dynamically downscaled atmospheric model output for hydrologic simulations in three mountainous basins in the western United States. J Hydrol 282(1–4):56–75. https://doi.org/10.1016/s0022-1694(03)00252-x

Hetze F, Vaessen V, Himmelsbach T, Struckmeier W, Villholth K (2008) Groundwater and climate change: challenges and possibilities. BGR http://www.bgr.bund.de/

Hobbins MT, Ramirez JA (2001) The complementary relationship in estimation of regional evapotranspiration: an enhanced advection-aridity model. Water Resour Res 37(5):1389–1403

Huang J, Zhang J, Zhang Z, Xu C, Wang B, Yao J (2011) Estimation of future precipitation change in the Yangtze River basin by using statistical downscaling method. Stoch Env Res Risk A 25 (6):781–792. https://doi.org/10.1007/s00477-010-0441-9

Huth R (2002) Statistical downscaling of daily temperature in Central Europe. J Clim 15 (13):1731–1742. https://doi.org/10.1175/1520-0442(2002)015<1731:sdodti>2.0.CO;2

IPCC (2007) Climate change 2007: the physical science basis. In: Solomon S, Qin D, Manning M, Chen Z, Marquis M, Averyt KB, Tignor M, Miller HL (eds) Contribution of working Group I to the fourth assessment report of the intergovernmental panel on climate change. Cambridge University Press, Cambridge/New York, 996 pp

IPCC (2008) Climate change and water. IPCC technical paper VI. Available at: http://www.ipcc.ch/pdf/technical-papers/climate-change-water-en.pdf

Irrigation Department (2008) Canal Atlas of Rohtak district. Irrigation Department, Office of the Executive Engineer, Rohtak (Haryana), India

IWASRI (International Waterlogging and Salinity Research Institute) (1998) Integrated water resources management program for Pakistan. Economic, social and environmental matters. Internal Report 98/5, Lahore, Pakistan

Jalota SK, Arora VK (2002) Model-based assessment of water balance components under different cropping systems in North – West India. Agric Water Manag 57:75–87

Jurriens R, Mollinga PP (1996) Scarcity by design: protective irrigation in India and Pakistan. ICID J 45(2):31–53

Kazmi SI, Ertsen MW, Asi MR (2012) The impact of conjunctive use of canal and tube well water in Lagar irrigated area, Pakistan. Phys Chem Earth Parts A/B/C 47-48:86–98. https://doi.org/10.1016/j.pce.2012.01.001

Lerner DN, Issar AS, Simmers I (1990) Groundwater recharge. A guide to understanding and estimating natural recharge. Int Contrib Hydrogeol Verlang Heinz Heise 8:345

Liu S, Bai J, Jia Z, Jia L, Zhou H, Lu L (2010) Estimation of evapotranspiration in the land of China. Hydrol Earth Syst Sci 14(3):573–584

Llamas MR, Martinez-Santos P (2005) Intensive groundwater use: silent revolution and potential source of social conflicts. ASCE J Water Resour Plann Manag 131:337–341

Madramootoo CA (2012) Sustainable groundwater use in agriculture. Irrig Drain 61:26–33. https://doi.org/10.1002/ird.1658

Mahmood R, Babel MS (2013) Evaluation of SDSM developed by annual and monthly sub-models for downscaling temperature and precipitation in the Jhelum basin, Pakistan and India. Theor Appl Climatol 113:27–44. https://doi.org/10.1007/s00704-012-0765-0

Malmberg GT (1975) Reclamation by tubewell drainage in Rechna doab and adjacent areas, Punjab region, Pakistan. Geological Survey water supply paper 1608-O. United States Government Printing Office, Washington, DC, 1975

Maréchal JC, Dewandel B, Ahmed S, Galeazzi L, Zaidi FK (2006) Combined estimation of specific yield and natural recharge in a semi-arid groundwater basin with irrigated agriculture. J Hydrol 329(1-2):281–293

Mukherji A, Shah T (2005) Socio-ecology of groundwater irrigation in South Asia: an overview of issues and evidence. In: Selected papers of the symposium on intensive use of groundwater, held in Valencia (Spain), 10–14 December 2002, IAH Hydrogeology Selected Papers. Balkema Publishers

Patwardhan A, Nieber J, Johns E (1990) Effective rainfall estimation methods. J Irrig Drain Eng 116 (2):182–193

Postel SL (2003) Securing water for people, crops and ecosystems: new mindset and new priorities. Nat Res Forum 27:89–98

Punjab Development Statistics (2015) Bureau of statistics. Government of Punjab, Lahore. http://www.bos.gop.pk

Punjab Private Sector Groundwater Development Consultants (1998) Canal seepage analysis for calculation of recharge to groundwater, Technical Report No. 14, PPSGDP, PMU, P&DD, Government of Pakistan

Qureshi AS, Shah T, Mujeeb A (2003) The groundwater economy of Pakistan. IWMI working paper No. 19

Qureshi AS, McCornick PG, Sarwar A, Sharma BR (2010) Challenges and Prospects of Sustainable Groundwater Management in the Indus Basin, Pakistan. Water Resour Manag 24(8):1551–1569

Salzmann N, Frei C, Vidale PL, Hoelzle M (2007) The application of Regional Climate Model output for the simulation of high-mountain permafrost scenarios. Global Planet Chang 56 (1–2):188–202. https://doi.org/10.1016/j.gloplacha.2006.07.006

Sarwar A (2000) A transient model approach to improve on-farm irrigation and drainage in semi-arid zones. PhD dissertation, Wageningen University, The Netherlands

Sarwar A, Eggers H (2006) Development of a conjunctive use model to evaluate alternative management options for surface and groundwater resources. Hydrogeol J 14(8):1676–1687. https://doi.org/10.1007/s10040-006-0066-8

Scanlon BR, Healy RW, Cook PG (2002) Choosing appropriate techniques for quantifying groundwater recharge. Hydrogeol J 10:18–39

Schicht RJ, Walton WC (1961) Hydrologic budgets for three small watersheds in Illinois. Illinois state water survey report invest. 40:40

Shah T (2003) Governing the groundwater economy: comparative analysis of national institutions and policies in South Asia, China and Mexico. Water Perspect 1(1):2–27

Shah T, Burke J, Villholth K (2007) Groundwater: a global assessment of scale and significance. In: Molden D (ed) Water for food, water for life. Earthscan/IWMI, London/Colombo, pp 395–423

Siebert S, Burke J, Faures JM, Frenken K, Hoogeveen J, Döll P, Portmann FT (2010) Groundwater use for irrigation – a global inventory. Hydrol Earth Syst Sci 14:1863–1880. https://doi.org/10.5194/hess-14-1863-2010

Singh A (2011) Estimating long term regional groundwater recharge for the evaluation of potential solution alternatives to waterlogging and salinization. J Hydrol 406(3–4):245–255

Soomro AB (1975) A review of problem of waterlogging and salinity in Pakistan. Master Thesis No. 915, Asian Institute of Technology, Bangkok, Thailand

Sunyer MA, Madsen H, Ang PH (2011) A comparison of different regional climate models and statistical downscaling methods for extreme rainfall estimation under climate change. Atmos Res. https://doi.org/10.1016/j.atmosres.2011.06.011

Taylor AB, Martin NA, Everard E, Kelly TJ (2012) Modelling the Vale of St Albans: parameter estimation and dual storage. In: Shepley MG, Whiteman MI, Hulme PJ, Grout MW (eds) Groundwater resources modelling: a case study from the UK. Geological Society, London, Special Publications (364): 193–204

Thi T, Nguyen H, De Bie CAJM, Amjad A, Smaling EMA, Chu TH (2012) Mapping the irrigated rice cropping patterns of the Mekong delta, Vietnam through hyper-temporal SPOT NDVI image analysis. Int J Remote Sens 33(2):415–434

Tyagi NK, Sharma DK, Luthra SK (2000a) Determination of evapotranspiration and crop coefficients of rice and sunflower. Agric Water Manag 45:41–54

Tyagi NK, Sharma DK, Luthra SK (2000b) Evapotranspiration and crop coefficients of wheat and sorghum. J Irrig Drain Eng 126:215–222

UNCED (2002) Conservation and management of resources for development. United Nations conference on environment and development. Johannesburg Summit, 26th Aug-04th Sep, Agenda 21, Ch: 14-18

UNDESA (United Nations Department of Economic and Social Affairs, Population Division) (2009) World population prospects: the 2008 Revision, Highlights, Working Paper No. ESA/P/WP.210. New York, UN

Usman M, Liedl R, Shahid MA (2014) Managing irrigation water by yield and water productivity assessment of a Rice-Wheat system using remote sensing. J Irrig Drain Eng. https://doi.org/10.1061/(ASCE)IR.1943-4774.0000732

Usman M, Liedl R, Awan UK (2015a) Spatio-temporal estimation of consumptive water use for assessment of irrigation system performance and management of water resources in irrigated Indus Basin, Pakistan. J Hydrol. https://doi.org/10.1016/j.jhydrol.2015.03.031

Usman M, Liedl R, Kavousi A (2015b) Estimation of distributed seasonal net recharge by modern satellite data in irrigated agricultural regions of Pakistan. Environ Earth Sci. https://doi.org/10.1007/s12665-015-4139-7

Usman M, Liedl R, Shahid MA, Abbas A (2015c) Land use/land cover classification and its change detection using multi-temporal MODIS NDVI data. J Geogr Sci 25:1479–1506. https://doi.org/10.1007/s11442-015-1247-y

Van der Velde EJ, Kijne JW (1992) Salinity and irrigation operations in Punjab. Pakistan: are there management options? Paper at Workshop on IIMI-India Collaborative Research in Irrigation Management (Lahore, IIMI)

Wardlow BD, Egbert S, Kastens JH (2007) Analysis of time-series MODIS 250 m vegetation index data for crop classification in the U.S. Central Great Plains (2007). Drought Mitigation Center Faculty Publications. Paper 2. http://digitalcommons.unl.edu/droughtfacpub/2

Wetterhall FA, Bárdossy D, Chen SH, Xu CY (2006) Daily precipitation-downscaling techniques in three Chinese regions. Water Resour Res 42:W11423. https://doi.org/10.1029/2005WR004573

Wilby RL, Dawson CW (2013) The statistical downscaling model: Insights from one decade of application. Int J Climatol 33:1707–1719. https://doi.org/10.1002/joc.3544

Wilby RL, Harris I (2006) A framework for assessing uncertainties in climate change impacts: low-flow scenarios. Water Resour Res 42:W02419. https://doi.org/10.1029/2005WR004065

Wilby RL, Hay LE, Gutowski WJ, Arritt RW, Takle ES, Pan Z, Leavesley GH, Martyn PC (2000) Hydrological responses to dynamically and statistically downscaled climate model output. Geophys Res Lett 27(8):1199. https://doi.org/10.1029/1999GL006078

Wilson K, Goldstein A, Falge E, Aubinet M, Baldocchi D, Berbigier P (2002) Energy balance closure at FLUXNET sites. Agric For Meteorol 113:223–243

Xu CY (1999) Climate change and hydrologic models: a review of existing gaps and recent research developments. Water Resour Manag 13(5):369–382. https://doi.org/10.1023/a:1008190900459

Yang W, Bárdossy A, Caspary HJ (2010) Downscaling daily precipitation time series using a combined circulation- and regression- based approach. Theor Appl Climatol 102(3–4):439–454. https://doi.org/10.1007/s00704-010-0272-0

Yin L, Guangcheng H, Jinting H, Dongguang W, Jiaqiu D, Wang X, Li H (2011) Groundwater recharge estimation in the Ordos Plateau, China: Comparison of methods. Hydrogeol J 19 (8):1563–1575

Yuan F, Sawaya KE, Loeffelholz BC, Bauer ME (2005) Land cover classification and change analysis of the Twin Cities (Minnesota) metropolitan area by multi-temporal Landsat remote sensing. Remote Sens Environ 98:317–328

Chapter 10
Closed and Semi-closed Systems in Agriculture

Ebrahim Hadavi and Noushin Ghazijahani

Abstract Agriculture is the endeavor that served for thousands of years as a cradle for human civilizations. However, it may also act as a brake. Indeed, climate change is increasingly constraining the available pool of land and water resources. The current agriculture has become highly dependent on a flow of chemical input from pesticide and insecticide industry, itself remaining as a source of a problem both for human and global health. Protected agriculture is among the answers to the looming crisis in water and climate change.

We reviewed the current systems of protected agriculture as they migrate from 'open' to 'semi-closed' and 'fully-closed' systems of plant production from the sustainability point of view. Not a long time ago, the problems related to the soil-borne diseases acted as a trigger for a remarkable shift to use of simpler growth media. Now the plant factories and similar approaches are applying the state of the art techniques of fertilization and lighting to deliver a clean product; but still they remain dependent on a large flow of external inputs, as well as an intensive monitoring and management platform. On the other side are the emerging agro-ecological perspectives that are in support of a concept of disease management based on higher biodiversity in the plant culture media and the surrounding environment. Such systems try to address the pest and disease issue by use of a much more complex media, and a high biodiversity in the production environment that could reduce the management intensity and thus reduce the cost. As these systems are more self-sufficient, they have the ability to create more 'closed' systems of production. However, there is no a consensus in the scientific society in this regard, therefore a clear dilemma between a selection of simple or complex growing media exists, which determines the technologies to be developed and used. The same problem persists in the studies of bio-regenerative life support system where sophisticated designs for plant production units destined for potential use in outer space are

E. Hadavi (✉)
Department of Horticultural Science, Karaj Branch, Islamic Azad University, Karaj, Iran
e-mail: hadavi@kiau.ac.ir

N. Ghazijahani
Department of Microbiology, Karaj Branch, Islamic Azad University, Karaj, Iran

E. Lichtfouse (ed.), *Sustainable Agriculture Reviews 33*, Sustainable Agriculture Reviews 33, https://doi.org/10.1007/978-3-319-99076-7_10

attempted. This review tries to shed light on the future image of both approaches from the sustainability outlook based on the degree of inherent self-sufficiency or recyclability of inputs, which could be further interpreted as the level of 'closed-ness' of the systems.

Keywords Food security · Agroecology · Soil rehabilitation · Livestock · Innovation systems · Transitions

10.1 Introduction

Ignoring the small amounts of matter that enter or leave the earth, our planet is a system that is closed to matter, but open to energy. The vast quantities of radiant energy, mostly from the Sun, drive dynamic processes in the earth's atmosphere, hydrosphere, biosphere e.g. photosynthesis, and even lithosphere. Most of this energy is eventually emitted back into space as thermal radiation while a fraction is stored as chemical-bond energy in coal, oil, and natural gas (Salisbury et al. 1997). While in fact we are living in the biosphere of a gigantic closed terrarium, but our ordinary agriculture systems are considered 'open' because of inputs we have from the outside of our field ecosystems. The greenhouse production systems might be considered the most notable examples of semi-closed systems of agricultural production, even though still a large amount of natural or synthetic inputs are needed to sustain their production.

However, the importance of answer to the 'closed-ness' of a production system remains questioned itself; the technology and commercial sectors don't consider it an important feature while from the sustainability point of view and results of space experiments it appears to be crucial. As we realize that we are running out of some key plant nutrients in future, as well as the environmental consequences of our conventional high-input production systems, we find out that the input-intensive system of agricultural production is marching forward in a dead-end. Thus, we may seek ways to make our production/consumption cycles more closed and thus more efficient and environmentally friendly.

In the current review, we will cover the contemporary semi-closed and closed systems of agricultural production and later we discuss their sustainability issues and future use.

10.2 The Background

A definition is provided for closed and semi-closed plant production systems, which included the plant parts and organs as well as the plant culture systems (Goto et al. 2013). By confining that definition to plant production, we can define the 'Closed and semi-closed agricultural systems' as any type of environment (closed or semi-closed) in which plants are cultured and/or maintained in a restricted space, where free exchange of mass and/or energy between the system's interior and exterior are restricted. Examples are Controlled Ecological Life Support Systems, space farming, greenhouses, plastic tunnels and factory-style plant production systems.

10.2.1 Evolution and the Research Trend

The beginning of greenhouse growing structures dates back to more than 2000 years ago (Jiang et al. 2004a). The food production in greenhouses often termed 'controlled environment agriculture' (CEA); it was started by the growing of off-season fruit crops and later for vegetable production (Dalrymple 1973; Jensen 1997b). However, the glass structures remained somehow luxury until the introduction of polyethylene in 1948 (Jensen 1997a).

While initially, the greenhouses served as plant protector, nowadays, they can best be seen as plant or vegetable factories. The automation level is increasing rapidly using computers and once a system is totally enclosed and an artificial light source is used, the setup is called a 'phytotron' in an experimental scale or a 'plant factory' in a commercial scale (Jensen 1997a).

Several distinctive driving forces are propelling the research on the closed and semi-closed agricultural systems. The first is the most important upcoming challenge of food security, which is the ever-limiting nature of available water resources for agricultural production in areas with a large availability of land and sunlight. Elaboration of sophisticated closed and semi-closed production systems in such areas could enable us to make the best use of our available water resources (Kozai et al. 1997; van Kooten et al. 2006). The second is the year-round market demand for many horticultural products that turn the off-season production to a profitable endeavor (Castilla and Hernandez 2006). The use of a greenhouse, if the energy issues are considered (passive greenhouses) would be a reasonable way to increase global food production with lower inputs including water. The third is the increasing need for transplants as another driving force of extension of such systems. The last but maybe the most sophisticated research is aimed to the creation of closed agricultural production systems for use in outer space or even other planets (Blüm et al. 1994; MacElroy et al. 1989; Mitchell 1994; Tibbits and Alford 1982).

The consumption of fossil fuel and water and the emission of waste are reduced considerably in closed systems of agriculture production (Kozai and Fujiwara 2016). The studies on controlled ecological life support systems have acted as the driving engine of studies on closed and semi-closed agricultural systems since late years of the earlier century. While missions in which the robotic technology is used for probing outer space and planets don't need such arrangements; in a case considering long-term missions of humankind, then there would be a need for functioning and sustainable closed systems of crop production. In fact, possessing of such a system is the only way that could realize such missions in the future.

"However, to develop a closed plant production system as a sustainable plant production system, electricity consumption for lighting must also be decreased significantly, for example, with intelligent lighting. Such systems will require innovative methodology incorporating advanced global technologies under an appropriate clear vision, mission, goals, and concepts leading to sustainability" (Kozai and Fujiwara 2016). The advances in covering and lighting material could increase its popularity in the future (Kozai et al. 1997). If it could be kept running at a low input level, the system could be more environmentally friendly and could benefit many aspects of urban life as discussed earlier by Eigenbrod and Gruda (2015) as being complementary to rural agriculture.

10.2.2 The Question of Growing Media and Fertilization Strategy Selection

Choice of the growing media and fertilization technique remains as the key in determining the 'closed-ness' of the systems. In the field production, the growing process is not only naturally interrupted between seasons, but also a crop rotation is practiced. On the other side is the continuous plant growing, which is usually the case in greenhouse production that can lead to an excessive buildup of soil pathogens. Thus, a soil replacement becomes a required routine practice in soil based greenhouses. In addition, the soil still needs ample fertilization in greenhouse production systems (Jensen 1997a; Reddy 2016). Therefore, to address these issues, research turned to nutrient solution culture or soilless culture systems (Withrow and Withrow 1948). Soilless culture is defined by FAO as "any method of growing plants without the use of soil as a rooting medium, in which the inorganic nutrients absorbed by the roots are supplied via the irrigation water" (FAO 2013).

By banning of methyl bromide which was used as the only popular soil fumigant before the 1990s, the soilless methods were appreciated more than before (Ristaino and Thomas 1997). This caused for an increased interest in soilless culture methods in the current century, while the high capital requirement and inherent technical complications remain as burdens for the popularity of hydroponic systems (Jiang and Yu 2006). However, the shift toward soilless methods usually means more dependence on imported inputs and thus the system becomes more 'open' materially. To address this, the starting years of current century has been also characterized by a shift from 'open' to 'closed-cycle' cultivation systems by reuse of drainage solution. While on one hand, the cultivation of greenhouse crops in closed hydroponic systems can greatly reduce the pollution of water resources, on the other hand, closed cultivation systems have more inherent complications like an accumulation of salt ions in the recycled nutrient solution (FAO 2013). Furthermore, the increased risk of disease spread via the recycled leachate demands the installation of a solution disinfection system (Wohanka 2002). A much-sophisticated instrumentation like the systems needed for control of individual ions, is another constraint created by switching to closed systems of soilless culture (Van Os et al. 2008). The recycling of nutrient solution has been shown to reduce the use of potable water by 33% and considerable amounts of N, P and K are saved from being released to the environment (Grewal et al. 2011). Integration of aquaculture and hydroponic vegetable production systems as 'Aquaponics', has proved as an effective way for more closure of the production cycle, yielding both large reduction in water use and operation costs (Diver 2000; Enduta et al. 2011; Klinger and Naylor 2012; Li and Li 2009).

However, there is a documented history suggesting that soil-based systems have operated for virtually unlimited periods if they are replenished with compost or organic nutrients, on the contrary, there is no evidence for the indefinite operation of hydroponic systems (Nelson et al. 2008). On the other side, there is evidence that the chemical hydroponic is at risk of an end. In fact, phosphate rock is a finite resource that cannot be manufactured, therefore any measure that increases the pressure on earth phosphorus reserves will just reduce the time to the eruption of the loomed potential phosphate crisis (Cordell et al. 2009; Neset and Cordell 2012; Vaccari 2009; Vance 2001). The same question is valid for N as well. The other problem is that the maintenance of chemical hydroponics system in which the water and nutrient supply are supposed to be equal to the uptake by the crop, demands a water of very high quality that is of limited availability worldwide (Voogt and Sonneveld 1997). While the soil-borne diseases caused a remarkable shift to simpler media that offer an easier maintenance and management in technical terms, but these remain highly dependent on external inputs. By contrary, the more complex media remain simpler in management and production and if their disease management is addressed, they could find their popularity again due to the less reliance on external chemical inputs.

10.3 Variation in Semi-closed to Closed Production Systems

All agricultural production systems are essentially 'artificial ecosystems'. However, the 'protected' production systems that are used mainly for horticultural crops are considered as better descriptions for the artificial ecosystems. Though, they largely rely on material input and so are categorized as 'open'. However, it is a loose description and practically there is a large variation among protected production systems in term of the degree of reliance on the external inputs. Closed artificial ecosystems are closed ecosystems designed and controlled by humans ranging from agricultural systems, bio-regenerative life support systems to microcosms, and aquaria, which may be widely useful in practical or research applications (Sun et al. 2016). In practice, these artificial experimental ecosystems have different degrees of material closure and are of great value for theoretical and applied ecology development as simplified but representative models of natural biosystems (Pechurkin and Shirobokova 2001). While both agriculture ecosystems as well as plant factories are 'artificial ecosystems' in essence but 'closed artificial ecosystems' and 'bio-regenerative life support systems' are distinguished from them by the characteristic of 'closed-ness'. This closed-ness in the matter is usually partial and a total closedness is less common. However, some waste products may not be recycled easily and even an artificial biosphere needs energy input from the environment. So the basics of bio-regenerative life support systems research are more or less the same as the closed artificial ecosystems (Blüm et al. 1994). In current review, we made a categorization mainly based on the technology and growing media complexity of the production systems. First, we will cover the systems which are developed for conventional agriculture production, and later the systems that are used to study the production of food in outer space are covered.

10.3.1 Conventional Protected Agriculture

10.3.1.1 Plant Factory

The term "plant factory" first was used in the literature by Tsuruka et al., for a factory-like artificial lighting system (Tsuruoka et al. 1984). However, the concept of phytotrons and producing plants with artificial lighting was established years before (Hashimoto 1991). Later, many studies have been done on lighting, air conditioning, growth management, nutrient solution control, mechanization, and automation. Plant factories are promising systems for plant production in buildings, underground, and in outer space. They are samples of technology-intensive closed plant production systems in which an artificial lighting system is used for growing plants instead of the sunlight. The environmental conditions, such as light, air temperature, humidity, and CO2 concentration, can be controlled for optimum growing conditions, vegetable production in such systems is not affected by weather conditions (Yoshida et al. 2016). The technologies which have been developed for use in plant factories are expected to be introduced into plant production systems in controlled ecological life support systems (Goto 1997). However, the basic concept of plant factories is usually based on chemical hydroponic, which is far from a closed agricultural system, even though its concepts and technology may help to elaborate more refined closed systems.

10.3.1.2 The Eco–organic Soilless Culture System

This system of culture is a widespread practice in China. They invented the system to bypass the high capital investment and technical complications of classic soilless culture. Even though, the system is named 'soilless' but this naming may be controversial as the nutritive media consisting a mix of locally available substrates (coal cinder, peat moss, vermiculite, coir, sawdust, sand, rice husks, sunflower stems, maize stems, and mushroom waste) bears the nutrition task and fresh water is used to irrigate the trough via flow pipe. In addition, solid manure is used for fertilization (Jiang and Yu 2006). However, this culture system is a great way to cultivate crops with both a low cost and environmental impact (Jiang et al. 2004b). As the fertilizer input is limited, the system would be more closed when compared to true soilless systems.

10.3.1.3 Organoponic

An organoponic system, consists of a mix of soil and organic matter in raised beds that can be constructed on almost any plot of land, which is a common urban cultivation practice in Cuba (Lovell 2010). This system shares the basic idea of raised beds and use of organic fertilizers with the prevalent eco-organic soilless culture system of China, which is mentioned before. However, in this system, the use of soil in beds is practiced as the difference with the previous method.

Organic farming practices often lead to more closed agricultural systems. The system is characterized with high utilization of local resources and extremely limited external input aiming to improve the efficiency of resources (Muktamar et al. 2016). The concept and 'closedness' of this system makes it similar to that of the eco-organic soilless culture system.

10.3.1.4 Organic Hydroponic

While "organic hydroponics" appears to be an oxymoron at first sight (Atkin and Nichols 2003), but there are studies showing that it could be a reality in near future. "The coupling of plant growth and waste recycling systems is an important step toward the development of bio-regenerative life support systems"; This is what was aimed by Garland et al., whom after a series of studies on the recycling methods of organic residues as the source of nutrients for hydroponic production concluded that 'the direct use of organic fertilizer is deleterious to plant growth' (1997). As a solution, an anaerobic decomposition process was proposed and developed for inclusion in controlled ecological life support systems (Schwartzkopf et al. 1993). The feasibility for utilization of anaerobic degradative processes for mineral recycling and secondary food production from crop residues in such a system was investigated subsequently by Schwingel and Sager (1996). Later, Strayer et al. (1997) integrated the anaerobic bioreactor of Schwingel and Sager (1996) with two other aerobic components, one further converting the residues into an edible yeast biomass followed by another aerobic nitrification component that also functioned to remove biodegradable soluble organic compounds which are remained in the liquid output from the yeast production stage. They concluded the array to be a sustainable recycling system for crop residues in controlled ecological life support systems (Strayer et al. 1997). The relatively low efficiency of generating nitrate from organic nitrogen in the organic fertilizer (less than 30%) in the earlier studies is reported to be enhanced substantially to 97.6% by regulating the amounts of organic fertilizer and inoculum, with moderate aeration (Shinohara et al. 2011). Later works on the system revealed that in contrary to the conventional hydroponic system based on inorganic fertilizers, the cultivated plants have unique rhizosphere characteristics including the formation of a biofilm in the rhizosphere, and development of root hairs leading even to better disease control (Fujiwara et al. 2012). Another method is introduced in which the soil leachates remained as the main source of nutrients and the nitrogen is supplied by a bioreactor containing nitrogen-fixing bacteria (Farajollahzadeh et al. 2013). This method is considered as a variation of soil culture in which the plant and soil are connected by a cycling liquid system. They believed that, by use of organic soil as the source of nutrients, it could be considered within the set definitions of 'organic agriculture'. Such systems, which both utilize hydroponic methods and meet the organic standards are being further developed so as to allow the use of the certified "organic" label on hydroponically grown produce (Brentlinger 2005). In the 'organic hydroponic', complete recycling of water, nutrient and media are needed in order to become eligible to receive the "certified hydroponic" label (Brooke 2000). Interestingly, these are the same concepts which are followed in controlled ecological life support systems research, in which a more sophisticated closed system of culture is aimed.

10.3.2 Food Production Systems for Outer Space

In the 1950s, Folsome sealed aquatic ecosystems consisting of algae, brine shrimp, and other organisms in closed 1–5 l glass flasks. It was shown that these mini-communities, if contain sufficient metabolic diversity at the time of closure and are provided with an adequate energy flow, will remain viable for prolonged periods (Nelson et al. 1993). Controlled Ecological Life Support System is a variation of bio-regenerative life-support system which is dedicated to the production of food for astronauts based on the controlled growth of higher plants (Blüm et al. 1994; MacElroy et al. 1989). The concept of using biological systems for life support in space has been studied as early as the 1950s (Wheeler and Sager 2006). Later, an integrated regenerative life support systems concept was introduced in which the by-products of one system become a useful material for another (Jones 1975). However, the system was not considered to be closed because of the need for elimination or disposition of the waste products.

These systems are designed for the maximum potential of recycling and sustainability in mind. As mentioned before, these are developed to enable long-run space missions by mankind. Human exploration of the solar system will include missions lasting years at a time. Longer duration missions will require regenerable human life support systems with a maximized degree of self-sufficiency (Barta and Henninger 1994). Space agencies are considering long-term space journeys for upcoming decades. That is why they are pioneering the research on closed agricultural systems.

Materially closed micro-biospheres, including humans, were created by ecologists which included both sealed microcosms, as well as open, but boundary-defined, microcosms to study ecosystem processes. On the other hand, the experimental life-support systems were designed for use in spacecrafts and as prototypes for space habitations (Salisbury et al. 1997).

Early closed ecological system facilities developed for space application focused on providing human life support and included green algae. In former USSR a series of projects on controlled ecological life support systems were carried out starting from Bios-1, then Bios-2 and Bios-3. In Bios-1 which was constructed in 1965, the atmosphere for one human was regenerated in a sealed 12 m3 chamber connected through air ducts with an 18 L algal cultivator containing *Chlorella vulgaris* (Salisbury et al. 1997). Despite the fact that the 'continuous production algal systems' use growing space efficiently, produce oxygen, are rich in protein, may be used as food supplements, and are also efficient in processing metabolic wastes, but on the other side, the productive algal systems have been difficult to maintain for long periods, do not provide a balanced tasty diet and require intensive maintenance and harvesting. (Tibbits and Alford 1982). Even though, these research efforts enabled the first human to breath one day in a bio-regenerative life-support system that supplied all of his required air. However, later studies revealed that relying on only one companion species to support humans life is impossible apparently because Chlorella was not palatable as a human food (Nelson et al. 1993; Salisbury et al. 1997). That's why higher plants were a candidate for food production for large scale or long duration closed facilities (Tibbits and Alford 1982). In 1968, a chamber of higher plants was attached to Bios-1 sealed chambers and renamed to 'Bios-2'. Wheat and a set of vegetables (e.g., beetroots, carrots, cucumbers, and dill) were cultured in phytotron and air purification was provided by both higher plants (approximately 25%) and algae (approximately 75%). This three-component system demonstrated the feasibility of direct gas exchange between humans and higher plants (Salisbury et al. 1997). In the next stage, Bios-3 was a completely underground structure constructed of welded stainless steel plates to provide a hermetic seal (Salisbury et al. 1997). "The test facility had a volume of more than 300 cubic meters divided among a chamber for algae tanks, a hydroponic cropping area, and a human living area that included food processing, medical, and control rooms, kitchen and dining areas, and separate apartments for the three crew members" (Nelson et al. 1993). As the food production didn't sustain the crew of three, later a phytotron was substituted with the algal cultivators (Salisbury et al. 1997). wheat, chufa (sedge nuts), and vegetable crops including lettuce, potatoes, radishes, and beets were grown hydroponically to provide approximately half the nutrition for crews (Nelson et al. 1993; Salisbury et al. 1997). "Experiments lasted as long as 6 months and almost all air was regenerated, although catalytic burners were used to oxidize trace gas buildups. More than 90% of water was recycled, the main mechanism being purification by plant transpiration and then condensation to provide drinking and irrigation water. Human wastes, aside from a portion of the urine, were not processed inside the facility, but were exported; some food, including dried meat for needed protein, was imported. Overall, the health of the crew of Bios-3 was good, although some simplification of their intestinal microbiota occurred" (Nelson et al. 1993). These line of studies later came to effect in a space greenhouse, that realized the growth of the first "space" vegetables (radish and Chinese cabbage) onboard the orbital Mir complex since June 1990 which used up human waste (Ivanova et al. 1993). Apparently, while these simple semi-closed ecosystems could be kept in an affordable size, but the limited diversity didn't allow to reach a fully closed ecosystem that is needed for longer outer space journeys and possible settlement on other planets. That is possibly the reason encouraging the US scientists to start the Biosphere 2 facility, which covered 1.2 ha of the desert in Oracle, Arizona, which stands "in stark contrast to these relatively simple systems" as described by Nelson (Nelson et al. 1993). Seven so-called biomes (ocean, freshwater and saltwater marshes, tropical rain forest, savanna, desert, intensive agriculture, and human habitat) were included to mimic the biomes of Earth, or Biosphere 1. More than 3000 species of plants and animals lived inside Biosphere 2, in which eight "biospherians" were sealed for 2 years (Andre and Chagvardieff 1997; Salisbury et al. 1997). This arrangement was the first successful experience of a fully closed life system which maintained a high diversity of participants including human. This diversity mimicked that of the earth biosphere which keeps working by holding the equilibrium by means of interactions among the present ecosystems and food webs. Perhaps the most interesting observation was the unexpected decrease in oxygen concentration, much of which occurred as oxygen was used in respiration and in the decay of organic matter sealed in the structure. This decay used much oxygen and produced much carbon dioxide. Some of the carbon dioxide reacted with structural concrete inside the structure, and the result was a net loss of oxygen without an equivalent buildup of carbon dioxide (Andre and Chagvardieff 1997; Salisbury et al. 1997). Nelson reports on this first experiment in which human was part of the experiment; "On September 26, 1991 a crew of eight people passed through the airlock beginning the experimental habitation of Biosphere 2, a closed ecological system built in the Arizona desert north of Tucson. Two years later they emerged – somewhat thinner but against considerable odds in overall good health and with a viable life support system" (1995).

Tibbitts and Henninger conclude that to support a single human in space, an area of 25–40 m^2 is needed. This Area can provide significantly more food and clean water that is needed by one person (1997). Showing that some plants like wheat could produce mass effectively in irradiance level that is triple of that of sun together with higher planting densities (5X) and increased light periods, an average production of 60 g m^{-2} day^{-1} through a year has been realized which could reduce the area required to feed a person from 178 m^2 in real world to 12 m^2 (Andre and Chagvardieff 1997). These type of studies continue with the state of the art manner (Graham and Bamsey 2016), and the reader is referred to a recent comprehensive review by Wheeler (2017).

The concept of 'artificial ecosystems' even thought was developed for space projects, now it is more and more taken into account by the scientific community as a new tool to study basic and applied problems related to evolution and the functioning of natural ecosystems. In fact, it serves as a complementary to open field studies.

10.3.2.1 The Selection of Culture Media

The ultimate goal of controlled ecological life support systems is to attain the highest practical level of mass recycle and deliver self-sufficiency and safety. (Bubenheim and Wydeven 1994). In the existing controlled ecological life support system models, both hydroponics, as well as soil culture, are tested. While the hydroponics systems are easier to be automated, it is difficult to elaborate self-regenerating as there is a constant demand for fresh nutrients while in the case of the soil-based culture of higher plants, a complicated waste processing, and recycling system is needed (Blüm et al. 1994). Some have tried adding an aquatic compartment to the hydroponic system in which fish could live sustainably by use of algal products (Blüm 1992).

From a stability point of view, using the soil as culture medium is considered intermediate between intensive agroecosystems and soilless controlled ecological life support systems and so is more stable (Thiéry 1994). Based on a review of the literature, Haeuplik-Meusburger concludes that in outer space missions, many researchers favor a soil-based system because of its capability to use in situ resources on other planets and the possibilities of recycling waste products (2014). Trace gases and other compounds could build up in a tightly sealed environment; however, a multitude and diversity of microbes in a soil medium create a metabolic safety net which could metabolize most compounds of potential toxic effect. During the 3 years that Biosphere 2 functioned as a materially closed ecological system from 1991 to 1994, it was demonstrated that the soil bed reactor approach was quite effective at controlling trace gases such as methane, ethylene, ethane, and propane. Facilitation of waste recycling and the return of nutrients to the soil is considered the other benefit of soil as the medium (Nelson et al. 2008). By use of soil medium, in addition to the reduction in inputs and consumables such as nutrient solutions for hydroponics, Nelson states that "Composting of inedible crop wastes is a simple and effective method of building soil; this can be combined with worm-beds which might provide food for fish if aquaculture is part of the agriculture system, and starting and stopping compost operations offer a method of manipulating carbon dioxide generation. Soil-based agriculture also offers the easy integration of constructed wetlands as a low-energy method of treating and recycling human sewage, growing edible crops in the wetland and sending remaining nutrients in output water back into the irrigation supply. Thus many resources can be readily utilized and recycled without high-energy or complex technological requirements" (2008). Use of soil as the medium means an "increased utilization of low-energy, natural mechanisms which have successfully operated over geologic time frames in Earth's biosphere rather than energy-expensive, high technology protocols" as Nelson quotes Schwartzkopf (1990). The other benefits are "Increased buffering capacities and improved system stability" (Glenn and Frye 1990). For instance, we learned from the Biosphere 2 that "a soil ecosystem containing a vigorous consortium of microfauna and flora is less likely to succumb to rapid colonization by a pathogenic species, as the invasive species of microbiota are generally kept in check by competition and predation" (Andre and Chagvardieff 1997).

10.4 Conclusion

While the earth is materially a closed system, our ordinary agriculture systems are considered 'open' because of the inputs we have from the outside of our field ecosystems. However, there are other artificial agricultural ecosystems that are more 'closed' like the greenhouse production systems, plant factories and so on. The importance of answer to the 'closed-ness' of a production system could predict its sustainability. Thus, making our production/consumption cycles more 'closed', makes them more efficient, sustainable and environmentally friendly.

There is a dilemma for future studies: on the one hand is a technology-intensive soilless culture approach with chemical nutrient, and on the other hand is a technology-assisted approach with emphasis on organic materials. In the first approach, we try to control all the playing factors by use of hi-tech devices while in the second approach we try to find out a design which could reach to an optimum internal equilibrium or homeostasis in ecological and biological cycles. Up to now, however, the available inputs have been determining, in western countries with a lot of companies available for technological assistance to the grower a mutual relationship between the two has been in benefit of the both. On the other hand in other places like China, relying on foreign companies contributes to a huge financial burden both for startup and maintenance phases so the research on organic ways of nutrition has created an affordable type of soilless culture as eco-organic type soilless culture system. The other driving knowledge comes from a constructive two-way interaction between controlled ecological life support systems studies for space applications and the science of terrestrial closed and semi-closed agricultural systems. New knowledge and techniques from either side could find an application in the other.

On the other side, more self-sufficient production systems, which depend less on foreign input, fall in contrast with the current business model of agroforestry-related companies as they will continue to push for more input-dependent models of production to guaranty their income. Therefore, maybe we need to address this paradox by rethinking around the future model of a sustainable agri-business model which could support a migration toward sustainability in agricultural production systems. In fact, "The striking increase in the use of nitrogen (N) and phosphorus (P) fertilizers between 1960 and 2000 by intensive agricultural practices have led to degradation of air and water quality" (Vance 2001). The future conversion of our agriculture practices toward semi-closed and closed systems will help to avoid the looming crisis for our planet. In the other word, we may describe the sustainability of a given production system by the maximum of attainable closed-ness to matter that it can realize during the long-term run.

10.4.1 The Prospect

High input concepts like "vertical farming" will remain as a matter of controversy. While they try to give the cities a 'green' color but mostly remain as energy and fertilizer intensive approaches and thus open in nature. A similar critical approach may be applied to current organic agriculture that characterizes it as more driven by attitude or creed rather than the science. There is a need for a scientific solution that could convince both sides. The current review suggests that the closed-ness of agricultural production systems could create a common scientific platform which could yield to a third definition of sustainable agriculture or 'recyclable agriculture', which could satisfy the reasonable concerns of both parties. In fact, the knowledge of controlled ecological life support systems, which is collected with a fully practical point of view could be applied here as an agreed platform for prediction of sustainability in our earth ecosystems based on their anticipated self-sufficiency in outer space. This would let us keep more of our precious inputs for the generations to come.

If the human is going to colonize the outer space he/she should accumulate the necessary knowledge and approach, which could be used to realize the "space villages" concept. In fact, the same knowledge could pave the way for us if we plan to reside in our big shared spaceship, the planet earth, for a long future. Moving toward the development of agricultural systems which are more 'closed-compatible' and not essentially titled 'organic' or 'industrial' could provide us a better decision-making criteria for further studies and policies and related technology development that will yield us a sustainability not only in regional or national but also in a global level.

References

Andre M, Chagvardieff P (1997) CELSS research: interaction between space and terrestrial approaches in plant science. In: Plant production in closed ecosystems. Springer, Dordrecht, pp 245–261

Atkin K, Nichols M (2003) Organic hydroponics. South Pacific Soilless Culture Conference-SPSCC, vol 648, ISHS, Leuven, pp 121–127

Barta D, Henninger D (1994) Regenerative life support systems—why do we need them? Adv Space Res 14:403–410

Blüm V (1992) C.E.B.A.S., a closed equilibrated biological aquatic system as a possible precursor for a long-term life support system? Adv Space Res 12:193–204

Blüm V, Gitelson J, Horneck G, Kreuzberg K (1994) Opportunities and constraints of closed man-made ecological systems on the moon. Adv Space Res 14:271–280

Brentlinger D (2005) New trends in hydroponic crop production in the US. In: International conference and exhibition on soilless culture: ICESC, vol 2005742, pp 31–33

Brooke LL (2000) The organic-hydroponic debate. In: Best of growing edge: popular hydroponics and gardening for small commercial growers, vol 2, New Moon Publishing, Corvallis, p 204

Bubenheim D, Wydeven T (1994) Approaches to resource recovery in controlled ecological life support systems. Adv Space Res 14:113–123

Castilla N, Hernandez J (2006) Greenhouse technological packages for high-quality crop production. In: XXVII international horticultural congress-IHC2006: international symposium on advances in environmental control, automation, vol 761, pp 285–297
Cordell D, Drangert JO, White S (2009) The story of phosphorus: global food security and food for thought. Glob Environ Chang 19(2):292–305
Dalrymple D (1973) A global review of greenhouse food production. United States Department of Agriculture, Boletín
Diver S (2000) Aquaponics-integration of hydroponics with aquaculture. Attra, Fayetteville
Eigenbrod C, Gruda N (2015) Urban vegetable for food security in cities. A review. Agron Sustain Dev 35(2):483–498
Enduta A, Jusoh A, Ali N, Wan Nik W (2011) Nutrient removal from aquaculture wastewater by vegetable production in aquaponics recirculation system. Desalin Water Treat 32:422–430
FAO (2013) Good agricultural practices for greenhouse vegetable crops. Principles for mediterranean climate areas. FAO, Rome
Farajollahzadeh Z, Hadavi E, Khandan-Mirkohi A (2013) Introducing a potentially organic hydroponics system in production of pot gerbera flowers. Acta Hortic 1037:1075–1082
Fujiwara K, Aoyama C, Takano M, Shinohara M (2012) Suppression of Ralstonia solanacearum bacterial wilt disease by an organic hydroponic system. J Gen Plant Pathol 78:217–220
Garland J, Mackowiak C, Strayer R, Finger B (1997) Integration of waste processing and biomass production systems as part of the KSC breadboard project. Adv Space Res 20:1821–1826
Glenn E, Frye R (1990) Soil bed reactors as endogenous control systems of CELSS. In: Workshop on artificial ecological systems, pp 41–58
Goto E (1997) Environmental control for plant production in space CELSS. In: Plant production in closed ecosystems. Springer, Dordrecht, pp 279–296
Goto E, Kurata K, Hayashi M, Sase S (2013) Plant production in closed ecosystems. In: The international symposium on plant production in closed ecosystems held in Narita, Japan, August 26–29, 1996. Springer Science & Business Media, Dordrecht
Graham T, Bamsey M (2016) Editor's note for the topical issue 'agriculture in space'. Open Agric 1
Grewal HS, Maheshwari B, Parks SE (2011) Water and nutrient use efficiency of a low-cost hydroponic greenhouse for a cucumber crop: an Australian case study. Agric Water Manag 98:841–846
Haeuplik-Meusburger S, Paterson C, Schubert D, Zabel P (2014) Greenhouses and their humanizing synergies. Acta Astronaut 96:138–150
Hashimoto Y (1991) Computer integrated plant growth factory for agriculture and horticulture. In: Hashimoto Y, Day W (eds) Mathematical and control applications in agriculture and horticulture. Pergamon, Oxford, pp 105–110
Ivanova TN, Bercovich YA, Mashinskiy AL, Meleshko GI (1993) The first "space" vegetables have been grown in the "SVET" greenhouse using controlled environmental conditions. Acta Astronaut 29:639–644
Jensen MH (1997a) Food production in greenhouses. In: Goto E, Kurata K, Hayashi M, Sase S (eds) Plant production in closed ecosystems: the international symposium on plant production in closed ecosystems held in Narita, Japan, August 26–29, 1996. Springer, Dordrecht, pp 1–14
Jensen MH (1997b) Hydroponics worldwide. In: International symposium on growing media and hydroponics, vol 481, pp 719–730
Jiang W-J, Yu H (2006) Twenty years development of soilless culture in mainland China, In: XXVII international horticultural congress-IHC2006: global horticulture: diversity and harmony, an introduction to IHC2006, vol 759, pp 181–186
Jiang W, Qu D, Mu D, Wang L (2004a) Protected cultivation of horticultural crops in China. In: Horticultural reviews. Wiley, New York, pp 115–162
Jiang W, Qu D, Mu D, Wang L (2004b) Protected cultivation of horticultural crops in China. Hortic Rev 30:115–162

Jones WL (1975) Life-support systems for interplanetary spacecraft and space stations for long-term use. In: Calvin M, Gazenko OG (eds) Fondations of space biology and medicine. NASA/HAYKA, Washington/Moscow, pp 247–273
Klinger D, Naylor R (2012) Searching for solutions in aquaculture: charting a sustainable course. Annu Rev Environ Resour 37:247–276
Kozai T, Fujiwara K (2016) Moving toward self-learning closed plant production systems. In: LED lighting for urban agriculture. Springer, Singapore, pp 445–448
Kozai T, Kubota C, Kitaya Y (1997) Greenhouse technology for saving the earth in the 21st century. In: Plant production in closed ecosystems. Springer, Dordrecht, pp 139–152
Li W, Li Z (2009) In situ nutrient removal from aquaculture wastewater by aquatic vegetable Ipomoea aquatica on floating beds. Water Sci Technol 59:1937–1943
Lovell ST (2010) Multifunctional urban agriculture for sustainable land use planning in the United States. Sustainability 2:2499–2522
Macelroy RD, Tremor J, Bubenheim DL, Gale J (1989) The CELSS research program – a brief review of recent activities. In: Lunar base agriculture: soils for plant growth. American Society of Agronomy, Inc., Crop Science Society of America, Inc., and Soil Science Society of America, Inc., Madison, pp 165–172
Mitchell CA (1994) Bioregenerative life-support systems. Am J Clin Nutr 60:820S–824S
Muktamar Z, Setyowati N, Sudjatmiko S, Chozin M (2016) Selected macronutrient uptake by sweet corn under different rates of liquid organic fertilizer in closed agriculture system. Int J Adv Sci Eng Inf Technol 6:258–261
Nelson M, Dempster WF (1995) Living in space: results from biosphere 2's initial closure, an early testbed for closed ecological systems on Mars. Life Support Biosph Sci 2:81–102
Nelson M, Burgess TL, Alling A, Alvarez-Romo N, Dempster WF, Walford RL, Allen JP (1993) Using a closed ecological system to study earth's biosphere. Bioscience 43:225–236
Nelson M, Dempster W, Allen J (2008) Integration of lessons from recent research for "Earth to Mars" life support systems. Adv Space Res 41:675–683
Neset TSS, Cordell D (2012) Global phosphorus scarcity: identifying synergies for a sustainable future. J Sci Food Agric 92(1):2–6
Pechurkin N, Shirobokova I (2001) Closed artificial ecosystems as a means of ecosystem studies for earth and space needs. Adv Space Res 27:1497–1504
Reddy PP (2016) Greenhouse technology. In: Sustainable crop protection under protected cultivation, Springer Singapore, Singapore, pp 13–22
Ristaino JB, Thomas W (1997) Agriculture, methyl bromide, and the ozone hole: can we fill the gaps? Plant Dis 81:964–977
Salisbury FB, Gitelson JI, Lisovsky GM (1997) Bios-3: Siberian experiments in bioregenerative life support attempts to purify air and grow food for space exploration in a sealed environment began in 1972. Bioscience 47:575–585
Schwartzkopf SH, Cullingford HS (1990) Conceptual design for a lunar-base CELSS
Schwartzkopf SH, Stroup TL, Williams DW (1993) Anaerobically-processed waste as a nutrient source for higher plants in a controlled ecological life support system. SAE Tech Pap 102, 1464(1)
Schwingel W, Sager J (1996) Anaerobic degradation of inedible crop residues produced in a controlled ecological life support system. Adv Space Res 18:293–297
Shinohara M, Aoyama C, Fujiwara K, Watanabe A, Ohmori H, Uehara Y, Takano M (2011) Microbial mineralization of organic nitrogen into nitrate to allow the use of organic fertilizer in hydroponics. Soil Sci Plant Nutr 57:190–203
Strayer R, Finger B, Alazraki M (1997) Evaluation of an anaerobic digestion system for processing CELSS crop residues for resource recovery. Adv Space Res 20:2009–2015
Sun Y, Xie B, Wang M, Dong C, Du X, Fu Y, Liu H (2016) Microbial community structure and succession of airborne microbes in closed artificial ecosystem. Ecol Eng 88:165–176
Thiéry J (1994) Can technology make life-support systems more stable than isolated terrestrial ecosystems? Life Sci Res Space: 227

Tibbits T, Alford D (1982) Controlled ecological life support system: use of higher plants. NASA Scientific and Technical Information Branch, Washington, DC

Tibbitts T, Henninger D (1997) Food production in space: challenges and perspectives. In: Plant production in closed ecosystems. Springer, Dordrecht, pp 189–203

Tsuruoka H, Kanenko T, Takatsuji M (1984) Lighting design for a 3-dimensional plant factory. Trans Soc Instrum Control Eng 20:734–739

Vaccari DA (2009) Phosphorus: a looming crisis. Sci Am 300(6):54–59

Vance CP (2001) Symbiotic nitrogen fixation and phosphorus acquisition. Plant nutrition in a world of declining renewable resources. Plant Physiol 127(2):390–397

van Kooten O, Heuvelink E, Stanghellini C (2006) New developments in greenhouse technology can mitigate the water shortage problem of the 21st century. In: XXVII international horticultural congress-IHC2006: international symposium on sustainability through integrated and organic, vol 767, pp 45–52

Van Os E, Gieling TH, Lieth J (2008) Technical equipment in soilless production systems, soilless culture: theory and practice. Elsevier, Amsterdam, pp 157–207

Voogt W, Sonneveld C (1997) Nutrient management in closed growing systems for greenhouse production. In: Plant production in closed ecosystems. Springer, Dordrecht, pp 83–102

Wheeler RM (2017) Agriculture for space: people and places paving the way. Open Agric 2:14–32

Wheeler RM, Sager JC (2006) Crop production for advanced life support systems. Technical reports, 1

Withrow R, Withrow A (1948) Nutriculture. Purdue University Agricultural Experiment Station. Boletín 328

Wohanka W (2002) Nutrient solution disinfection. In: Savvas D, Passam HC (eds) Hydroponic production of vegetables and ornamentals. Embryo Publications, Athens, pp 345–372

Yoshida H, Mizuta D, Fukuda N, Hikosaka S, Goto E (2016) Effects of varying light quality from single-peak blue and red light-emitting diodes during nursery period on flowering, photosynthesis, growth, and fruit yield of everbearing strawberry. Plant Biotechnol 33(4):267–276

Chapter 11
Bioenergy and Sustainable Agriculture

Hossein Zahedi

Abstract Intensive use of non renewable fossil fuels hampers the development of the human society. Energy plays a central role in the global economy. Changes in energy costs have significant effects on economic growth. Bioenergy is an alternative energy source able to supply liquid transportation fuels. Sustainable development requires energy source stability and environmental maintenance. Biomass is the principal supporter of renewable energy, accounting for 59.2% of total renewable sources in 2015 in the European Union (EU). This chapter reviews renewable energy aspects focusing on bioenergy production, crops residues and algae to produce biofuel.

Keywords Biofuel · Algae · Crops residual · Sustainable development

11.1 Introduction

Progresses in the examination and utilization of energy sources over the past century have undeniably changed global lifestyle. Energy plays a significant role to the attainment of economic dominance and independence on the international scene and approach to adequate and sustainable energy is necessary for industrialization (Bailis 2011). Today, fossil fuel involving oil, natural gas, and coal supply 32.6%, 23.7% and 30% of the total world energy consuming (BP 2015) and are the main sources of the worldwide energy supply. Despite the extensive utilization and application of fossil resources, these are non-renewable and will not last eternally. Oil, natural gas and coal supplies will be consumed in about 45, 60, and 120 years, respectively (IEA 2013). Reliance on imported oil and shortage of fossil reserves can make political and economic dangers to stability. Most depend on gains from exports of fossil fuel for countries like Iran cause a monopoly and a one product economy.

Environmental issues and social costs of carbon are other great harms of fossil fuel (Mitchell 2002). Investment and attempts to utilize other energy sources were

H. Zahedi (✉)
Department of Agriculture, Eslamshahr Branch, Islamic Azad University, Eslamshahr, Iran

E. Lichtfouse (ed.), *Sustainable Agriculture Reviews 33*, Sustainable Agriculture Reviews 33, https://doi.org/10.1007/978-3-319-99076-7_11

made after the oil crisis in the 1970s (Oliveira et al. 2010). Global consideration of renewable energy sources has raised their percentage of consuming to about 19% of worldwide energy consumption in 2012. It is estimated that the renewable energy contribution of the European Union will raise 55–75% by 2050 (Scarlat et al. 2015a). Biomass is the principal supporter of renewable energy, accounting for 59.2% of sum up renewable sources in 2015 in the EU (Scarlat et al. 2015a). It is the fourth-largest energy source in back of oil, coal, and natural gas in the global (Onwudili 2014). The energy supplied annually for internationally produced biomass is about 8-times higher than the global total energy necessity (Akay et al. 2005) and it is appraised to supply about 10–20% by 2050 (Onwudili 2014). It is necessary to advance sustainable bioenergy production from biomass. First-generation biofuels can be generated from edible crude materials such as corn and soybeans, meaning that the production of like biofuel contests with the food stock. This has made second-generation biofuel acquired from non-edible sources such as agricultural residue of raising significance.

The annual worldwide production of agricultural biomass is about 11 billion tonnes (Scarlat et al. 2015b). The universal agricultural residue and waste from corn, barley, oats, rice, wheat, sorghum, and sugarcane can possibly produce 491 gallons of bioethanol each year (Kim and Dale 2004). North American countries, conducted by the US, are the chief biofuel manufacturers with a 44.1% portion of total production in 2014. The renewable fuels standard program in the US (EPA 2010), forecasts that about 44.5% of the 36 billion gallons of renewable fuel will be manufactured with cellulosic biofuels, that nearly 56.9% will be produced from agricultural residue by 2022. The staying shares belong to South and Central America, involving Brazil (28.7%), Europe and Eurasia, including the Netherlands (16.5%) and the Asia Pacific, including China (10.6%) (BP 2015).

The major technologies used to transform bioenergy stored in biomass are either biochemical or thermochemical. In biochemical alternation (e.g., fermentation and anaerobic digestion), enzymes and microorganisms are used to convert lignocellulosic matters into a particular biofuel. In thermochemical changing (e.g., pyrolysis and gasification), long chain biopolymers in the biomass are made hot in the absence of oxygen to produce a combination of short chain hydrocarbons, gaseous products such as hydrogen, methane, and solids (Basu 2013; Mohamed et al. 2014). Despite the fact that biochemical methods are very slower than thermochemical methods, they enable highly selective break down of biomass into products such as bioethanol. Moreover, biochemical methods are managed under mild situations and need less energy to produce biofuel (Chundawat et al. 2011). An influential possible restriction for biofuel development in global is the continuous discussion on criterion for sustainable production, energy stabilities and greenhouse gasses redeeming of biofuels, which could become protectionist obstacles distinguishing versus supplier countries. In addition, if carbon emissions from land use alternation are accounted for and high carbon compactness lands are cleared for biofuel production, most crop-derived biofuels will not qualify to be sold in Europe (Boerrigter 2006a).

The transportation sector accounts for 21% of common worldwide fossil fuel CO_2 emissions to the atmosphere, second only to emissions from power production. Universal economic growth supposed to average 3.2% per year to 2030, growth in energy and demand for transport is predicted to expand at an average annual rate of 2.1% over the same period. Transport sector contribution to the total anthropogenic greenhouse gas emissions is planned to rise to 23% in 2030 (Kheshgi et al. 2004). Bioenergy has been identified as an important component in many future's energy scenarios. Replacement of fossil fuels by biofuel seems to be an efficient strategy to meet not only the future world energy requires but also the necessity for diminishing carbon emissions from fossil fuels. Although there is an intensifying demand for fossil energy due to rising economic activities in the emerging markets, particularly China and India, rising oil prices have supported major consumers globally to sharply increase their use of "green" biofuel (Metzger and Huttermann 2008; IEA 2008).

11.2 Role of Land in Biofuel Source

In order to attain coming biofuel demands, forest lands can be either clarified to plant biofuel crops, or maintained as productive forests and harvested for sustainable wood-for-energy production. Leaving sustainability considerations separately, both alternatives are in a technical manner viable. Anyway, it is not yet obvious which one will be more beneficial in the emerging biofuels market. The oil crisis of the early 1970s initiated interest in the adoption of the land based agriculture-drawn from fuels known as biofuel in a bid to enhance the supply of fossils. Although, it was thought that the mass cultivation of these first generation biofuel resources such as sugarcane, corn, soybean, rapeseed, oil palm trees could solve both problems of edible oil and fuel at the same time, it became clear with time, that the expanding worldwide demand for fuel could not be met sustainably by these fuel sources (Schenk et al. 2008).

11.2.1 First Generation Biofuel

First-generation fuels hint to the biofuel a product of sugar, starch, vegetable oil, or animal fats applying routine technology (Boerrigter 2006b). These fluid biofuels comprise the approachable fuels such as pure plant oil from oil producing crops, biodiesel from esterification of pure plant oil or waste vegetable oils, bio-ethanol from sugar or starch crop maturation, and ethanol derived. The most well-common first generation, transport biofuel is recorded below.

11.2.1.1 Vegetable Oil

Vegetable oil can be used for either food or fuel. The possible to run engines on immediately run vegetable oils dates back to the nineteenth century, remarkably to try by the well-known German inventor, Rudolph Diesel directs to the fruitful development of his engine in 1895 (EPA 2002). In greatest cases, vegetable oil is used to produce biodiesel, which is good to most diesel motors when blended with unoriginal diesel fuel.

11.2.1.2 Biodiesel

Biodiesel states to a diversity of ester based fuels (fatty esters) normally defined as the mono alkyl esters prepared from several types of vegetable oils, such as soybean, canola or hemp oil, or occasionally from animal fats through a simple transesterification procedure. When oils are mixed with methanol and sodium hydroxide in the outcome biodiesel and glycerol are manufactured by the chemical response. One portion of glycerol is made for every ten portions biodiesel. Biodiesel can be effectively used in any diesel engine when it is synthesized with mineral diesel in an immaculate composition.

11.2.1.3 Bioalcohol

Ethanol, propanol and butanol are generally biologically formed alcohols that are created by the action of microcosm and enzymes through sugar stretches or cellulose. It is frequently requested that biodiesel offers a straight replacement for gasoline as it could be used openly in a gasoline engine Acetone–butanol–ethanol fermentation and testing changes form butane with the process presented in which butanol is the sole liquid product. It is probably that butanol is able to produce enough energy to be burnt "conventional" in the existing gasoline engines since it is less corrosive and water soluble than that of ethanol which could be dispersed through the existing system.

Ethanol Fuel

Ethanol is the most common biofuel global, chiefly in Brazil and USA, where it has been used in mixes with ethanol (gasohol) for nearly three decades. Ethanol fuels that are made by sugar fermentation taken from wheat, corn, sugar beet, sugarcane, molasses and any substance that alcohol beverages could be prepared from. The ethanol production procedures used are enzyme ingestion. The process needs noteworthy energy contribution for heat. Ethanol has a superior octane rating than petrol which can be composited with fossil fuel to any rate and consumed as a part of gasoline engines as a replacement for petrol (Girard and Fallot 2006). Nevertheless, this quality of the fuel could be misused if the engines density ratio is adjusted consequently. Thus that existing vehicle engine can run on mixtures of up to 15% bioethanol plus fuel gasoline. Ethanol oxygen content more leads to higher efficacy causing a clear ignition process at low temperature. Plastics or metals are well known suitability difficulties between ethanol and components of the engine. Car producers are lately inclining to produce more flexible fuel automobiles, which could run on any mixture of both bioethanol and petrol equal to 100%.

Methanol Fuel

Methanol might be produced from biomass as biomethanol but it is formed from natural gases. At the instant ethanol economy is a notable substitute to the hydrogen economy. If it is put in evaluation with today's hydrogen production from natural gas improbable hydrogen production and state-of-the-art clean solar thermal energy process.

11.2.2 Second Generation Biofuel

The second generation biofuel technologies have been established to overwhelm some important restrictions of the first generation biofuel, particularly their use as food. High agriculture inputs in the form of fertilizers are manufactured by the first generation to limit the greenhouse gas production which could be simply accomplished. They are neither cost competitive nor risky to the environment. A great deal of attention is felt to use tree biomass for the second generation. Also, it is sustainable source of supply whose producers are advanced to break down the plant material. Trees are said to comprise more carbohydrate and the raw material for biofuel than that of food crops. There is an inconceivable deal of interest in using tree biomass for second generation biofuel. Also to be an obvious wellspring of sustainable supply when techniques are produced for breaking down the plant substance economically and efficiently trees likewise comprise further carbohydrates and the raw matter for biofuel than food crops.

11.2.2.1 Cellulosic Ethanol

It is taken from nonfood crops or uneatable misuse products that have less impact on food such as switch gases, sawdust, rice hulls, paper pulp, and wood chips. Lignocelluloses are the "woody" structural substance of plants. This feedstock is plentiful and miscellaneous in some cases (similar to citrus peels or sawdust) which cause a considerable industry-specific disposal trouble. A complicated and additional pace is to make ethanol from cellulose and a technical problem to resolve. Domestic animals live on grass and consumes slow enzymatic digestive process to break into glucose. Lignocellulosic ethanol is made when sugar molecules are freed from cellulose by enzymes. Cellulose and lignin are complex carbohydrate molecules based on sugar, which are established in every plant. The sugar can be fermented to set up ethanol in a similar manner to the first generation bioethanol production. Lignocellulosic ethanol can reduce greenhouse gas emissions by approximately 90% when it is compared with fossil fuel (Demain et al. 2005).

11.3 Third Generation Biofuel

The rising demand for biofuels global increases the defy of originating from great areas of land for the production of feedstock. This is particularly the case where developing countries are seeking to utilize trade and investment occasions that may place significantly more pressure on finite land resources than if energy security was the only policy goal. In order to improve the difficulties often allied with land based biofuel feedstock, there have been calls for the acceptation of the third generation biofuel sources, which demand much less land and can be applied for diminishing CO_2 emissions into the ambiance. Specially the biofuel that is derived from Aquatic Microbial Oxygenic Photoautotroph (AMOPS), more usually indicated as cyanobacteria, algae and diatom (Dismukes et al. 2008) Has been promoted use more sustainable resource that could refer the worldwide fuel demands in the absence of influencing food provide in the developing countries. Of these, biofuel from algae shows to have greater forecasts being the only renewable energy source that could meet international require for fuel transport while addressing the carbon makeup and worldwide warming issues at the same time. This has created extraordinary interest in agriculture for the making of transportation biofuel.

11.3.1 Algae Biomass

As a possible resource for bioenergy industry. Algae is the fastest growing organism in the globe. Algae is very significant as a biomass source. Algae will sometime be competitive as a source for biofuel. Different species of algae may be better fitted for dissimilar types of fuel. Algae can be grown nearly anywhere, even on sewage or salt water, and do not need rich land or food crops, and processing needs less energy than the algae supply. Algae can be a substitution for oil-based fuels, one that is more useful and has no difficulties.

Algae are growing fast, and about 50% of their weight is oil. This lipid oil can be used as substitute oil for industry, mostly in the transportation sector (Cornell 2009). Lately, algae have become the latest possible source being aimed for biofuel production since they reveal numerous attractive features (Pandey 2017; Sialve et al. 2009). Studies demonstrate that they can make up to 60% of their biomass in the form of oil or carbohydrates, from which the biofuel and lots of other industrially significant products can be acquired. Most notably, algae require CO_2 to grow, which hints they can be used for biofixation and bioremediation. As it grows, the oil is reaped for fuel while the remaining green mass by-product can be utilized in fish and oyster farms. In fact, algae could produce up to 10,000 gallons per acre (about 94, 000 l per ha) of biofuel per year though corn would only do 60 gallons per acre (about 560 l per ha) yearly. The potential use of algae for CO_2 sequestration is depicted in Fig.11.1 (Benemann 2003). In evaluation with other renewable energy sources like wind, solar, geothermal, tidal energy, algae derived energy is more restricted and stable compared to land based biomass agriculture has the possible to produce larger amounts of biofuel with no productive land or good water use. In spite of the fact of all these, the most important hindrance militating versus the prevalent utilization of algae for biofuel production remnants its high cost of cultivation (Dismukes et al. 2008; Belay 2007).

The process of growing, harvesting and altering algae into fuel and other significant products in an economically competitive approach is still being perfected; the following rewards are often qualified to argue: (1) Algae can be grown nearly anywhere, even on sewage or saltwater which does not need fertile land or food crops. (2) Algae is very competent and can be made cost efficient with more attempt. (3) It is very energy and oil dense with a very high yield per acre and sequester CO_2 enduringly while growing. (4) It only needs sunlight and water, which are neither valued for farm use nor drinkable. (5) Algae only takes hours to reproduce, since they have high photon alteration efficiency. (6) Algae is very Eco friendly being non-poisonous, which do not have sulfur, and are extremely biodegradable.

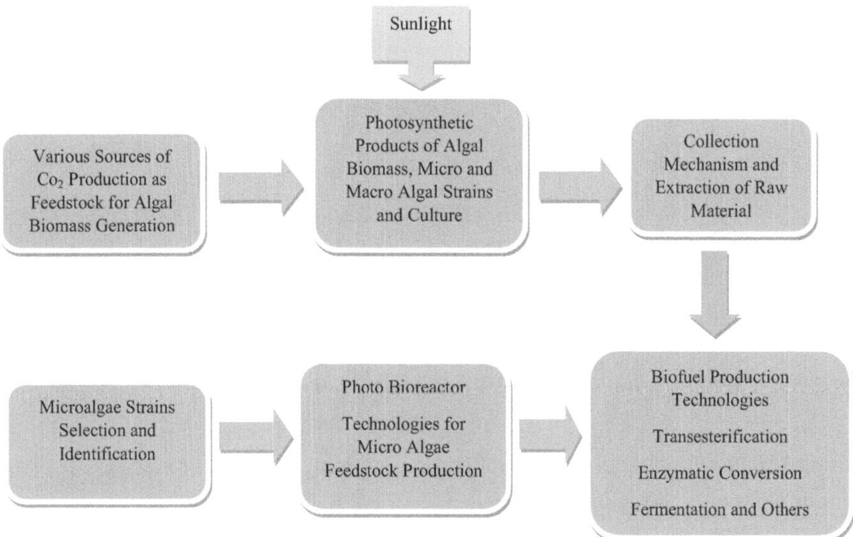

Fig. 11.1 CO_2 sequestration using algae

For more than three decades, researchers in the US Aquatic Species Program (ASP) studied the use of algae for the making of energy. Initially, the group focused its attention on the production of hydrogen, but later through the year 1982, their primary research modified to study the oil production (Benemann 2008). The 10 years attempt, included over 20 private companies and few government, research institutions, in similar efforts in order to expand closed photobioreactor technologies for the creation of high value products using power plant flue gas for CO_2. The program focused on the development of optical fiber photobioreactors that use concentrating mirrors to collect light that is installed into a director by method of light helps of different plans, despite the fact that other closed photobioreactors were also discovered. However, these researches and development attempts were not prolonged, partly due to the very critical economic projections for such advances, research beside similar lines continues currently somewhere else (Kremer et al. 2006; Nakamura and Senior 2005).

Latest commercial developments in microalgae biotechnology have been the mass growing of some original algal species, particularly *Haematococcus Pluvialis*, a source of the *carotenoid astaxanthin,* used in salmon aquaculture and also in food complements. Although all large-scale algal production systems utilize open ponds, a number of small-scale commercial production systems using closed photobioreactors have been started. A diagram of the microalgal biodiesel value chain is available as shown in Fig. 11.2; it begins with determination of the most proper species, depending upon local ecological conditions, and on the configuration proposed for cultivation; then it experiences gathering of biomass and extraction of oil accordingly and finishes with the biodiesel processing unit. Then, repercussions

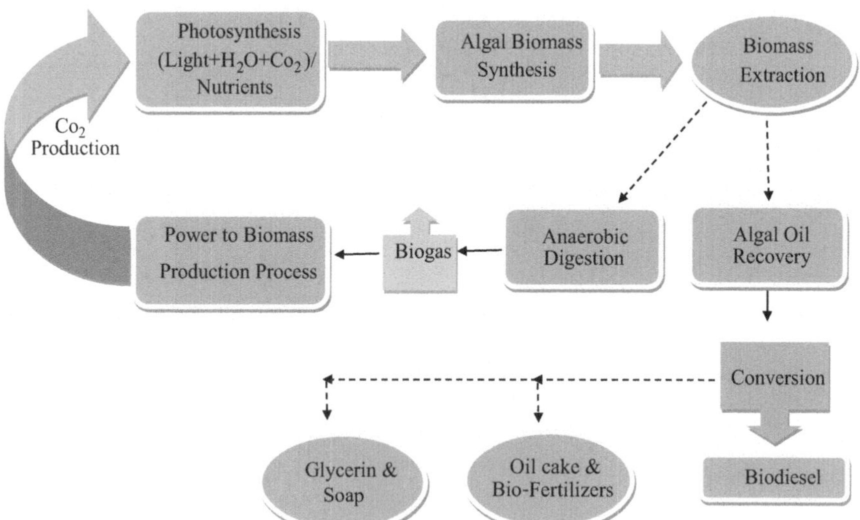

Fig. 11.2 Schematic presentation of various steps of algal biodiesel

created may be reused or utilized for energy production related with each of these steps as displayed and discoursed in the following subsections (Delucchi 1997).

11.4 Agricultural Residues as a Source of Bioenergy

Agricultural residue is mainly composed of cellulose and hemicellulose carbohydrates and lignin as the glue that binds plant fibers. Biochemical alteration of lignified tissue needs appropriate pretreatment and hydrolysis. Pretreating lignocellulosic substances interrupts the lignin and diminishes cellulose crystallinity. Through hydrolysis, the pretreated remains is transformed into sugar, including hexose and pentose. Microorganisms are capable to utilize the sugar to produce biofuel (Limayem and Ricke 2012). Crop waste and residue are cheap and plentiful and, more outstandingly, are not considered to be part of the human food supply. Fewer life cycle greenhouse gas (GHG) emissions are released from second-generation biofuel (Sims et al. 2010). A diversity of bioprocessing techniques, pretreatment methods, and microbial strain variations has been developed to commercialize progressed biofuel biorefineries through the past decade.

Corn ethanol production has raised gradually up to 57 gallons and will stay constant up to 2022, whereas cellulosic ethanol manufacture will increase to about 60.6 gallons with at least lessen by 60% in GHG emissions relative to the 2005 fuel baseline (EPA 2010). Data presented by the IEA Bioenergy Task 39 (2015), shows that 87 projects have been planned to produce second-generation biofuel during the biochemical pathway, with 40 biorefineries placed in North America. Corn residue gives the most share (51.28%) of internationally-distributed corn, wheat, rice, barley, and sugarcane residue for usage as the feedstock of these plants. A number of research efforts have surveyed cellulosic bioenergy feedstocks such as switchgrass, miscanthus, energy cane, energy sorghum, willow, hybrid poplar, forest residues, and agricultural rests, and the alteration technologies that can use these feedstocks (Kumar and Gayen 2011; Suntana et al. 2009).

Of these feedstocks, the resource with the maximum close term potential (1–5 years) for attaining national targets is agricultural residues (U.S. Department of Energy 2011). Recognizing a sustainable and trustworthy agricultural rest resource base has been a significant challenge for the developing cellulosic biofuels industry (Wilhelm et al. 2010). Agricultural residue elimination must be managed cautiously to be sustainable, and spatial and temporal changeability (soil, climate, and managing practices) influence the dependability of the supply. Residues play a number of serious roles in an agronomic system comprising direct and indirect effects on physical, chemical, and biological processes within the soil (Wilhelm et al. 2007, 2010). Unnecessary residue elimination can degrade the long period productive capacity of soil resources (Sheehan et al. 2003; Mann et al. 2002). Wilhelm et al. (2010) performed an extensive review of sustainability pointers for agricultural remains removal. The result of this study was the documentation of six environmental factors that potentially bound agricultural residue removal soil erosion from wind and water; soil organic carbon; plant nutrient balances; soil, water, and temperature dynamics; soil compaction; and off-site environmental impacts.

11.5 Sustainable Agricultural Crops and Residues

In this section, major agricultural crops for production of second-generation biofuels are resolved based on the accessibility and production rate.

11.5.1 Wheat

The use of cereal residue for making of biofuel has been enhanced both academically and industrially. Cereal establishes nearly 21% of the world's total agricultural biomass (Scarlat et al. 2015b). Each year, the large quantity of wheat straw made on farms is a potential sustainable lignocellulosic biomass for biofuel production (Novy et al. 2015). In the EU, approximately 18.8% of the total feedstock used for bioethanol production is provided by wheat and the European Council has resolved to rise residue utilization instead of the grain exploitation by 2020 (Scarlat et al. 2015b). On average, 1.3 kg of straw is produced per kg of wheat grain (Schnitzer et al. 2013).

11.5.2 Barley

Barley causes about 2.2% of worldwide manufacture and 14.7% of Asian production (FAOSTAT 2013). About 0.4 million tonnes of barley are adopted to produce bioethanol in the EU, involving 1.8% of the total feedstock (Scarlat et al. 2015b).

11.5.3 Rice

Its remains are suitable potential crude substance for second-generation biofuel production. Almost 90.6% of the glob's rice is produced in Asia (FAOSTAT 2013); therefore, efforts have been prepared to generate energy in the form of heat, electricity, and biofuel from rice straw in Asian countries (Ranjan et al. 2013).

11.5.4 Corn

Another sustainable nominee for second-generation biofuels manufacture is corn residue. Corn is the primary crop produced in the US (35.37 Mt in 2013) (FAOSTAT 2013) and US first- and Second-generation biofuel biorefineries are chiefly located in the Midwest, particularly in Iowa, where corn is widely grown (Office of Energy Efficiency and Renewable Energy 2013; RFF 2015). Investigators have assessed the sustainability and environmental and economic features of corn residue as biorefinery feedstock (Jin et al. 2014; Khanal et al. 2014). The removal of corn Stover generally reduces soil greenhouse gas emissions (Jin et al. 2014). Assumptions about a plant placed in Greece show better environmental presentation for ethanol acquired from corn stover and a better economic performance for ethanol manufactured from cotton stalks. The lower percentage of biomass loss through carrying and storage makes corn stove a most promising feedstock (Petrou and Pappis 2014). Corn supplies 19.8%of feedstock used for first-generation bioethanol production in the EU (Scarlat et al. 2015b).

11.5.5 Potato

Potato is widely cultivated and is the fourth-largest crop after corn, rice, and wheat internationally (Liang and McDonald 2014). Separately from residue produced on farms, a significant amount of waste is produced in food industries that process potatoes. Depending on the peeling process, 15–40% of a potato could be wasted. In a usual potato processing plant, 6–10% of potato peel waste would outcome from peeling. It is improved with carbohydrates, comprising starch (25%), cellulose and hemicellulose (30%) and fermentable sugars. It's beside contains protein (18%) and lipids (1%) (Arapoglou et al. 2010; Liang and McDonald 2014).

11.5.6 Alfalfa

Alfalfa has great protein content and carbohydrates are contained in the leaves and stems. Alfalfa leaves is a source of protein for animal feed, but the low digestibility of the stems, which constitute about 50% of the plant means that the carbohydrates in the stems could be changed to biofuel (González-García et al. 2010; Zhou and Runge 2015). Alfalfa is an agronomically useful and effective plant. It is a perennial and has nitrogen fixing features; therefore, it can fix atmospheric nitrogen, which can reduce the need for nitrogen fertilizer. Soil erosion prevention and water retention are the other remarkable profits of alfalfa (Sarkar 2009; Small 2011).

11.5.7 Sugarcane

Industrial crops such as soybeans and sugarcane are commonly utilized as raw substances for biofuel production, particularly, bioethanol and biodiesel. Sugar cane is a perennial crop that provisions as regards 75% of the globe's sugar (Swapna and Srivastava 2012). It's industrial remains, and waste (bagasse and molasses) is a sustainable resource for Bioenergy production. Brazil positions first in sugarcane production with regarding 40% of international production (FAOSTAT 2013). In Brazil, the marketable production of sugarcane bioethanol is grown-up. Burgess is typically engaged in boilers to generate electricity; still, strong attempts have been made to make bioethanol from bagasse. A techno-economic study reports that, with no subsidies, the production of bioethanol from sugarcane bagasse and lives in Brazil can struggle with first-generation bioethanol production from starch in Europe (Macrelli et al. 2012).

11.5.8 Sugar Beet

Sugar cane, sugar beets are the key providers of the globe's sugar (Swapna and Srivastava 2012). In the EU, sugar beets are the most important crop for bioethanol production, giving 57.9% of the total feedstock (Scarlat et al. 2015b). Notable residual resources, involving molasses and pulp, outcome from sugar beet processing that can be transformed into biofuel (Tukacs-Hájos et al. 2014). Sugar beet pulp includes 20–25% cellulose, 25–36% hemicellulose, 20–25% pectin, 10–15% protein, and 1–2% lignin. The pectin contents of sugar beet pulp is much higher than lignocellulosic biomass such as straw and can biochemically is changed into biofuel by bacteria (Zheng et al. 2013).

11.5.9 Oilseeds

At current, these resources are usually transformed into biodiesel. Soybeans are basically produced in the US and Argentina and theses countries exploits soybeans as a main feedstock for biodiesel production (Rincón et al. 2014). About 77% of the total feedstock applied for biodiesel is supplied by soybeans in Brazil (Castanheira et al. 2015). In the EU, biodiesel is the main transport biofuel and 65.9% of oleo chemical material engaged to produce biodiesel is supplied by rapeseed oil (Scarlat et al. 2015b). Non-edible plant-based resources such as jatropha and microalgae can effectively be used to produce biodiesel, but jatropha is rain reliant and requires a high quantity of water for optimal growth, which is not well-matched with the arid to semi-arid climate (Tabatabaei et al. 2011). Microbial oil formed by oleaginous

Table 11.1 Comparison of crop dependent biodiesel production efficiencies from plant oils

Plant source	Bio-diesel L/ha/year	Area to produce global oil demand (106 ha)	Area required as % of global land mass	Area as % of arable land mass
Cotton	325	15,002	100.7	756.9
Soybean	446	10,932	73.4	551.6
Mustard	572	8524	57.2	430.1
Seed	952	5121	34.4	258.4
Sunflower	1190	4097	27.5	206.7
Rapeseed	1892	2577	17.3	130
Jatropha	5950	819	5.5	41.3
Oil palm	12,000	406	2.7	20.5
Algae (50 gm-2 day-1 at 50% TAG)	98,500	49	0.3	2.5

microorganisms can be used for biodiesel production. Agricultural residue and waste is a low-cost substrate for these microorganisms, although this new technology needs more study for commercialization (Leiva-Candia et al. 2014). In addition biodiesel, oil crop biomass can provide as a substitute feedstock for biofuel, like bioethanol and biogas. The high carbohydrate content (60%) of rapeseed straw builds it a sustainable and possible biomass for biofuel production (López-Linares et al. 2014). At an average annual growth rate of 42%, the global biodiesel market is estimated to reach about 168 billion liters by 2016 (Gouveia and Oliveira 2008). In order to meet the rapid expansion in biodiesel production capacity observed not only in developed countries but also in developing countries such as China, Brazil, Argentina, Indonesia and Malaysia, other oil sources especially non-edible oils need to be explored (Li et al. 2008). A comparison of the oil yield of various crops with algae (Table 11.1) shows that microalgae seem to be the only source of renewable biodiesel that has the potential to completely displace petroleum-derived transport fuels without the controversial "food for fuel" conflicts (Benemann and Oswald 1996).

11.5.10 Horticultural Crops

Horticultural crops deposit and waste have lately obtained worldwide interest for the production of biofuel. Pruning residue involving the leaves and branches is regular biomass left on the ground (Buratti et al. 2015; García Martín et al. 2013). Waste and residue in the field, an important amount of agro-industrial waste is attained in the path of fruit processing.

11.5.10.1 Apple

Solid apple waste includes 20–30% of the weight and is produced as pomace in apple processing (Dhillon et al. 2013); regarding 13% of the apple fruit is the peel. Food fiber accounts for almost 65% of the apple pomace and peels and carbohydrates are the major ingredient of these fibers (Rabetafika et al. 2014).

11.5.10.2 Grape

Grape stalks and pomace accumulate in large amounts throughout harvesting and industrial processing of grapes. A total of 6–10% of the total weight of the processed grapes in the food industry is wasted (El Boushy and Poel 2000). Carbohydrates contained in the grape skin and stalks have potential for biofuel production (Fabbri et al. 2015; Mendes et al. 2013; Ping et al. 2011).

11.5.10.3 Date

Each year, significant waste is produced by date trees which can supply as lignocellulosic feedstock for biofuel production (Chandrasekaran and Bahkali 2013).

11.6 Conclusion

Biofuels as the primary products of the biochemical conversion of biomass can play a key role in world's energy outlook. Practical exploitation of bioenergy and bioproducts from agricultural residue and construction of biorefineries in world requires a more comprehensive economic evaluation, energy management, and policies for biofuel producers. The preliminary information covered by this research can increase awareness of fossil fuel limitations, potential crops to produce biomass-based biofuel, their applications, and their environmental impact. The limitations of the first and second generation biofuel resources show clearly that they are grossly inadequate to meet global demands for transport fuels in a sustainable way. Although, the use of microalgae for production of third generation biofuel has been studied for many years now, the fact remains that R&D activities still need to be undertaken to reduce the production cost of algal biomass to an acceptable level that could compete favorably with biomass from higher plants before commercial algae for energy cultivation can commence. Algae production technologies are quite mature but presently only its application for bio-fixation, especially wastewater treatment is economically feasible. Biofuel production from algae will become competitive in the medium term if considered along with production of higher value co-products such as bio-fertilizers, biopolymers.

References

Akay G, Dogru M, Calkan OF, Calkan B (2005) Biomass processing in biofuel applications. In: Lens P (ed) Biofuels for fuel cells: renewable energy from biomass fermentation. IWA, London, p 51

Arapoglou D, Varzakas T, Vlyssides A, Israilides C (2010) Ethanol production from potato peel waste (PPW). Waste Manag 30(10):1898–1902. https://doi.org/10.1016/j.wasman.2010.04.017

Bailis R (2011) Energy and poverty: perspective of poor countries. In: Galarraga I, Gonzlez-Eguino M, Markandya A (eds) Handbook of sustainable energy. Edward Elgar Publishing, Incorporated, pp 505–537. https://doi.org/10.4337/9780857936387.00035

Basu P (2013) Introduction. In: Basu P (ed) Biomass gasification, pyrolysis and torrefaction, 2nd edn. Academic, Boston, pp 1–27. https://doi.org/10.1016/c2011-0-07564-6

Belay A (2007) Spirulina (*Arthrospira*) production and quality assurance. In: Gershwin M, Belay A (eds) Spirulina in human nutrition and health. CRC Press, Boca Raton. https://doi.org/10.1201/9781420052572

Benemann JR (2003) Biofixation of CO2 and greenhouse gas abatement with microalgae-technology road map. Final Report submitted to the US Department of Energy, National Energy Technology Laboratory

Benemann JR (2008) Opportunities and challenges in algae biofuels production. Algae World Singapore, November 17–18

Benemann JR, Oswald WJ (1996) Systems and economic analysis of microalgae ponds for conversion of CO_2 to biomass, final report. United States Department of Energy, Washington, DC. https://doi.org/10.2172/493389

Boerrigter H (2006a) Economy of biomass-to-liquids (BTL) plants: an engineering assessment. ECN Biomass, Coal & Environmental Research, Petten

Boerrigter H (2006b) Economy of biomass-to-liquids (BTL) plants: an engineering assessment. Internal Report, Energy Commission of The Netherlands

BP (2015) BP Statistical Review of World Energy 2015. BP p.l.c. http://www.bp.com/content/dam/bp/pdf/Energy-economics/statistical-review-2015/bpstatistical-review-of-world-energy-2015-full-report.pdf, https://doi.org/10.1111/oets.00048

Buratti C, Barbanera M, Lascaro E (2015) Ethanol production from vineyard pruning residues with steam explosion pretreatment. Environ Prog Sustain Energy 34(3):802–809. https://doi.org/10.1002/ep.12043

Castanheira ÉG, Grisoli R, Coelho S, Anderi da Silva G, Freire F (2015) Life-cycle assessment of soybean-based biodiesel in Europe: comparing grain, oil and biodiesel import from Brazil. J Clean Prod 102:188–201. https://doi.org/10.1016/j.jclepro.2015.04.036

Chandrasekaran M, Bahkali AH (2013) Valorization of date palm (*Phoenix dactylifera*) fruit processing by-products and wastes using bioprocess technology – review. Saudi J Biol Sci 20 (2):105–120. https://doi.org/10.1016/j.sjbs.2012.12.004

Chundawat SPS, Beckham GT, Himmel ME, Dale BE (2011) Deconstruction of lignocellulosic biomass to fuels and chemicals. Annu Rev Chem Biomol Eng 2:121–145. https://doi.org/10.1146/annurev-chembioeng-061010-114205

Cornell CB (2009) First algae biodiesel plant goes online. Retrieved on 27th April, from http://gas2.org/2008/03/29/first-algae-biodiesel-plant-goesonline

Delucchi MA (1997) Emissions of non-CO2 greenhouse gases from the production and use of transportation fuels and electricity. Institute of Transportation Studies paper number UCD-ITS-RR-97-05

Demain A, Newcomb M, Wu D (2005) Cellulase, clostridia, and ethanol. Microbiol Mol Biol Rev 69:124–154. https://doi.org/10.1128/mmbr.69.1.124-154.2005

Dhillon GS, Kaur S, Brar SK (2013) Perspective of apple processing wastes as low-cost substrates for bioproduction of high value products: a review. Renew Sustain Energy Rev 27:789–805 10.1016/j.rser.2013.06.046

Dismukes GC, Carrieri D, Bennette N, Ananyev GM, Posewitz MC (2008) Aquatic prototroph: efficient alternatives to land-based crops for biofuels. Curr Opin Biotechnol 19:235–240. https://doi.org/10.1016/j.copbio.2008.05.007

El Boushy ARY, Poel AFBVD (2000) Fruit, vegetable and brewers' waste. In: Boushy ARYE, Poel AFBVD (eds) Handbook of poultry feed from waste. Springer, New York, pp 173–311. https://doi.org/10.1007/978-94-017-1750-2_6

EPA (2002) Clean alternative fuels: biodiesel. United States Environmental protection Agency; EPA 420-F-00-032. http://www.epa.gov

EPA (2010) Renewable fuel standard program (RFS2) regulatory impact analysis. U.S. Environmental Protection Agency, Ann Arbor http://www3.epa.gov/otaq/renewablefuels/420r10006.pdf

Fabbri A, Bonifazi G, Serranti S (2015) Micro-scale energy valorization of grape marcs in winery production plants. Waste Manag 36:156–165. https://doi.org/10.1016/j.wasman.2014.11.022

FAOSTAT (2013) Food and Agriculture Organization of the United Nations, Statistics Division

García Martín JF, Sánchez S, Cuevas M (2013) Evaluation of the effect of the dilute acid hydrolysis on sugars release from olive prunings. Renew Energy 51:382–387. https://doi.org/10.1016/j.renene.2012.10.002

Girard P, Fallot A (2006) Review of existing and emerging technologies for the production of biofuels in developing countries. Energy Sustain Dev 10(2):92–108. https://doi.org/10.1016/s0973-0826(08)60535-9

González-García S, Moreira MT, Feijoo G (2010) Environmental performance of lignocellulosic bioethanol production from Alfalfa stems. Biofuels Bioprod Biorefin 4(2):118–131. https://doi.org/10.1002/bbb.204

Gouveia L, Oliveira AC (2008) Microalgae as a raw material for biofuels production. J Indust Microbiol Biotechnol 36(2):269–274. https://doi.org/10.1007/s10295-008-0495-6

IEA (2008) Energy technology perspectives: scenarios and strategies to 2050. OECD/IEA, Paris. https://doi.org/10.1787/9789264041431-en

IEA (2013) Resources to reserves 2013. International Energy Agency, Paris http://www.iea.org/publications/freepublications/publication/Resources2013.pdf

IEA Bioenergy Task 39 (2015) Commercializing liquid biofuels from biomass, a database on facilities for the production of advanced liquid and gaseous biofuels for transport. J Microbiol Exp 4(4). https://doi.org/10.15406/jmen.2017.04.00117, http://demoplants.bioenergy2020.eu

Jin V, Baker J, Johnson JF, Karlen D, Lehman RM, Osborne S, Sauer T, Stott D, Varvel G, Venterea R, Schmer M, Wienhold B (2014) Soil greenhouse gas emissions in response to corn stover removal and tillage management across the US corn belt. Bioenergy Res 7(2):517–527. https://doi.org/10.1007/s12155-014-9421-0

Khanal S, Anex RP, Gelder BK, Wolter C (2014) Nitrogen balance in Iowa and the implications of corn-stover harvesting. Agric Ecosyst Environ 183:21–30. https://doi.org/10.1016/j.agee.2013.10.013

Kheshgi H, Akhurst M, Christensen D, Cox R, Greco R, Kempsell S, Kramer G, Organ R, Stileman T (2004) Transportation and climate change: opportunities, challenges and long term strategies. In: Summary brochure of the international petroleum industry, environment conservation association, Baltimore, USA, 12–13 October

Kim S, Dale BE (2004) Global potential bioethanol production from wasted crops and crop residues. Biomass Bioenergy 26:361–375. https://doi.org/10.1016/j.biombioe.2003.08.002

Kremer G, Bayless DJ, Vis M, Prudich M, Cooksey K, Muhs J (2006) Enhanced practical photosynthetic CO2 mitigation. Office of Scientific and Technical Information (OSTI). https://doi.org/10.2172/888741

Kumar M, Gayen K (2011) Developments in biobutanol production: new insights. Appl Energy 88 (6):1999–2012. https://doi.org/10.1016/j.apenergy.2010.12.055

Leiva-Candia DE, Pinzi S, Redel-Macías MD, Koutinas A, Webb C, Dorado MP (2014) The potential for agro-industrial waste utilization using oleaginous yeast for the production of biodiesel. Fuel 123:33–42. https://doi.org/10.1016/j.fuel.2014.01.054

Li Q, Du W, Liu D (2008) Perspectives of microbial oils for biodiesel production. Appl Microbiol Biotechnol 80(5):749–756. https://doi.org/10.1007/s00253-008-1625-9

Liang S, McDonald AG (2014) Chemical and thermal characterization of potato peel waste and its fermentation residue as potential resources for biofuel and bioproducts production. J Agric Food Chem 62(33):8421–8429. https://doi.org/10.1021/jf5019406

Limayem A, Ricke SC (2012) Lignocellulosic biomass for bioethanol production: current perspectives, potential issues and future prospects. Prog Energy Combust Sci 38(4):449–467. https://doi.org/10.1016/j.pecs.2012.03.002

López-Linares JC, Romero I, Cara C, Ruiz E, Moya M, Castro E (2014) Bioethanol production from rapeseed straw at high solids loading with different process configurations. Fuel 122:112–118. https://doi.org/10.1016/j.fuel.2014.01.024

Macrelli S, Mogensen J, Zacchi G (2012) Techno-economic evaluation of 2nd generation bioethanol production from sugar cane bagasse and leaves integrated with the sugar-based ethanol process. Biotechnol Biofuels 5(1):22. https://doi.org/10.1186/1754-6834-5-22

Mann L, Tolbert V, Cushman J (2002) Potential environmental effects of corn (*Zea mays* L.) stover removal with emphasis on soil organic matter and erosion. Agric Ecosyst Environ 89 (3):149–166. https://doi.org/10.1016/s0167-8809(01)00166-9

Mendes JAS, Xavier AMRB, Evtuguin DV, Lopes LPC (2013) Integrated utilization of grape skins from white grape pomaces. Ind Crop Prod 49:286–291. https://doi.org/10.1016/j.indcrop.2013.05.003

Metzger JO, Huttermann A (2008) Sustainable global energy supply based on lignocellulosic biomass from afforestation of degraded areas. Naturwissenschaften Springer 96:279–288. https://doi.org/10.1007/s00114-008-0479-4

Mitchell JV (2002) A new political economy of oil. Q Rev Econ Finance 42:251–272. https://doi.org/10.1787/9789264090705-en

Mohamed M, Yusup S, Machmudah S, Goto M, Uemura Y (2014) Upgrading of oil palm empty fruit bunch to value-added products. In: Hakeem KR, Jawaid M, Rashid U (eds) Biomass bioenergy. Springer International Publishing, pp 63–78. https://doi.org/10.1007/978-3-319-07578-5_3

Nakamura T, Senior CL (2005) Recovery and sequestration of CO2 from stationary combustion systems by photosynthesis of microalgae. Office of Scientific and Technical Information (OSTI), Pittsburgh. https://doi.org/10.2172/892742

Novy V, Longus K, Nidetzky B (2015) From wheat straw to bioethanol: integrative analysis of a separate hydrolysis and co-fermentation process with implemented enzyme production. Biotechnol Biofuels 8(1):46. https://doi.org/10.1186/s13068-015-0232-0

Office of Energy Efficiency & Renewable Energy (2013) Integrated biorefineries: biofuels, biopower, and bioproducts. U.S. Department of Energy, Washington, DC http://www.energy.gov/sites/prod/files/2014/06/f16/ibr_portfolio_overview.pdf

Oliveira AD, Ribeiro EP, Bone RB, Losekann L (2010) Energy restrictions to growth: the past, present and future energy supply in Brazil. In: Amann E, Baer W, Coes D (eds) Energy, bio fuels and development. Routledge, London, pp 51–64

Onwudili J (2014) Hydrothermal gasification of biomass for hydrogen production. In: Jin F (ed) Application of hydrothermal reactions to biomass conversion. Springer, Berlin/Heidelberg, pp 219–246. https://doi.org/10.1007/978-3-642-54458-3_10

Pandey A (2017) Microalgae biomass production for CO2 mitigation and biodiesel production. Journal of Microbiology & Experimentation 4(4). https://doi.org/10.15406/jmen.2017.04.00117

Petrou EC, Pappis CP (2014) Bioethanol production from cotton stalks or corn stover? A comparative study of their sustainability performance. ACS Sustain Chem Eng 2(8):2036–2041. https://doi.org/10.1021/sc500249d

Ping L, Brosse N, Sannigrahi P, Ragauskas A (2011) Evaluation of grape stalks as a bioresource. Ind Crop Prod 33(1):200–204. https://doi.org/10.1016/j.indcrop.2010.10.009

Rabetafika HN, Bchir B, Blecker C, Richel A (2014) Fractionation of apple by-products as source of new ingredients: current situation and perspectives. Trends Food Sci Technol 40(1):99–114. https://doi.org/10.1016/j.tifs.2014.08.004

Ranjan A, Khanna S, Moholkar VS (2013) Feasibility of rice straw as alternate substrate for biobutanol production. Appl Energy 103:32–38 https://doi.org/10.1016/j.apenergy.2012.10.035

RFF (2015) 2015 ethanol industry outlook. The Renewable Fuels Foundation. http:// www.ethanolrfa.org/wp-content/uploads/2015/09/c5088b8e8e6b427bb3_cwm626ws2.pdf

Rincón LE, Jaramillo JJ, Cardona CA (2014) Comparison of feedstocks and technologies for biodiesel production: an environmental and techno-economic evaluation. Renew Energy 69:479–487. https://doi.org/10.1016/j.renene.2014.03.058

Sarkar A (2009) Alfalfa in molecular farming. In: Sarkar A (ed) Molecular farming. Discovery Publishing House Pvt. Limited, New Delhi, pp 310–320

Scarlat N, Dallemand JF, Monforti-Ferrario F, Banja M, Motola V (2015a) Renewable energy policy framework and bioenergy contribution in the European Union – an overview from national renewable energy action plans and progress reports. Renew Sustain Energy Rev 51:969–985. https://doi.org/10.1016/j.envdev.2015.03.006

Scarlat N, Dallemand JF, Monforti-Ferrario F, Nita V (2015b) The role of biomass and bioenergy in a future bioeconomy: policies and facts. Environ Dev 15:3–34. https://doi.org/10.1016/j.envdev.2015.03.006

Schenk PM, Sky R, Thomas-Hall SR, Stephens E, Marx UC, Mussgnug JH, Posten C et al (2008) Second generation biofuels: high-efficiency microalgae for biodiesel production. Bioenergy Res 1:20–43. https://doi.org/10.1007/s12155-008-9008-8

Schnitzer M, Monreal CM, Powell EE (2013) Wheat straw biomass: a resource for high-value chemicals. J Environ Sci Health B 49(1):51–67. https://doi.org/10.1080/03601234.2013.836924

Sheehan J, Aden A, Paustian K, Killian K, Brenner J, Walsh M et al (2003) Energy and environmental aspects of using corn stover for fuel ethanol. J Ind Ecol 7(3–4):117–146. https://doi.org/10.1162/108819803323059433

Sialve B, Bernet N, Bernard O (2009) Anaerobic digestion of microalgae as a necessary step to make microalgal biodiesel sustainable. Biotechnol Adv 27:409–416. https://doi.org/10.1016/j.biotechadv.2009.03.001

Sims REH, Mabee W, Saddler JN, Taylor M (2010) An overview of second generation biofuel technologies. Bioresour Technol 101(6):1570–1580. https://doi.org/10.1016/j.biortech.2009.11.046

Small E (2011) The economic importance of medicago. In: Small E (ed) Alfalfa and relatives: evolution and classification of Medicago. NRC Press, Ottawa, pp 5–14

Suntana AS, Vogt KA, Turnblom EC, Upadhye R (2009) Bio-methanol potential in Indonesia: forest biomass as a source of bioenergy that reduces carbon emissions. Appl Energy 86:S215–S221. https://doi.org/10.1016/j.apenergy.2009.05.028

Swapna M, Srivastava S (2012) Molecular marker applications for improving sugar content in sugarcane. Springer Briefs Plant Sci:1–49. https://doi.org/10.1007/978-1-4614-2257-0_1

Tabatabaei M, Tohidfar M, Jouzani GS, Safarnejad M, Pazouki M (2011) Biodiesel production from genetically engineered microalgae: future of bioenergy in Iran. Renew Sustain Energy Rev 15(4):1918–1927. https://doi.org/10.1016/j.rser.2010.12.004

Tukacs-Hájos A, Pap B, Maróti G, Szendefy J, Szabó P, Rétfalvi T (2014) Monitoring of thermophilic adaptation of mesophilic anaerobe fermentation of sugar beet pressed pulp. Bioresour Technol 166:288–294. https://doi.org/10.1016/j.biortech.2014.05.059

U.S. Department of Energy (2011) U.S. Billion-Ton update. Biomass supply for a bioenergy and bioproducts industry. U.S. Department of Energy, Washington, DC. https://doi.org/10.2172/1219219

Wilhelm WW, Johnson JMF, Karlen DL, Lightle DT (2007) Corn stover to sustain soil organic carbon further constrains biomass supply. Agron J 99(6):1665. https://doi.org/10.2134/agronj2007.0150

Wilhelm WW, Hess JR, Karlen DL, Johnson JMF, Muth DJ, Baker JM et al (2010) Balancing limiting factors & economic drivers for sustainable Midwestern US agricultural residue feedstock supplies. Ind Biotechnol 6(5):271–287. https://doi.org/10.1089/ind.2010.6.271

Zheng Y, Lee C, Yu C, Cheng YS, Zhang R, Jenkins BM, VanderGheynst JS (2013) Dilute acid pretreatment and fermentation of sugar beet pulp to ethanol. Appl Energy 105:1–7. https://doi.org/10.1016/j.apenergy.2012.11.070

Zhou S, Runge TM (2015) Improving ethanol production from alfalfa stems via ambient-temperature acid pretreatment and washing. Bioresour Technol 193:286–292 https://doi.org/10.1016/j.biortech.2015.06.096

Index

E. Lichtfouse (ed.), *Sustainable Agriculture Reviews 33*, Sustainable Agriculture Reviews 33, https://doi.org/10.1007/978-3-319-99076-7

The manufacturer's authorised representative in the EU is Springer Nature Customer Service Centre GmbH, Europaplatz 3, 69115 Heidelberg, Germany. If you have any concerns regarding our products, please contact ProductSafety@springernature.com

Printed and bound by CPI Group (UK) Ltd, Croydon, CR0 4YY

15/07/2026

02167631-0005